Birkhäuser

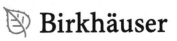

Nadir Jeevanjee

An Introduction to Tensors and Group Theory for Physicists

Second Edition

Birkhäuser

Nadir Jeevanjee
Department of Physics
University of California at Berkeley
Berkeley, CA, USA

ISBN 978-3-319-33089-1 ISBN 978-3-319-14794-9 (eBook)
DOI 10.1007/978-3-319-14794-9

Mathematics Subject Classification (2010): 15A-01, 81R-01, 22E-01, 20C-01

Springer Cham Heidelberg New York Dordrecht London

Softcover reprint of the hardcover 2nd edition 2015

Printed on acid-free paper

Springer International Publishing AG Switzerland is part of Springer Science+Business Media (www.springer.com)

To My Parents

Preface to the First Edition

This book is composed of two parts: Part I (Chaps. 1–3) is an introduction to tensors and their physical applications, and Part II (Chaps. 4–6) introduces group theory and intertwines it with the earlier material. Both parts are written at the advanced undergraduate/beginning graduate level, although in the course of Part II the sophistication level rises somewhat. Though the two parts differ somewhat in flavor, I have aimed in both to fill a (perceived) gap in the literature by connecting the component formalisms prevalent in physics calculations to the abstract but more conceptual formulations found in the math literature. My firm belief is that we need to see tensors and groups in coordinates to get a sense of how they work, but also need an abstract formulation to understand their essential nature and organize our thinking about them.

My original motivation for the book was to demystify tensors and provide a unified framework for understanding them in all the different contexts in which they arise in physics. The word tensor is ubiquitous in physics (stress tensor, moment of inertia tensor, field tensor, metric tensor, tensor product, etc.) and yet tensors are rarely defined carefully, and the definition usually has to do with transformation properties, making it difficult to get a feel for what these objects *are*. Furthermore, physics texts at the beginning graduate level usually only deal with tensors in their component form, so students wonder what the difference is between a second rank tensor and a matrix, and why new, enigmatic terminology is introduced for something they've already seen. All of this produces a lingering unease, which I believe can be alleviated by formulating tensors in a more abstract but conceptually much clearer way. This coordinate-free formulation is standard in the mathematical literature on differential geometry and in physics texts on General Relativity, but, as far as I can tell, is not accessible to undergraduates or beginning graduate students in physics who just want to learn what a tensor is *without* dealing with the full machinery of tensor analysis on manifolds.

The irony of this situation is that a proper understanding of tensors doesn't require much more mathematics than what you likely encountered as an undergraduate. In Chap. 1, I introduce this additional mathematics, which is just an extension of the linear algebra you probably saw in your lower division coursework.

This material sets the stage for tensors and hopefully also illuminates some of the more enigmatic objects from quantum mechanics and relativity, such as bras and kets, covariant and contravariant components of vectors, and spherical harmonics. After laying the necessary linear algebraic foundations, we give in Chap. 2 the modern (component-free) definition of tensors, all the while keeping contact with the coordinate and matrix representations of tensors and their transformation laws. Applications in classical and quantum physics follow.

In Part II of the book, I introduce group theory and its physical applications, which is a beautiful subject in its own right and also a nice application of the material in Part I. There are many good books on the market for group theory and physics (see the references), so rather than be exhaustive, I have just attempted to present those aspects of the subject most essential for upper division and graduate level physics courses. In Chap. 4, I introduce abstract groups but quickly illustrate that concept with myriad examples from physics. After all, there would be little point in making such an abstract definition if it didn't subsume many cases of interest! We then introduce Lie groups and their associated Lie algebras, making precise the nature of the symmetry "generators" that are so central in quantum mechanics. Much time is also spent on the groups of rotations and Lorentz transformations, since these are so ubiquitous in physics.

In Chap. 5, I introduce representation theory, which is a mathematical formalization of what we mean by the "transformation properties" of an object. This subject sews together the material from Chaps. 2 and 3 and is one of the most important applications of tensors, at least for physicists. Chapter 6 then applies and extends the results of Chap. 5 to a few specific topics: the perennially mysterious "spherical" tensors, the Wigner–Eckart theorem, and Dirac bilinears. These topics are unified by the introduction of the representation operator, which is admittedly somewhat abstract but neatly organizes these objects into a single mathematical framework.

This text aims (perhaps naively!) to be simultaneously intuitive and rigorous. Thus, although much of the language (especially in the examples) is informal, almost all the definitions given are precise and are the same as one would find in a pure math text. This may put you off if you feel less mathematically inclined; I hope, however, that you will work through your discomfort and develop the necessary mathematical sophistication, as the results will be well worth it. Furthermore, if you can work your way through the text (or at least most of Chap. 5), you will be well prepared to tackle graduate math texts in related areas.

As for prerequisites, it is assumed that you have been through the usual undergraduate physics curriculum, including a "mathematical methods for physicists" course (with at least a cursory treatment of vectors and matrices) as well as the standard upper division courses in classical mechanics, quantum mechanics, and relativity. Any undergraduate versed in those topics, as well as any graduate student in physics, should be able to read this text. To undergraduates who are eager to learn about tensors but haven't yet completed the standard curriculum, I apologize; many of the examples and practically all of the motivation for the text come from those courses, and to assume no knowledge of those topics would preclude discussion of the many "examples" that motivated me to write this book in the first place.

However, if you are motivated and willing to consult the references, you could certainly work through this text and would no doubt be in excellent shape for those upper division courses once you take them.

Exercises and problems are included in the text, with exercises occurring within the chapters and problems occurring at the end of each chapter. The exercises in particular should be done as they arise, or at least carefully considered, as they often flesh out the text and provide essential practice in using the definitions. Very few of the exercises are computationally intensive, and many of them can be done in a few lines. They are designed primarily to test your conceptual understanding and help you internalize the subject. Please don't ignore them!

Besides the aforementioned prerequisites, I've also indulged in the use of some very basic mathematical shorthand for brevity's sake; a guide is below. Also, be aware that for simplicity's sake, I've set all physical constants such as c and $\hbar$ equal to 1. Enjoy!

Berkeley, CA Nadir Jeevanjee

Preface to the Second Edition

While teaching courses based on the first edition, I found myself augmenting the material in the text in various ways and thought that these additions might make the text a bit more user friendly. This was the motivation to undertake a second edition, which in addition to these significant additions also includes the usual correction of typos and other minor improvements. The major new elements are as follows:

More motivation. One well-received feature of the first edition was the introductory chapter on tensors, which quickly and intuitively conveys some of the main points and sets the stage for the more detailed treatment that follows. Such motivation was conspicuously absent in Part II of the book, so I have added introductory sections to these later chapters, with a similar aim: to convey the take-home messages as directly as possible, leaving the details and secondary examples for the remainder of the chapter. I have also added an epilogue to Chap. 2 which takes stock of the various mathematical structures built up in that chapter, in the hope of coherently organizing them.

More figures and tables. The theory of Lie groups is at heart a geometric one, but making the geometric picture precise requires the machinery of differential geometry, which I very specifically wished to avoid in this book.[1] That, however, is no reason to omit the actual pictures one should have in mind when thinking about Lie groups, and so those figures, along with several others, are now included in the text. I've also included more tables in the representation theory chapter, for help in organizing the menagerie of representations that arise in physics.

More varied formatting. In teaching from the first edition, I also perceived an opportunity to make the visual format of the text more expressive of the content. To this end, I have taken some of the punch lines of various sections, which

[1] This is consonant with the approach taken by Hall [11]. There are, of course, many excellent books which *do* take the geometric approach, which for the committed theoretical physicist is undoubtedly the right one; see, e.g., Frankel [6] and Schutz [18].

previously had been bold but in-line in the text, and have separated them from the main text for emphasis. I have also introduced some boxed text for important side discussions to emphasize that they are a departure from the main storyline but still significant. Less important departures are relegated to footnotes (which are frequent).

In addition to the above, there have been many smaller improvements and corrections, many of which were pointed out or suggested by students and other readers. Thanks are due to all the Phys 198 students at UC Berkeley for these and for enduring my various pedagogical experiments, both with this book and in the classroom. Thanks are also due to Roger Berlind and Prof. John Colarusso, who read the text extremely closely and provided detailed feedback through extended correspondence. Their interest and attention have been gratifying, as well as extremely helpful in improving the book.

Valuable and detailed feedback on the first edition of this book were given by Hal Haggard, Mark Moriarty, Albert Shieh, Felicitas Hernandez, and Emily Rauscher. Early mentorship and support were given by professors Robert Penner and Ko Honda of the U.S.C. mathematics department, whose encouragement was instrumental in my pursuing mathematics and physics to the degree that I have. Thanks are also due to my colleagues past and present at Birkhauser, for taking a chance on this young author with the first edition and supporting his ambitions for a second.

Finally, I must again thank my family, which has now grown to include my partner Erika, and our children Seamus and Samina; they are the reason.

Berkeley, CA, USA Nadir Jeevanjee

Notation

Some Mathematical Shorthand

$\mathbb{N}$	The set of natural numbers (positive integers)
$\mathbb{Z}$	The set of positive and negative integers
$\mathbb{R}$	The set of real numbers
$\mathbb{C}$	The set of complex numbers
$\in$	"is an element of", "an element of", i.e. $2 \in \mathbb{R}$ reads "2 is an element of the real numbers"
$\notin$	"is not an element of"
$\forall$	"for all"
$\subset$	"is a subset of", "a subset of"
$\equiv$	Denotes a definition
$f : A \to B$	Denotes a map f that takes elements of the set A into elements of the set B
$f : a \mapsto b$	Indicates that the map f sends the element a to the element b
$\circ$	Denotes a composition of maps, i.e. if $f : A \to B$ and $g : B \to C$, then $f \circ g : A \to C$ is given by $(f \circ g)(a) \equiv f(g(a))$
$A \times B$	The set $\{(a, b)\}$ of all ordered pairs where $a \in A$, $b \in B$. Referred to as the *cartesian product* of sets A and B. Extends in the obvious way to n-fold products $A_1 \times \ldots \times A_n$.
$\mathbb{R}^n$	$\underbrace{\mathbb{R} \times \ldots \times \mathbb{R}}_{n \text{ times}}$
$\mathbb{C}^n$	$\underbrace{\mathbb{C} \times \ldots \times \mathbb{C}}_{n \text{ times}}$
$\{A \mid Q\}$	Denotes a set A subject to condition Q. For instance, the set of all even integers can be written as $\{x \in \mathbb{R} \mid x/2 \in \mathbb{Z}\}$
$\square$	Denotes the end of a proof or example

Dirac Dictionary

We summarize here all of the translations given in the text between quantum mechanical Dirac notation and standard mathematical notation.

Standard Notation	Dirac Notation
Vector $\psi \in \mathcal{H}$	$\|\psi\rangle$
Dual vector $L(\psi)$	$\langle\psi\|$
Inner product $(\psi\|\phi)$	$\langle\psi\|\phi\rangle$
$A(\psi),\ A \in \mathcal{L}(\mathcal{H})$	$A\|\psi\rangle$
$(\psi, A\phi)$,	$\langle\psi\|A\|\phi\rangle$
$T_i{}^j e^i \otimes e_j$	$\sum_{i,j} T_{ij}\|j\rangle\langle i\|$
$e_i \otimes e_j$	$\|i\rangle\,\|j\rangle$ or $\|i, j\rangle$

Contents

Part I Linear Algebra and Tensors

1 A Quick Introduction to Tensors 3

2 Vector Spaces 11
2.1 Definition and Examples 11
2.2 Span, Linear Independence, and Bases 17
2.3 Components 21
2.4 Linear Operators 25
2.5 Dual Spaces 30
2.6 Non-degenerate Hermitian Forms 33
2.7 Non-degenerate Hermitian Forms and Dual Spaces 40
Epilogue: Tiers of Structure in Linear Algebra 43
Chapter 2 Problems 46

3 Tensors 51
3.1 Definition and Examples 52
3.2 Change of Basis 57
3.3 Active and Passive Transformations 65
3.4 The Tensor Product: Definition and Properties 70
3.5 Tensor Products of V and V^* 71
3.6 Applications of the Tensor Product in Classical Physics 75
3.7 Applications of the Tensor Product in Quantum Physics 77
3.8 Symmetric Tensors 85
3.9 Antisymmetric Tensors 88
3.10 Pseudovectors 96
Chapter 3 Problems 102

Part II Group Theory

4 Groups, Lie Groups, and Lie Algebras 109
4.1 Invitation: Lie Groups and Infinitesimal Generators 110
4.2 Groups: Definition and Examples 115

4.3 The Groups of Classical and Quantum Physics 125
4.4 Homomorphism and Isomorphism 135
4.5 From Lie Groups to Lie Algebras 145
4.6 Lie Algebras: Definition, Properties, and Examples 150
4.7 The Lie Algebras of Classical and Quantum Physics 158
4.8 Abstract Lie Algebras 165
4.9 Homomorphism and Isomorphism Revisited 171
Chapter 4 Problems 180

5 Basic Representation Theory 187
5.1 Invitation: Symmetry Groups and Quantum Mechanics 187
5.2 Representations: Definitions and Basic Examples 192
5.3 Further Examples 197
5.4 Tensor Product Representations 208
5.5 Symmetric and Antisymmetric Tensor Product Representations 215
5.6 Equivalence of Representations 220
5.7 Direct Sums and Irreducibility 229
5.8 More on Irreducibility 237
5.9 The Irreducible Representations of $\mathfrak{su}(2)$, $SU(2)$, and $SO(3)$ 242
5.10 Real Representations and Complexifications 248
5.11 The Irreducible Representations of $\mathfrak{sl}(2,\mathbb{C})_{\mathbb{R}}$, $SL(2,\mathbb{C})$, and $SO(3,1)_o$ 251
5.12 Irreducibility and the Representations of $O(3,1)$ and Its Double Covers 259
Chapter 5 Problems 264

6 The Representation Operator and Its Applications 271
6.1 Invitation: Tensor Operators, Spherical Tensors, and Wigner–Eckart 271
6.2 Representation Operators, Selection Rules, and the Wigner–Eckart Theorem 276
6.3 Gamma Matrices and Dirac Bilinears 282
Chapter 6 Problems 286

A Complexifications of Real Lie Algebras and the Tensor Product Decomposition of $\mathfrak{sl}(2,\mathbb{C})_{\mathbb{R}}$ Representations 289
A.1 Direct Sums and Complexifications of Lie Algebras 289
A.2 Representations of Complexified Lie Algebras and the Tensor Product Decomposition of $\mathfrak{sl}(2,\mathbb{C})_{\mathbb{R}}$ Representations 292

References 297

Index 299

Part I
Linear Algebra and Tensors

Chapter 1
A Quick Introduction to Tensors

The reason tensors are introduced in a somewhat ad-hoc manner in most physics courses is twofold: first, a detailed and proper understanding of tensors requires mathematics that is slightly more abstract than the standard linear algebra and vector calculus that physics students use everyday. Second, students don't necessarily *need* such an understanding to be able to manipulate tensors and solve problems with them. The drawback, of course, is that many students feel uneasy with tensors; they can use them for computation but don't have an intuitive feel for what they're doing. One of the primary aims of this book is to alleviate that unease. Doing that, however, requires a modest investment (about 30 pages) in some abstract linear algebra, so before diving into the details we'll begin with a rough overview of what a tensor is, which hopefully will whet your appetite and tide you over until we can discuss tensors in full detail in Chap. 3.

Many older books define a tensor as a collection of objects which carry indices and which "transform" in a particular way specified by those indices. Unfortunately, this definition usually doesn't yield much insight into what a tensor *is*. One of the main purposes of the present text is to promulgate the more modern definition of a tensor, which is equivalent to the old one but is more conceptual and is in fact already standard in the mathematics literature. This definition takes a tensor to be a *function* which eats a certain number of vectors (known as the ***rank*** r of the tensor) and produces a number. The distinguishing characteristic of a tensor is a special property called ***multilinearity***, which means that it must be linear in each of its r arguments (recall that linearity for a function with a single argument just means that $T(v + cw) = T(v) + cT(w)$ for all vectors v and w and numbers c). As we will explain in a moment, this multilinearity enables us to express the value of the function on an *arbitrary* set of r vectors in terms of the values of the function on r *basis* vectors like $\hat{\mathbf{x}}$, $\hat{\mathbf{y}}$, and $\hat{\mathbf{z}}$. These values of the function on basis vectors are nothing but the familiar ***components*** of the tensor, which in older treatments are usually introduced first as part of the definition of the tensor.

We'll make this concrete by considering a couple of extended examples.

N. Jeevanjee, *An Introduction to Tensors and Group Theory for Physicists*,
DOI 10.1007/978-3-319-14794-9_1

Example 1.1. *The Levi–Civita symbol and the volume tensor ϵ on $\mathbb{R}^3$*

You may have encountered the Levi–Civita symbol in coursework in classical or quantum mechanics. We will denote it here[1] as $\bar{\epsilon}_{ijk}$, where the indices i, j, and k range from 1 to 3. The symbol takes on different numerical values depending on the values of the indices, as follows:

$$\bar{\epsilon}_{ijk} \equiv \begin{cases} 0 & \text{unless } i \neq j \neq k \\ +1 & \text{if } \{i,j,k\} = \{1,2,3\},\ \{2,3,1\},\ \text{or } \{3,1,2\} \\ -1 & \text{if } \{i,j,k\} = \{3,2,1\},\ \{1,3,2\},\ \text{or } \{2,1,3\}. \end{cases} \tag{1.1}$$

Sometimes one sees this defined in words, as follows: $\bar{\epsilon}_{ijk} = 1$ if $\{i,j,k\}$ is a "cyclic permutation" of $\{1,2,3\}$, -1 if $\{i,j,k\}$ is an "anti-cyclic permutation" of $\{1,2,3\}$, and 0 otherwise.

The Levi–Civita symbol is usually introduced to physicists as a convenient shorthand that simplifies expressions and calculations; for instance, it allows one to write a simple expression for the components of the cross product of two vectors v and w:

$$(v \times w)^i = \sum_{j,k=1}^{3} \bar{\epsilon}_{ijk} v^j w^k .$$

It also allows for a compact expression of the quantum-mechanical angular momentum commutation relations:

$$[L_i, L_j] = \sum_{k=1}^{3} i\bar{\epsilon}_{ijk} L_k .$$

Despite its utility, however, the Levi–Civita symbol is rarely given any mathematical or physical interpretation, and (like tensors more generally) ends up being something that students know how to *use* but don't have a feel for. In this example we'll show how our new point of view on tensors sheds considerable light on both the mathematical nature and the geometric interpretation of the Levi–Civita symbol.

To begin, let's define a rank-three tensor, denoted ϵ, where ϵ eats three vectors u, v, and w and produces a number $\epsilon(u,v,w)$. We'd like to interpret $\epsilon(u,v,w)$ as the (oriented) volume of the parallelepiped spanned by u, v, and w; see Fig. 1.1. From vector calculus we know that we can accomplish this by defining

$$\epsilon(u,v,w) \equiv (u \times v) \cdot w. \tag{1.2}$$

[1] Usually the "–" across the top is omitted, but we will need it to conceptually distinguish the Levi–Civita symbol from the epsilon tensor defined below.

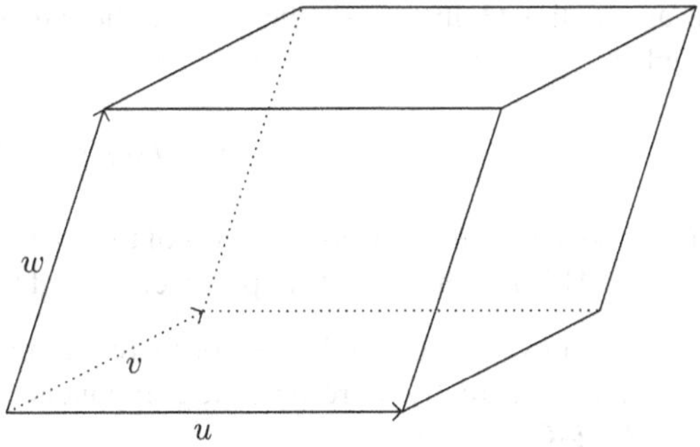

Fig. 1.1 The parallelepiped spanned by u, v, and w

For ϵ to really be a tensor, however, it must be ***multilinear***, i.e. linear in each argument. This means

$$\epsilon(u_1 + cu_2, v, w) = \epsilon(u_1, v, w) + c\epsilon(u_2, v, w) \tag{1.3a}$$

$$\epsilon(u, v_1 + cv_2, w) = \epsilon(u, v_1, w) + c\epsilon(u, v_2, w) \tag{1.3b}$$

$$\epsilon(u, v, w_1 + cw_2) = \epsilon(u, v, w_1) + c\epsilon(u, v, w_2) \tag{1.3c}$$

for all numbers c and vectors u, v, w, v_1, etc. Let's check that (1.3a) holds:

$$\begin{aligned} \epsilon(u_1 + cu_2, v, w) &= ((u_1 + cu_2) \times v) \cdot w \\ &= (u_1 \times v + cu_2 \times v) \cdot w \\ &= (u_1 \times v) \cdot w + c(u_2 \times v) \cdot w \\ &= \epsilon(u_1, v, w) + c\epsilon(u_2, v, w). \end{aligned}$$

Thus ϵ really is linear in the first argument. The check for (1.3b) and (1.3c) proceeds similarly and is left as an exercise. Thus, ϵ satisfies our definition of a tensor as a multilinear function. But, how does this square with our usual notion of a tensor as a set of numbers with some specified transformation properties? We claimed above that these numbers, known as the *components* of a tensor, are nothing but the tensor evaluated on sets of basis vectors. So, let's evaluate ϵ on three arbitrary basis vectors. Ordinarily, a basis vector is one of either $\hat{\mathbf{x}}$, $\hat{\mathbf{y}}$, or $\hat{\mathbf{z}}$, but for the purposes of this example it will be easier to call these e_1, e_2, and e_3, respectively. We'll arbitrarily choose three of them (since ϵ is rank three) and call these choices e_i, e_j, and e_k, where it's possible that i, j, and k are not all distinct. Then we leave it as an exercise for you to check, using (1.2), that

$$\epsilon(e_i, e_j, e_k) = \begin{cases} 0 & \text{unless } i \neq j \neq k \\ +1 & \text{if } \{i, j, k\} = \{1, 2, 3\},\ \{2, 3, 1\},\ \text{or } \{3, 1, 2\} \\ -1 & \text{if } \{i, j, k\} = \{3, 2, 1\},\ \{1, 3, 2\},\ \text{or } \{2, 1, 3\}. \end{cases} \tag{1.4}$$

If, as mentioned above, we then define the ***components*** of the ϵ tensor to be the numbers

$$\epsilon_{ijk} \equiv \epsilon(e_i, e_j, e_k), \tag{1.5}$$

then (1.4) tells us that the components of the ϵ tensor are nothing but the Levi–Civita symbol! This is a major shift in perspective, and tells us several things. First:

1. The Levi–Civita symbol is not merely a mathematical convenience or shorthand; it actually represents the components of a tensor, the volume tensor (or ***Levi–Civita tensor***).

Furthermore, Eq. (1.5) tells us that:

2. **The components of a tensor are just the values of the tensor evaluated on a corresponding set of basis vectors.**

Combining 1 and 2 above then gives the following geometric interpretation of the Levi–Civita symbol:

3. ϵ_{ijk} **is the volume of the oriented parallelepiped spanned by** e_i, e_j, **and** e_k.

Another property worth noting is that the definition (1.2) of ϵ does not require us to choose a basis for our vector space. This makes sense, because ϵ computes volumes of parallelepipeds, which are geometrical quantities which exist independently of any basis. We can thus add a fourth observation to our list:

4. **The** ϵ **tensor exists independently of any basis**. This is in contrast to its components, which by (1.5) are manifestly basis-dependent.

While all this may illuminate the nature of the Levi–Civita symbol, and tensors more generally, we still don't know that the ϵ_{ijk} as defined here "transform" in the manner specified by the usual definition of a tensor. We'll see how this works in our next example, that of a generic rank-two tensor.

Exercise 1.1. Complete the proof of multilinearity by verifying (1.3b) and (1.3c), using the definition (1.2). Also use (1.2) to verify (1.4).

Example 1.2. *A generic rank-two tensor*

In this example we'll analyze a generic rank-two tensor, using our modern definition of a tensor as a multilinear function. This new viewpoint will clear up some of the pervasive and perennial confusion related to tensors, as we'll see.

Consider a rank-two tensor T, whose job it is to eat two vectors v and w and produce a number $T(v, w)$. In analogy to (1.3), multilinearity for this tensor means

$$\begin{aligned} T(v_1 + cv_2, w) &= T(v_1, w) + cT(v_2, w) \\ T(v, w_1 + cw_2) &= T(v, w_1) + cT(v, w_2) \end{aligned} \tag{1.6}$$

for any number c and all vectors v and w. An important consequence of multilinearity is that if we have a coordinate basis for our vector space, say $\hat{\mathbf{x}}$, $\hat{\mathbf{y}}$, and $\hat{\mathbf{z}}$, then

T is determined entirely by its components, i.e. its values on the basis vectors, as follows: first, expand v and w in the coordinate basis as

$$v = v_x\hat{\mathbf{x}} + v_y\hat{\mathbf{y}} + v_z\hat{\mathbf{z}}$$
$$w = w_x\hat{\mathbf{x}} + w_y\hat{\mathbf{y}} + w_z\hat{\mathbf{z}}.$$

Then we have

$$\begin{aligned}
T(v,w) &= T(v_x\hat{\mathbf{x}} + v_y\hat{\mathbf{y}} + v_z\hat{\mathbf{z}}\,,\ w_x\hat{\mathbf{x}} + w_y\hat{\mathbf{y}} + w_z\hat{\mathbf{z}})\\
&= v_xT(\hat{\mathbf{x}},\ w_x\hat{\mathbf{x}} + w_y\hat{\mathbf{y}} + w_z\hat{\mathbf{z}}) + v_yT(\hat{\mathbf{y}},\ w_x\hat{\mathbf{x}} + w_y\hat{\mathbf{y}} + w_z\hat{\mathbf{z}})\\
&\quad + v_zT(\hat{\mathbf{z}},\ w_x\hat{\mathbf{x}} + w_y\hat{\mathbf{y}} + w_z\hat{\mathbf{z}})\\
&= v_xw_xT(\hat{\mathbf{x}},\hat{\mathbf{x}}) + v_xw_yT(\hat{\mathbf{x}},\hat{\mathbf{y}}) + v_xw_zT(\hat{\mathbf{x}},\hat{\mathbf{z}}) + v_yw_xT(\hat{\mathbf{y}},\hat{\mathbf{x}}) + v_yw_yT(\hat{\mathbf{y}},\hat{\mathbf{y}})\\
&\quad + v_yw_zT(\hat{\mathbf{y}},\hat{\mathbf{z}}) + v_zw_xT(\hat{\mathbf{z}},\hat{\mathbf{x}}) + v_zw_yT(\hat{\mathbf{z}},\hat{\mathbf{y}}) + v_zw_zT(\hat{\mathbf{z}},\hat{\mathbf{z}}).
\end{aligned}$$

As in the previous example, we define the components of T as

$$T_{xx} \equiv T(\hat{\mathbf{x}},\hat{\mathbf{x}}), \quad T_{xy} \equiv T(\hat{\mathbf{x}},\hat{\mathbf{y}}), \quad T_{yx} \equiv T(\hat{\mathbf{y}},\hat{\mathbf{x}}), \tag{1.7}$$

and so on. This then gives

$$\begin{aligned}
T(v,w) &= v_xw_xT_{xx} + v_xw_yT_{xy} + v_xw_zT_{xz} + v_yw_xT_{yx} + v_yw_yT_{yy}\\
&\quad + v_yw_zT_{yz} + v_zw_xT_{zx} + v_zw_yT_{zy} + v_zw_zT_{zz},
\end{aligned} \tag{1.8}$$

which may look familiar from discussion of tensors in the physics literature. In that literature, the above equation is often part of the *definition* of a 2nd rank tensor; here, though, we see that its form is really just a consequence of multilinearity. Another advantage of our approach is that the components $\{T_{xx}, T_{xy}, T_{xz}\ldots\}$ of T have a meaning beyond that of just being coefficients that appear in expressions like (1.8); Eq. (1.7) again shows that **components are the values of the tensor when evaluated on a given set of basis vectors**. We re-emphasize this fact because it is crucial in getting a feel for tensors and what they mean.

Another nice feature of our definition of a tensor is that it allows us to *derive* the tensor transformation laws which historically were taken as the definition of a tensor. Say we switch to a new set of basis vectors $\{\hat{\mathbf{x}}', \hat{\mathbf{y}}', \hat{\mathbf{z}}'\}$ which are related to the old basis vectors by

$$\begin{aligned}
\hat{\mathbf{x}}' &= A_{x'x}\hat{\mathbf{x}} + A_{x'y}\hat{\mathbf{y}} + A_{x'z}\hat{\mathbf{z}}\\
\hat{\mathbf{y}}' &= A_{y'x}\hat{\mathbf{x}} + A_{y'y}\hat{\mathbf{y}} + A_{y'z}\hat{\mathbf{z}}\\
\hat{\mathbf{z}}' &= A_{z'x}\hat{\mathbf{x}} + A_{z'y}\hat{\mathbf{y}} + A_{z'z}\hat{\mathbf{z}}.
\end{aligned} \tag{1.9}$$

This does not affect the action of T, since T exists independently of any basis, but if we'd like to compute the value of $T(v,w)$ in terms of the new components

(v'_x, v'_y, v'_z) and (w'_x, w'_y, w'_z) of v and w, then we'll need to know what the new components $\{T_{x'x'}, T_{x'y'}, T_{x'z'}, \ldots\}$ look like. Computing $T_{x'x'}$, for instance, gives

$$\begin{aligned} T_{x'x'} &= T(\hat{\mathbf{x}}', \hat{\mathbf{x}}') \\ &= T(A_{x'x}\hat{\mathbf{x}} + A_{x'y}\hat{\mathbf{y}} + A_{x'z}\hat{\mathbf{z}},\ A_{x'x}\hat{\mathbf{x}} + A_{x'y}\hat{\mathbf{y}} + A_{x'z}\hat{\mathbf{z}}) \\ &= A_{x'x}A_{x'x}T_{xx} + A_{x'y}A_{x'x}T_{yx} + A_{x'z}A_{x'x}T_{zx} + A_{x'x}A_{x'y}T_{xy} + \\ &\quad A_{x'y}A_{x'y}T_{yy} + A_{x'z}A_{x'y}T_{zy} + A_{x'x}A_{x'z}T_{xz} + A_{x'y}A_{x'z}T_{yz} + A_{x'z}A_{x'z}T_{zz}. \end{aligned} \tag{1.10}$$

You probably recognize this as an instance of the standard tensor transformation law, which used to be taken as the definition of a tensor. Here, the transformation law is another consequence of multilinearity. In Chap. 3 we will introduce more convenient notation which will allow us to use the Einstein summation convention, so that we can write the general form of (1.10) as

$$T_{i'j'} = A^k_{i'} A^l_{j'} T_{kl},$$

a form which may be more familiar to some readers.

One common source of confusion is that in physics textbooks, tensors (usually of the 2nd rank) are often represented as matrices, so then the student wonders what the difference is between a matrix and a tensor. Above, we have defined a tensor as a multilinear function on a vector space. What does that have to do with a matrix? Well, *if* we choose a basis $\{\hat{\mathbf{x}}, \hat{\mathbf{y}}, \hat{\mathbf{z}}\}$, we can then write the corresponding components $\{T_{xx}, T_{xy}, T_{xz} \ldots\}$ in the form of a matrix $[T]$ as

$$[T] \equiv \begin{pmatrix} T_{xx} & T_{xy} & T_{xz} \\ T_{yx} & T_{yy} & T_{yz} \\ T_{zx} & T_{zy} & T_{zz} \end{pmatrix}.$$

Equation (1.8) can then be written compactly as

$$T(v, w) = (v_x, v_y, v_z) \begin{pmatrix} T_{xx} & T_{xy} & T_{xz} \\ T_{yx} & T_{yy} & T_{yz} \\ T_{zx} & T_{zy} & T_{zz} \end{pmatrix} \begin{pmatrix} w_x \\ w_y \\ w_z \end{pmatrix}, \tag{1.11}$$

where the usual matrix multiplication is implied. Thus, once a basis is chosen, the action of T can be neatly expressed using the corresponding matrix $[T]$. It is crucial to keep in mind, though, that this association between a tensor and a matrix depends entirely on a *choice of basis*, and that $[T]$ is useful mainly as a computational tool, not a conceptual handle. T is best thought of abstractly as a multilinear function, and $[T]$ **as its representation in a particular coordinate system**.

One possible objection to our approach is that matrices and tensors are often thought of as linear operators which take vectors into vectors, as opposed to objects which eat vectors and spit out numbers. It turns out, though, that for a 2nd rank tensor these two notions are equivalent. Say we have a linear operator R; then we can turn R into a 2nd rank tensor T_R by

$$T_R(v, w) \equiv v \cdot Rw, \tag{1.12}$$

where $\cdot$ denotes the usual dot product of vectors. You can roughly interpret this tensor as taking in v and w and giving back the "component of Rw lying along v". You can also easily check that T_R is multilinear, i.e. that it satisfies (1.6). If we compute the components of T_R we find that, for instance,

$$\begin{aligned}(T_R)_{xx} &= T_R(\hat{\mathbf{x}}, \hat{\mathbf{x}}) \\ &= \hat{\mathbf{x}} \cdot R\hat{\mathbf{x}} \\ &= \hat{\mathbf{x}} \cdot (R_{xx}\hat{\mathbf{x}} + R_{yx}\hat{\mathbf{y}} + R_{zx}\hat{\mathbf{z}}) \\ &= R_{xx}\end{aligned}$$

so the components of the tensor T_R are the same as the components of the linear operator R! In components, the action of T_R then looks like

$$T_R(v, w) = (v_x, v_y, v_z) \begin{pmatrix} R_{xx} & R_{xy} & R_{xz} \\ R_{yx} & R_{yy} & R_{yz} \\ R_{zx} & R_{zy} & R_{zz} \end{pmatrix} \begin{pmatrix} w_x \\ w_y \\ w_z \end{pmatrix},$$

which is identical to (1.11). This makes it obvious how to turn a linear operator R into a 2nd rank tensor—just sandwich the component matrix in between two vectors! This whole process is also reversible: we can turn a 2nd rank tensor T into a linear operator R_T by defining the components of the vector $R_T(v)$ as

$$(R_T(v))_x \equiv T(\hat{\mathbf{x}}, v), \tag{1.13}$$

and similarly for the other components. One can check (see Exercise 1.2 below) that these processes are inverses of each other, so this sets up a one-to-one correspondence between linear operators and 2nd rank tensors and we can thus regard them as equivalent. Since the matrix form of both is identical, one often does not bother trying to clarify exactly which one is at hand, and often times it is just a matter of interpretation.

How does all this work in a physical context? One nice example is the rank 2 moment-of-inertia tensor $\mathcal{I}$, familiar from rigid body dynamics in classical mechanics. In most textbooks, this tensor usually arises when one considers the relationship between the angular velocity vector $\boldsymbol{\omega}$ and the angular momentum $\mathbf{L}$ or kinetic energy KE of a rigid body. If one chooses a basis $\{\hat{\mathbf{x}}, \hat{\mathbf{y}}, \hat{\mathbf{z}}\}$ and then expresses, say, KE in terms of $\boldsymbol{\omega}$, one gets an expression like (1.8) with $v = w = \boldsymbol{\omega}$ and $T_{ij} = \mathcal{I}_{ij}$. From this expression, most textbooks then figure out how the $\mathcal{I}_{ij}$ must transform under a change of coordinates, and this behavior under coordinate change is then what identifies the $\mathcal{I}_{ij}$ as a tensor. In this text, though, we will take a different point of view. We will *define* $\mathcal{I}$ to be the function which, for a given rigid body, assigns to a state of rotation (described by the angular momentum vector $\boldsymbol{\omega}$) twice the corresponding kinetic energy KE. One can then calculate KE in terms of $\boldsymbol{\omega}$, which yields an expression of the form (1.8); this then shows that $\mathcal{I}$ is a tensor in the old sense of the term, which also means that it's a tensor in the modern sense of being a multilinear function. We can then think of $\mathcal{I}$ as a 2nd rank tensor defined by

$$KE = \frac{1}{2}\mathcal{I}(\boldsymbol{\omega}, \boldsymbol{\omega}).$$

From our point of view, we see that the components of $\mathcal{I}$ have a physical meaning; for instance, $\mathcal{I}_{xx} = \mathcal{I}(\hat{\mathbf{x}}, \hat{\mathbf{x}})$ is just twice the kinetic energy of the rigid body when $\boldsymbol{\omega} = \hat{\mathbf{x}}$. When one actually needs to compute the kinetic energy for a rigid body, one usually picks a convenient coordinate system, computes the components of $\mathcal{I}$ using the standard formulae,[2] and then uses the convenient matrix representation to compute

$$KE = \frac{1}{2}(\omega_x, \omega_y, \omega_z)\begin{pmatrix} \mathcal{I}_{xx} & \mathcal{I}_{xy} & \mathcal{I}_{xz} \\ \mathcal{I}_{yx} & \mathcal{I}_{yy} & \mathcal{I}_{yz} \\ \mathcal{I}_{zx} & \mathcal{I}_{zy} & \mathcal{I}_{zz} \end{pmatrix}\begin{pmatrix} \omega_x \\ \omega_y \\ \omega_z \end{pmatrix},$$

which is the familiar expression often found in mechanics texts.

As mentioned above, one can also interpret $\mathcal{I}$ as a linear operator which takes vectors into vectors. If we let $\mathcal{I}$ act on the angular velocity vector, we get

$$\begin{pmatrix} \mathcal{I}_{xx} & \mathcal{I}_{xy} & \mathcal{I}_{xz} \\ \mathcal{I}_{yx} & \mathcal{I}_{yy} & \mathcal{I}_{yz} \\ \mathcal{I}_{zx} & \mathcal{I}_{zy} & \mathcal{I}_{zz} \end{pmatrix}\begin{pmatrix} \omega_x \\ \omega_y \\ \omega_z \end{pmatrix},$$

which you probably recognize as the coordinate expression for the angular momentum $\mathbf{L}$. Thus, $\mathcal{I}$ can be interpreted *either* as the rank 2 tensor which eats two copies of the angular velocity vector $\boldsymbol{\omega}$ and produces the kinetic energy, or as the linear operator which takes $\boldsymbol{\omega}$ into $\mathbf{L}$. The two definitions are equivalent.

Many other tensors which we'll consider in this text are also nicely understood as multilinear functions which happen to have convenient component and matrix representations. These include the electromagnetic field tensor of electrodynamics, as well as the metric tensors of Newtonian and relativistic mechanics. As we progress we'll see how we can use our new point of view to get a handle on these usually somewhat enigmatic objects.

Exercise 1.2. Verify that the definitions in (1.12) and (1.13) invert each other. Do this by considering the tensor T_R corresponding to a linear operator R, and then the linear operator R_{T_R} corresponding to T_R. Show that $R_{T_R} = R$ by feeding in a vector v on both sides and taking a particular (say, x) component of the result.

[2]See Example 3.14 or any standard textbook such as Goldstein [8].

Chapter 2
Vector Spaces

Since tensors are a special class of functions defined on vector spaces, we must have a good foundation in linear algebra before discussing them. In particular, you'll need a little bit more linear algebra than is covered in most sophomore or junior level linear algebra/ODE courses. This chapter starts with the familiar material about vectors, bases, linear operators, etc., but eventually moves on to slightly more sophisticated topics such as dual vectors and non-degenerate Hermitian forms, which are essential for understanding tensors in physics. Along the way we'll also find that our slightly more abstract viewpoint clarifies the nature of many familiar but enigmatic objects, such as spherical harmonics, bras and kets, contravariant and covariant indices, and the Dirac delta function.

2.1 Definition and Examples

We begin with the definition of an abstract vector space. We're taught as undergraduates to think of vectors as arrows with a head and a tail, or as ordered triples of real numbers. However, physics (and especially quantum mechanics) requires a more abstract notion of vectors. Before reading the definition of an abstract vector space, keep in mind that the definition is supposed to distill all the essential features of vectors as we know them (like addition and scalar multiplication) while detaching the notion of a vector space from specific constructs, like ordered n-tuples of real or complex numbers (denoted as $\mathbb{R}^n$ and $\mathbb{C}^n$ respectively). The mathematical utility of this is that much of what we know about vector spaces depends only on the essential properties of addition and scalar multiplication, not on other properties particular to $\mathbb{R}^n$ or $\mathbb{C}^n$. If we work in the abstract framework and then come across other mathematical objects that don't look like $\mathbb{R}^n$ or $\mathbb{C}^n$ but that are abstract vector spaces, then most everything we know about $\mathbb{R}^n$ and $\mathbb{C}^n$ will apply to these spaces as well. Physics also forces us to use the abstract definition since many

N. Jeevanjee, *An Introduction to Tensors and Group Theory for Physicists*,
DOI 10.1007/978-3-319-14794-9_2

quantum-mechanical vector spaces are infinite-dimensional and cannot be viewed as $\mathbb{C}^n$ or $\mathbb{R}^n$ for any n. An added dividend of the abstract approach is that we will learn to think about vector spaces independently of any basis, which will prove very useful.

That said, an ***(abstract) vector space*** consists of the following:

- A set V (whose elements are called vectors)
- A set of scalars C (which for us will always be either $\mathbb{R}$ or $\mathbb{C}$)
- Operations of addition and scalar multiplication under which the vector space is closed,[1] and which satisfy the following axioms:

1. $v + w = w + v$ for all v, w in V (Commutativity)
2. $v + (w + x) = (v + w) + x$ for all v, w, x in V (Associativity)
3. There exists a vector 0 in V such that $v + 0 = v$ for all v in V
4. For all v in V there is a vector $-v$ such that $v + (-v) = 0$
5. $c(v + w) = cv + cw$ for all v and w in V and scalars c (Distributivity)
6. $1v = v$ for all v in V
7. $(c_1 + c_2)v = c_1 v + c_2 v$ for all scalars c_1, c_2 and vectors v
8. $(c_1 c_2)v = c_1(c_2 v)$ for all scalars c_1, c_2 and vectors v

Some parts of the definition may seem tedious or trivial, but they are just meant to ensure that the addition and scalar multiplication operations behave the way we expect them to. In determining whether a set is a vector space or not, one is usually most concerned with defining addition in such a way that the set is closed under addition and that axioms 3 and 4 are satisfied; most of the other axioms are so natural and obviously satisfied that one, in practice, rarely bothers to check them.[2] That said, let's look at some examples from physics, most of which will recur throughout the text.

Example 2.1. $\mathbb{R}^n$

This is the most basic example of a vector space, and the one on which the abstract definition is modeled. Addition and scalar multiplication are defined in the usual way: for $v = (v^1, v^2, \ldots, v^n)$, $w = (w^1, w^2, \ldots, w^n)$ in $\mathbb{R}^n$, we have

$$(v^1, v^2, \ldots, v^n) + (w^1, w^2, \ldots, w^n) \equiv (v^1 + w^1, v^2 + w^2, \ldots, v^n + w^n), \tag{2.1}$$

[1] Meaning that these operations always produce another member of the set V, i.e. a vector.

[2] Another word about axioms 3 and 4, for the mathematically inclined (feel free to skip this if you like): the axioms don't demand that the zero element and inverses are unique, but this actually follows easily *from* the axioms. If 0 and $0'$ are two zero elements, then

$$0 = 0 + 0' = 0',$$

and so the zero element is unique. Similarly, if $-v$ and $-v'$ are both inverse to some vector v, then

$$-v' = -v' + 0 = -v' + (v - v) = (-v' + v) - v = 0 - v = -v,$$

and so inverses are unique as well.

and

$$c(v^1, v^2, \ldots, v^n) \equiv (cv^1, cv^2, \ldots, cv^n), \tag{2.2}$$

and you should check that the axioms are satisfied. These spaces, of course, are basic in physics; $\mathbb{R}^3$ is our usual three-dimensional cartesian space, $\mathbb{R}^4$ is spacetime in special relativity, and $\mathbb{R}^n$ for higher n occurs in classical physics as configuration spaces for multiparticle systems (e.g., $\mathbb{R}^6$ is the configuration space in the classic two-body problem, as you need six coordinates to specify the position of two particles in three-dimensional space).

Example 2.2. $\mathbb{C}^n$

This is another basic example—addition and scalar multiplication are defined by (2.1) and (2.2), just as for $\mathbb{R}^n$, and the axioms are again straightforward to verify. Note, however, that we can take $\mathbb{C}^n$ to be a ***complex*** vector space (i.e., we may take the set C in the definition to be $\mathbb{C}$), since the right-hand side of (2.2) is guaranteed to be in $\mathbb{C}^n$ even when c is complex. The same is *not* true for $\mathbb{R}^n$, which is why $\mathbb{R}^n$ is only a ***real*** vector space. This seemingly pedantic distinction can often end up being significant, as we'll see.

As for physical applications, $\mathbb{C}^n$ occurs in physics primarily as the ket space for finite-dimensional quantum-mechanical systems, such as particles with spin but without translational degrees of freedom. For instance, a spin 1/2 particle fixed in space has ket space identifiable with $\mathbb{C}^2$, and a more general fixed particle with spin s has ket space identifiable with $\mathbb{C}^{2s+1}$.

Box 2.1 $\mathbb{C}^n$ *as a Real Vector Space*

Note that for $\mathbb{C}^n$ we were not *forced* to take $C = \mathbb{C}$; in fact, we could have taken $C = \mathbb{R}$ since (2.2) certainly makes sense when $c \in \mathbb{R}$. In this case we would write the vector space as $\mathbb{C}^n_{\mathbb{R}}$, to remind us that we are considering $\mathbb{C}^n$ as a real vector space. It may not be obvious what the difference between $\mathbb{C}^n$ and $\mathbb{C}^n_{\mathbb{R}}$ is, since both consist of n-tuples of complex numbers, nor may it be obvious *why* anyone would take $C = \mathbb{R}$ when one could take $C = \mathbb{C}$. We will answer both these questions when we consider bases and dimension in the next section. As a matter of course, though, you should assume that we take $C = \mathbb{C}$ if possible, unless explicitly stated otherwise.

Example 2.3. $M_n(\mathbb{R})$ *and* $M_n(\mathbb{C})$*,* $n \times n$ *matrices with real or complex entries*

The vector space structure of $M_n(\mathbb{R})$ and $M_n(\mathbb{C})$ is similar to that of $\mathbb{R}^n$ and $\mathbb{C}^n$: denoting the entry in the ith row and jth column of a matrix A as A_{ij}, we define addition and (real) scalar multiplication for $A, B \in M_n(\mathbb{R})$ by

$$(A + B)_{ij} \equiv A_{ij} + B_{ij}$$
$$(cA)_{ij} \equiv cA_{ij}$$

i.e. addition and scalar multiplication are done component-wise. The same definitions are used for $M_n(\mathbb{C})$, which can of course be taken to be a complex vector space. You can again check that the axioms are satisfied. Though these vector spaces don't appear explicitly in physics very often, they have many important subspaces, one of which we consider in the next example.

Example 2.4. *$H_n(\mathbb{C})$, $n \times n$ Hermitian matrices with complex entries*

$H_n(\mathbb{C})$, the set of all $n \times n$ Hermitian matrices,[3] is obviously a subset of $M_n(\mathbb{C})$, and in fact it is a sub*space* of $M_n(\mathbb{C})$ in that it forms a vector space itself.[4] To show this it is not necessary to verify all of the axioms, since most of them are satisfied by virtue of $H_n(\mathbb{C})$ being a subset of $M_n(\mathbb{C})$; for instance, addition and scalar multiplication in $H_n(\mathbb{C})$ are just given by the restriction of those operations in $M_n(\mathbb{C})$ to $H_n(\mathbb{C})$, so the commutativity of addition and the distributivity of scalar multiplication over addition follow immediately. What does remain to be checked is that $H_n(\mathbb{C})$ is closed under addition and contains the zero "vector" (in this case, the zero matrix), both of which are easily verified.

As far as physical applications go, we know that physical observables in quantum mechanics are represented by Hermitian operators, and if we are dealing with a finite-dimensional ket space such as those mentioned in Example 2.2 then observables can be represented as elements of $H_n(\mathbb{C})$. As an example one can take a fixed spin 1/2 particle whose ket space is $\mathbb{C}^2$; the angular momentum operators are then represented as $L_i = \frac{1}{2}\sigma_i$, where the σ_i are the Hermitian ***Pauli matrices***

$$\sigma_x \equiv \begin{pmatrix} 0 & 1 \\ 1 & 0 \end{pmatrix}, \quad \sigma_y \equiv \begin{pmatrix} 0 & -i \\ i & 0 \end{pmatrix}, \quad \sigma_z \equiv \begin{pmatrix} 1 & 0 \\ 0 & -1 \end{pmatrix}. \tag{2.3}$$

Box 2.2 *$H_n(\mathbb{C})$ is* not *a Complex Vector Space*

One interesting thing about $H_n(\mathbb{C})$ is that even though the entries of its matrices can be complex, it does **not** form a complex vector space; multiplying a Hermitian matrix by i yields an anti-Hermitian matrix, as you can check, so $H_n(\mathbb{C})$ is not closed under complex scalar multiplication. Thus $H_n(\mathbb{C})$ is only a *real* vector space, even though its matrices contain complex numbers. This point is subtle but worth understanding!

Example 2.5. *$L^2([a,b])$, Square-integrable complex-valued functions on an interval*

---

[3] Hermitian matrices being those which satisfy $A^\dagger \equiv (A^T)^* = A$ where superscript T denotes the transpose and superscript $*$ denotes complex conjugation of the entries.

[4] Another footnote for the mathematically inclined: as discussed later in this example, though, $H_n(\mathbb{C})$ is only a *real* vector space, so it is only a subspace of $M_n(\mathbb{C})$ when $M_n(\mathbb{C})$ is considered as a real vector space.

This example is fundamental in quantum mechanics. A complex-valued function f on $[a,b] \subset \mathbb{R}$ is said to be ***square-integrable*** if

$$\int_a^b |f(x)|^2 dx < \infty. \tag{2.4}$$

Defining addition and scalar multiplication in the obvious way,

$$(f+g)(x) = f(x) + g(x)$$
$$(cf)(x) = cf(x),$$

and taking the zero element to be the function which is identically zero (i.e., $f(x) = 0$ for all x) yields a complex vector space. (Note that if we considered only real-valued functions then we would only have a real vector space.) Verifying the axioms is straightforward though not entirely trivial, as one must show that the sum of two square-integrable functions is again square-integrable (Problem 2-2). This vector space arises in quantum mechanics as the set of *normalizable* wavefunctions for a particle in a one-dimensional infinite potential well. Later on we'll consider the more general scenario where the particle may be unbound, in which case $a = -\infty$ and $b = \infty$ and the above definitions are otherwise unchanged. This vector space is denoted as $L^2(\mathbb{R})$.

Example 2.6. *$\mathcal{H}_l(\mathbb{R}^3)$ and $\tilde{\mathcal{H}}_l$, The Harmonic Polynomials and the Spherical Harmonics*

Consider the set $P_l(\mathbb{R}^3)$ of all complex-coefficient polynomial functions on $\mathbb{R}^3$ of fixed degree l, i.e. all linear combinations of functions of the form $x^i y^j z^k$ where $i + j + k = l$. Addition and (complex) scalar multiplication are defined in the usual way and the axioms are again easily verified, so $P_l(\mathbb{R}^3)$ is a vector space. Now consider the vector subspace $\mathcal{H}_l(\mathbb{R}^3) \subset P_l(\mathbb{R}^3)$ of ***harmonic*** degree l polynomials, i.e. degree l polynomials satisfying $\Delta f = 0$, where Δ is the usual three-dimensional Laplacian. You may be surprised to learn that the spherical harmonics of degree l are essentially elements of $\mathcal{H}_l(\mathbb{R}^3)$! To see the connection, note that if we write a degree l polynomial

$$f(x,y,z) = \sum_{\substack{i,j,k \\ i+j+k=l}} c_{ijk}\, x^i y^j z^k$$

in spherical coordinates with polar angle θ and azimuthal angle ϕ, we'll get

$$f(r,\theta,\phi) = r^l Y(\theta,\phi)$$

for some function $Y(\theta,\phi)$, which is just the restriction of f to the unit sphere. If we write the Laplacian out in spherical coordinates, we get

$$\Delta = \frac{\partial^2}{\partial r^2} + \frac{2}{r}\frac{\partial}{\partial r} + \frac{1}{r^2}\Delta_{S^2}, \tag{2.5}$$

where Δ_{S^2} is shorthand for the angular part of the Laplacian.[5] You will show below that applying this to $f(r,\theta,\phi) = r^l Y(\theta,\phi)$ and demanding that f be harmonic yields

$$\Delta_{S^2} Y(\theta,\phi) = -l(l+1)\, Y(\theta,\phi), \tag{2.7}$$

which is the definition of a spherical harmonic of degree l ! Conversely, if we take a degree l spherical harmonic and multiply it by r^l, the result is a harmonic function. If we let $\tilde{\mathcal{H}}_l$ denote the set of all spherical harmonics of degree l, then we have established a one-to-one correspondence between $\tilde{\mathcal{H}}_l$ and $\mathcal{H}_l(\mathbb{R}^3)$! The correspondence is given by restricting functions in $\mathcal{H}_l(\mathbb{R}^3)$ to the unit sphere, and conversely by multiplying functions in $\tilde{\mathcal{H}}_l$ by r^l, i.e.

$$\begin{aligned} \mathcal{H}_l(\mathbb{R}^3) &\overset{1-1}{\longleftrightarrow} \tilde{\mathcal{H}}_l \\ f &\longrightarrow f(r=1,\theta,\phi) \\ r^l Y(\theta,\phi) &\longleftarrow Y(\theta,\phi). \end{aligned}$$

In particular, this means that:

> **The familiar spherical harmonics $Y^l_m(\theta,\phi)$ are just the restriction of particular harmonic degree l polynomials to the unit sphere.**

For instance, consider the case $l = 1$. Clearly $\mathcal{H}_1(\mathbb{R}^3) = P_1(\mathbb{R}^3)$ since all first-degree (linear) functions are harmonic. If we write the functions

$$\frac{x+iy}{\sqrt{2}},\ z,\ -\frac{x-iy}{\sqrt{2}} \in \mathcal{H}_1(\mathbb{R}^3)$$

in spherical coordinates, we get

$$\frac{1}{\sqrt{2}} re^{i\phi}\sin\theta,\ r\cos\theta,\ -\frac{1}{\sqrt{2}} re^{-i\phi}\sin\theta \in \mathcal{H}_1(\mathbb{R}^3).$$

---

[5] The differential operator Δ_{S^2} is also sometimes known as the ***spherical Laplacian***, and is given explicitly by

$$\Delta_{S^2} = \frac{\partial^2}{\partial\theta^2} + \cot\theta\frac{\partial}{\partial\theta} + \frac{1}{\sin^2\theta}\frac{\partial^2}{\partial\phi^2}. \tag{2.6}$$

We won't need the explicit form of Δ_{S^2} here. A derivation and further discussion can be found in any electrodynamics or quantum mechanics book, like Sakurai [17].

Restricting these to the unit sphere yields

$$\frac{1}{\sqrt{2}}e^{i\phi}\sin\theta,\ \cos\theta,\ -\frac{1}{\sqrt{2}}e^{-i\phi}\sin\theta \in \tilde{\mathcal{H}}_1.$$

Up to (overall) normalization, these are the usual degree 1 spherical harmonics Y^1_m, $-1 \leq m \leq 1$. The $l = 2$ case is treated in Exercise 2.2 below. Spherical harmonics are discussed further throughout this text; for a complete discussion, see Sternberg [19].

Exercise 2.1. Verify (2.7).

Exercise 2.2. Consider the functions

$$\tfrac{1}{2}(x+iy)^2,\ -z(x+iy),\ \frac{1}{\sqrt{2}}(2z^2-x^2-y^2),\ -z(x-iy),\ \tfrac{1}{2}(x-iy)^2 \in P_2(\mathbb{R}^3). \qquad (2.8)$$

Verify that they are in fact harmonic, and then write them in spherical coordinates and restrict to the unit sphere to obtain, up to normalization, the familiar degree 2 spherical harmonics Y^2_m, $-2 \leq m \leq 2$.

Non-example $GL(n,\mathbb{R})$, *invertible* $n \times n$ *matrices*
The "general linear group" $GL(n,\mathbb{R})$, defined to be the subset of $M_n(\mathbb{R})$ consisting of invertible $n \times n$ matrices, is not a vector space though it seems like it could be. Why not?

2.2 Span, Linear Independence, and Bases

The notion of a basis is probably familiar to most readers, at least intuitively: it's a set of vectors out of which we can "make" all the other vectors in a given vector space V. In this section we'll make this idea precise and describe bases for some of the examples in the previous section.

First, we need the notion of the span of a set of vectors. If $S = \{v_1, v_2, \ldots, v_k\} \subset V$ is a set of k vectors in V, then the ***span*** of S, denoted Span $\{v_1, v_2, \ldots, v_k\}$ or Span S, is defined to be just the set of all vectors of the form

$$c^1 v_1 + c^2 v_2 + \ldots + c^k v_k, \quad c^i \in C.$$

Such vectors are known as ***linear combinations*** of the v_i, so Span S is just the set of all linear combinations of the vectors in S. For instance, if $S = \{(1,0,0),(0,1,0)\} \subset \mathbb{R}^3$, then Span S is just the set of all vectors of the form $(c^1, c^2, 0)$ with $c^1, c^2 \in \mathbb{R}$. If S has infinitely many elements, then the span of S

is again all the linear combinations of vectors in S, though in this case the linear combinations can have an arbitrarily large (but finite) number of terms.[6]

Next we need the notion of linear dependence: a (not necessarily finite) set of vectors S is said to be ***linearly dependent*** if there exists distinct vectors $v_1, v_2, \ldots, v_m$ in S and scalars $c^1, c^2, \ldots, c^m$, not all of which are 0, such that

$$c^1 v_1 + c^2 v_2 + \cdots + c^m v_m = 0. \tag{2.9}$$

What this definition really means is that at least one vector in S can be written as a linear combination of the others, and in that sense is dependent (you should take a second to convince yourself of this). If S is not linearly dependent, then we say it is ***linearly independent***, and in this case no vector in S can be written as a linear combination of any others. For instance, the set $S = \{(1,0,0), (0,1,0), (1,1,0)\} \subset \mathbb{R}^3$ is linearly dependent, whereas the set $S' = \{(1,0,0), (0,1,0), (0,1,1)\}$ is linearly *in*dependent, as you should check.

With these definitions in place we can now define a ***basis*** for a vector space V as an *ordered* linearly independent set $\mathcal{B} \subset V$ whose span is all of V. This means, roughly speaking, that a basis has enough vectors to "make" all of V, but no more than that. When we say that $\mathcal{B} = \{v_1, \ldots, v_k\}$ is an *ordered* set we mean that the order of the v_i is part of the definition of $\mathcal{B}$, so another basis with the same vectors but a different order is considered distinct. The reasons for this will become clear as we progress.

One can show[7] that all finite bases must have the same number of elements, so we define the ***dimension*** of a vector space V, denoted dim V, to be the number of elements of any finite basis. If no finite basis exists, then we say that V is ***infinite-dimensional***.

Basis vectors are often denoted e_i, rather than v_i, and we will use this notation from now on.

Exercise 2.3. Given a vector v and a finite basis $\mathcal{B} = \{e_i\}_{i=1\ldots n}$, show that the expression of v as a linear combination of the e_i is unique.

Example 2.7. *The complex plane $\mathbb{C}$ as both a real and complex vector space*

We begin with a trivial example, as it illustrates the role of the scalars C. Consider the vector spaces $\mathbb{C}^n$ and $\mathbb{C}^n_{\mathbb{R}}$ from Example 2.2 and Box 2.1, where the first space is a complex vector space but the second is only real. Now let $n = 1$. Then the first space is just $\mathbb{C}$, and the number 1 (or any other nonzero complex number, for that matter) spans the space since we can write any $z \in \mathbb{C}$ as $z = z \cdot 1 \in \text{Span}\{1\}$. Since a set with just one nonzero vector is trivially linearly independent (check!), 1 is a basis for $\mathbb{C}$ and $\mathbb{C}$ is one-dimensional.

---

[6]We don't generally consider infinite linear combinations like $\sum_{i=1}^{\infty} c^i v_i = \lim_{N\to\infty} \sum_{i=1}^{N} c^i v_i$ because in that case we would need to consider whether the limit exists, i.e. whether the sum converges in some sense. More on this later.

[7]See Hoffman and Kunze [13].

For the second space $\mathbb{C}_{\mathbb{R}}$, however, 1 does not span the space: scalar multiplying 1 by only real numbers cannot yield arbitrary complex numbers. To span the space we need a complex number such as i. The set $\{1, i\}$ then spans $\mathbb{C}_{\mathbb{R}}$ since any $z \in \mathbb{C}$ can be written as $z = a \cdot 1 + b \cdot i$, where $a, b \in \mathbb{R}$. Thus, $\mathbb{C}_{\mathbb{R}}$ is two-dimensional.

What we see, then, is that even though $\mathbb{C}$ and $\mathbb{C}_{\mathbb{R}}$ are identical as *sets*, they differ in vector space properties such as their dimensionality. We sometimes express this by saying that $\mathbb{C}$ has ***complex dimension*** one but ***real dimension*** two.

Furthermore, we usually think about complex numbers as $z = a + bi$ and visualize them as the "complex plane"; when we do this we are really thinking about $\mathbb{C}_{\mathbb{R}}$ rather than $\mathbb{C}$. So, even though it may seem strange to take $C = \mathbb{R}$ when we could take $C = \mathbb{C}$, this is exactly what we usually do with complex numbers!

Example 2.8. *$\mathbb{R}^n$ and $\mathbb{C}^n$*

$\mathbb{R}^n$ has the following natural basis, also known as the ***standard basis***:

$$\begin{gathered} (1, 0, \ldots, 0), \\ (0, 1, \ldots, 0), \\ \vdots \\ (0, \ldots, 1, 0), \\ (0, \ldots, 0, 1). \end{gathered} \tag{2.10}$$

You should check that this is indeed a basis, and thus that the dimension of $\mathbb{R}^n$ is, unsurprisingly, n. The same set serves as a basis for $\mathbb{C}^n$, provided of course that you take $C = \mathbb{C}$.[8]

Note that although the standard basis is the most natural one for $\mathbb{R}^n$ and $\mathbb{C}^n$, there are infinitely many other perfectly respectable bases out there; you should check, for instance, that $\{(1, 1, 0, \ldots, 0),\ (0, 1, 1, 0, \ldots, 0),\ \ldots\ ,\ (0, \ldots, 1, 1),\ (1, 0, \ldots, 0, 1)\}$ is also a basis when $n > 2$.

Example 2.9. *$M_n(\mathbb{R})$ and $M_n(\mathbb{C})$*

Let E_{ij} be the $n \times n$ matrix with a 1 in the ith row, jth column and zeros everywhere else. Then you can check that $\{E_{ij}\}_{i,j=1,\ldots,n}$ is a basis for both $M_n(\mathbb{R})$ and $M_n(\mathbb{C})$, and that both spaces have dimension n^2. Again, there are other nice bases out there; for instance, the symmetric matrices $S_{ij} \equiv E_{ij} + E_{ji},\ i \leq j$, and antisymmetric matrices $A_{ij} \equiv E_{ij} - E_{ji}, i < j$ taken together also form a basis for both $M_n(\mathbb{R})$ and $M_n(\mathbb{C})$.

[8] If you take $C = \mathbb{R}$, then you need to multiply the basis vectors in (2.11) by i and add them to the basis set, giving $\mathbb{C}^n$ a real dimension of $2n$.

Exercise 2.4. Let $S_n(\mathbb{R})$, $A_n(\mathbb{R})$ be the sets of $n \times n$ symmetric and antisymmetric matrices, respectively. Show that both are real vector spaces, compute their dimensions, and check that $\dim S_n(\mathbb{R}) + \dim A_n(\mathbb{R}) = \dim M_n(\mathbb{R})$, as expected.

Exercise 2.5. What is the real dimension of $M_n(\mathbb{C})$?

Example 2.10. $H_2(\mathbb{C})$

Let's find a basis for $H_2(\mathbb{C})$. First, we need to know what a general element of $H_2(\mathbb{C})$ looks like. In terms of complex components, the condition $A = A^\dagger$ reads

$$\begin{pmatrix} a & b \\ c & d \end{pmatrix} = \begin{pmatrix} \bar{a} & \bar{c} \\ \bar{b} & \bar{d} \end{pmatrix} \tag{2.11}$$

where the bar denotes complex conjugation. This means that $a, d \in \mathbb{R}$ and $b = \bar{c}$, so in terms of real numbers we can write a general element of $H_2(\mathbb{C})$ as

$$\begin{pmatrix} t+z & x-iy \\ x+iy & t-z \end{pmatrix} = tI + x\sigma_x + y\sigma_y + z\sigma_z \tag{2.12}$$

where I is the identity matrix and $\sigma_x, \sigma_y, \sigma_z$ are the Pauli matrices defined in (2.3). You can easily check that the set $\mathcal{B} = \{I, \sigma_x, \sigma_y, \sigma_z\}$ is linearly independent, and since (2.12) shows that $\mathcal{B}$ spans $H_2(\mathbb{C})$, $\mathcal{B}$ is a basis for $H_2(\mathbb{C})$. We also see that $\dim H_2(\mathbb{C}) = 4$.

Exercise 2.6. Using the matrices S_{ij} and A_{ij} from Example 2.9, construct a basis for $H_n(\mathbb{C})$ and compute its dimension.

Exercise 2.7. $H_2(\mathbb{C})$ and $M_2(\mathbb{C})$ are both four-dimensional, yet $H_2(\mathbb{C})$ is clearly a proper subset of $M_n(\mathbb{C})$, and hence should have lower dimensionality. Explain the apparent paradox.

Example 2.11. $Y^l_m(\theta, \phi)$

We saw in the previous section that the Y^l_m are elements of $\tilde{\mathcal{H}}_l$, which can be obtained from $\mathcal{H}_l(\mathbb{R}^3)$ by restricting to the unit sphere. What's more is that the set $\{Y^l_m\}_{-l \le m \le l}$ is actually a basis for $\tilde{\mathcal{H}}_l$. In the case $l = 1$ this is clear: we have $\mathcal{H}_1(\mathbb{R}^3) = P_1(\mathbb{R}^3)$ and clearly $\{\frac{1}{\sqrt{2}}(x+iy), z, -1\frac{1}{\sqrt{2}}(x-iy)\}$ is a basis, and restricting this basis to the unit sphere gives the $l = 1$ spherical harmonics. For $l > 1$ proving our claim requires a little more effort; see Problem 2-3.

Another, simpler basis for $\mathcal{H}_1(\mathbb{R}^3)$ would be the ***cartesian basis*** $\{x, y, z\}$; physicists use the spherical harmonic basis because those functions are eigenfunctions of the orbital angular momentum operator L_z, which on $\mathcal{H}_l(\mathbb{R}^3)$ is represented[9] by $L_z = -i(x\frac{\partial}{\partial y} - y\frac{\partial}{\partial x})$. We shall discuss the relationship between the two bases in detail later.

[9]As mentioned in the preface, the $\hbar$ which would normally appear in this expression has been set to 1.

Not Quite Example $L^2([-a,a])$

From doing 1-D problems in quantum mechanics one already "knows" that the set $\{e^{i\frac{n\pi x}{a}}\}_{n\in\mathbb{Z}}$ is a basis for $L^2([-a,a])$. There's a problem, however; we're used to taking infinite linear combinations of these basis vectors, but our definition above only allows for *finite* linear combinations. What's going on here? It turns out that $L^2([-a,a])$ has more structure than your average vector space: it is an infinite-dimensional ***Hilbert space***, and for such spaces we have a generalized definition of a basis, one that allows for infinite linear combinations. We will discuss Hilbert spaces in Sect. 2.6.

2.3 Components

One of the most useful things about introducing a basis for a vector space is that it allows us to write elements of the vector space as n-tuples, in the form of either column or row vectors, as follows: Given $v \in V$ and a basis $\mathcal{B} = \{e_i\}_{i=1\ldots n}$ for V, we can write

$$\boxed{v = \sum_{i=1}^{n} v^i e_i}$$

for some numbers v^i, called the ***components of*** v ***with respect to*** $\mathcal{B}$. We can then represent v by the column vector, denoted $[v]_{\mathcal{B}}$, as

$$[v]_{\mathcal{B}} = \begin{pmatrix} v^1 \\ v^2 \\ \cdot \\ \cdot \\ \cdot \\ v^n \end{pmatrix}$$

or the row vector

$$[v]_{\mathcal{B}}^T = (v^1, v^2, \ldots, v^n),$$

where the superscript T denotes the usual transpose of a vector. The subscript $\mathcal{B}$ just reminds us which basis the components are referred to, and will be dropped if there is no ambiguity. With a choice of basis, then, every n-dimensional vector space can be made to "look like" $\mathbb{R}^n$ or $\mathbb{C}^n$. Writing vectors in this way greatly facilitates computation, as we'll see. One must keep in mind, however, that **vectors exist independently of any chosen basis**, and that their expressions as row or column vectors depend very much on the *choice* of basis $\mathcal{B}$.

To make this last point explicit, let's consider two bases for $\mathbb{R}^3$: the standard basis

$$\begin{aligned} e_1 &= (1,0,0) \\ e_2 &= (0,1,0) \\ e_3 &= (0,0,1), \end{aligned}$$

and the alternate basis we introduced in Example 2.8

$$\begin{aligned} e'_1 &= (1,1,0) \\ e'_2 &= (0,1,1) \\ e'_3 &= (1,0,1). \end{aligned}$$

We'll refer to these bases as $\mathcal{B}$ and $\mathcal{B}'$, respectively. Let's consider the components of the vector e'_1 in both bases. If we expand in the basis $\mathcal{B}$, we have

$$e'_1 = 1 \cdot e_1 + 1 \cdot e_2 + 0 \cdot e_3$$

so

$$[e'_1]_{\mathcal{B}} = \begin{pmatrix} 1 \\ 1 \\ 0 \end{pmatrix} \tag{2.13}$$

as expected. If we expand e'_1 in the basis $\mathcal{B}'$, however, we get

$$e'_1 = 1 \cdot e'_1 + 0 \cdot e'_2 + 0 \cdot e'_3,$$

and so

$$[e'_1]_{\mathcal{B}'} = \begin{pmatrix} 1 \\ 0 \\ 0 \end{pmatrix},$$

which is of course not the same as (2.13).[10] For fun, we can also express the standard basis vector e_1 in the alternate basis $\mathcal{B}'$; you can check that

$$e_1 = \tfrac{1}{2}e'_1 - \tfrac{1}{2}e'_2 + \tfrac{1}{2}e'_3,$$

[10]The simple form of $[e'_1]_{\mathcal{B}'}$ is no accident; you can easily check that if you express *any* set of basis vectors *in the basis that they define*, the resulting column vectors will just look like the standard basis.

and so

$$[e_1]_{\mathcal{B}'} = \begin{pmatrix} 1/2 \\ -1/2 \\ 1/2 \end{pmatrix}.$$

The following examples will explore the notion of components in a few other contexts.

Example 2.12. *Rigid Body Motion*

One area of physics where the distinction between a vector and its expression as an ordered triple is crucial is rigid body motion. In this setting we usually deal with two bases, an arbitrary but fixed *space axes* $K' = \{\hat{\mathbf{x}}', \hat{\mathbf{y}}', \hat{\mathbf{z}}'\}$ and a time-dependent *body axes* $K = \{\hat{\mathbf{x}}(t), \hat{\mathbf{y}}(t), \hat{\mathbf{z}}(t)\}$ which is fixed relative to the rigid body. These are illustrated in Fig. 2.1. When we write down vectors like the angular momentum vector L or the angular velocity vector $\boldsymbol{\omega}$, we must keep in mind what basis we are using, as the component expressions will differ drastically depending on the choice of basis. For example, if there is no external torque on a rigid body, $[L]_{K'}$ will be constant whereas $[L]_K$ will in general be time-dependent.

Example 2.13. *Different bases for* $\mathbb{C}^2$

As mentioned in Example 2.4, the vector space for a spin 1/2 particle is identifiable with $\mathbb{C}^2$ and the angular momentum operators are given by $L_i = \frac{1}{2}\sigma_i$. In particular, this means that

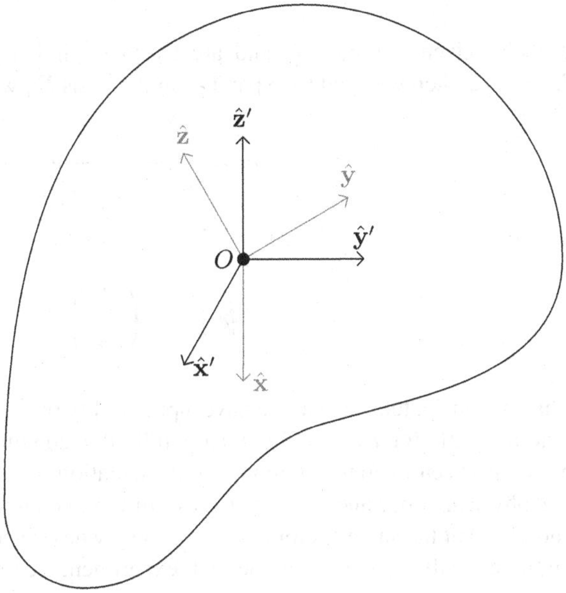

Fig. 2.1 Depiction of the fixed space axes K' and the time-dependent body axes K, in *gray*. K is attached to the rigid body and its basis vectors will change in time as the rigid body rotates

$$L_z = \begin{pmatrix} \frac{1}{2} & 0 \\ 0 & -\frac{1}{2} \end{pmatrix},$$

and so the standard basis vectors

$$e_1 = \begin{pmatrix} 1 \\ 0 \end{pmatrix}$$

$$e_2 = \begin{pmatrix} 0 \\ 1 \end{pmatrix} \tag{2.14}$$

are eigenvectors of L_z with eigenvalues of 1/2 and −1/2 respectively. Let's say, however, that we were interested in measuring L_x, where

$$L_x = \begin{pmatrix} 0 & \frac{1}{2} \\ \frac{1}{2} & 0 \end{pmatrix}.$$

Then we would need to use a basis $\mathcal{B}'$ of L_x eigenvectors, which you can check are given by

$$e'_1 = \frac{1}{\sqrt{2}} \begin{pmatrix} 1 \\ 1 \end{pmatrix}$$

$$e'_2 = \frac{1}{\sqrt{2}} \begin{pmatrix} 1 \\ -1 \end{pmatrix}.$$

If we're given the state e_1 and are asked to find the probability of measuring $L_x = 1/2$, then we need to expand e_1 in the basis $\mathcal{B}'$, which gives

$$e_1 = \frac{1}{\sqrt{2}} e'_1 + \frac{1}{\sqrt{2}} e'_2,$$

and so

$$[e_1]_{\mathcal{B}'} = \begin{pmatrix} \frac{1}{\sqrt{2}} \\ \frac{1}{\sqrt{2}} \end{pmatrix}. \tag{2.15}$$

This, of course, tells us that we have a probability of 1/2 for measuring $L_x = +1/2$ (and the same for $L_x = -1/2$). Hopefully this convinces you that the distinction between a vector and its component representation is not just pedantic, but can be of real physical importance. In fact, the two different component representations (2.14) and (2.15) of the same vector e_1 are precisely what is needed to understand the non-intuitive results of the Stern–Gerlach experiment; see Chap. 1 of Sakurai [17] for details.

Example 2.14. $L^2([-a,a])$

We know from experience in quantum mechanics that all square-integrable functions on an interval $[-a,a]$ have an expansion[11]

$$f = \sum_{m=-\infty}^{\infty} c_m e^{i\frac{m\pi x}{a}}$$

in terms of the "basis" $\{\exp(i\,\frac{m\pi x}{a})\}_{m\in\mathbb{Z}}$. This expansion is known as the ***Fourier series*** of f, and we see that the c_n, commonly known as the ***Fourier coefficients***, are nothing but the components of the vector f in the basis $\{e^{i\frac{m\pi x}{a}}\}_{m\in\mathbb{Z}}$.

2.4 Linear Operators

One of the basic notions in linear algebra, fundamental in quantum mechanics, is that of a linear operator. A ***linear operator*** on a vector space V is a function T from V to itself satisfying the ***linearity*** condition

$$T(cv + w) = cT(v) + T(w). \tag{2.16}$$

Sometimes we write Tv instead of $T(v)$. You should check that the set of all linear operators on V, with the obvious definitions of addition and scalar multiplication, forms a vector space, denoted $\mathcal{L}(V)$. You have doubtless seen many examples of linear operators: for instance, we can interpret a real $n\times n$ matrix as a linear operator on $\mathbb{R}^n$ that acts on column vectors by matrix multiplication. Thus $M_n(\mathbb{R})$ (and, similarly, $M_n(\mathbb{C})$) can be viewed as vector spaces whose elements are themselves linear operators. In fact, that was exactly how we interpreted the vector subspace $H_2(\mathbb{C}) \subset M_2(\mathbb{C})$ in Example 2.4; in that case, we identified elements of $H_2(\mathbb{C})$ as the quantum-mechanical angular momentum operators. There are numerous other examples of quantum-mechanical linear operators, for instance the familiar position and momentum operators $\hat{x}$ and $\hat{p}$ that act on $L^2([-a,a])$ by

$$(\hat{x}f)(x) = xf(x)$$
$$(\hat{p}f)(x) = -i\frac{\partial f}{\partial x}(x)$$

[11] This fact is proved in most real analysis books; see Rudin [16].

as well as the angular momentum operators L_x, L_y, and L_z which act on $P_l(\mathbb{R}^3)$ by

$$\begin{aligned} L_x(f) &= -i\left(y\frac{\partial f}{\partial z} - z\frac{\partial f}{\partial y}\right) \\ L_y(f) &= -i\left(z\frac{\partial f}{\partial x} - x\frac{\partial f}{\partial z}\right) \\ L_z(f) &= -i\left(x\frac{\partial f}{\partial y} - y\frac{\partial f}{\partial x}\right). \end{aligned} \tag{2.17}$$

Another class of less familiar examples is given below.

Example 2.15. *$\mathcal{L}(V)$ acting on $\mathcal{L}(V)$*

We are familiar with linear operators taking vectors into vectors, but they can also be used to take linear operators into linear operators, as follows: Given $A, B \in \mathcal{L}(V)$, we can define a linear operator $\text{ad}_A \in \mathcal{L}(\mathcal{L}(V))$ acting on B by

$$\text{ad}_A(B) \equiv [A, B],$$

where $[\cdot, \cdot]$ indicates the commutator. Note that ad_A is a linear operator on a (vector) space of linear operators! This action of A on $\mathcal{L}(V)$ is called the ***adjoint action*** or ***adjoint representation***.

The adjoint representation has important applications in quantum mechanics; for instance, the Heisenberg picture emphasizes $\mathcal{L}(V)$ rather than V and interprets the Hamiltonian as an operator in the adjoint representation. In fact, for any observable A the Heisenberg equation of motion reads

$$\frac{dA}{dt} = i\,\text{ad}_H(A). \tag{2.18}$$

The adjoint representation is also essential in understanding spherical tensors, which we'll discuss in detail in Chap. 6. □

Box 2.3 *Invertibility for Linear Operators*
One important property of a linear operator T is whether or not it is ***invertible***, i.e. whether there exists a linear operator T^{-1} such that $TT^{-1} = T^{-1}T = I$, where I is the identity operator.[12] You may recall that, in general, an inverse for

[12]Throughout this text I will denote the identity operator or identity matrix; it will be clear from context which is meant.

a map F between two arbitrary sets A and B (not necessarily vector spaces!) exists if and only if F is both ***one-to-one***, meaning

$$F(a_1) = F(a_2) \implies a_1 = a_2 \quad \forall\, a_1, a_2 \in A,$$

and ***onto***, meaning that

$$\forall\, b \in B \text{ there exists } a \in A \text{ such that } F(a) = b.$$

If this is unfamiliar, you should take a moment to convince yourself of this. In the particular case of a linear operator T on a vector space V (so that now instead of considering a generic map $F : A \to B$, we're considering a *linear* map $T : V \to V$), these two conditions are actually equivalent. You'll prove this in Exercise 2.8 below. Furthermore, these two conditions turn out also to be equivalent to the statement

$$T(v) = 0 \implies v = 0, \tag{2.19}$$

as you'll show in Exercise 2.9. Equation (2.19) thus gives us a necessary and sufficient criterion for invertibility of a linear operator, which we may express as follows:

***T* is invertible if and only if the only vector it sends to 0 is the zero vector.**

Exercise 2.8. Suppose V is finite-dimensional and let $T \in \mathcal{L}(V)$. Show that T being one-to-one is equivalent to T being onto. Feel free to introduce a basis to assist you in the proof.

Exercise 2.9. Suppose $T(v) = 0 \implies v = 0$. Show that this is equivalent to T being one-to-one, which by the previous exercise is equivalent to T being one-to-one *and* onto, which is then equivalent to T being invertible.

An important point to keep in mind is that **a linear operator is *not* the same thing as a matrix**; just as with vectors, the identification can only be made *once a basis is chosen*. For operators on finite-dimensional spaces this is done as follows: choose a basis $\mathcal{B} = \{e_i\}_{i=1\ldots n}$. Then the action of T is determined by its action on the basis vectors,

$$T(v) = T\left(\sum_{i=1}^{n} v^i e_i\right) = \sum_{i=1}^{n} v^i T(e_i) = \sum_{i,j=1}^{n} v^i T_i{}^j e_j, \tag{2.20}$$

where the numbers $T_i^{\ j}$, again called the ***components of*** T ***with respect to*** $\mathcal{B}$,[13] are defined by

$$\boxed{T(e_i) = \sum_{j=1}^{n} T_i^{\ j} e_j.} \tag{2.21}$$

Note that $T_i^{\ j} = T(e_i)^j$, the jth component of the vector $T(e_i)$. We now have

$$[v]_{\mathcal{B}} = \begin{pmatrix} v^1 \\ v^2 \\ \cdot \\ \cdot \\ \cdot \\ v^n \end{pmatrix} \quad \text{and} \quad [T(v)]_{\mathcal{B}} = \begin{pmatrix} \sum_{i=1}^n v^i T_i^{\ 1} \\ \sum_{i=1}^n v^i T_i^{\ 2} \\ \cdot \\ \cdot \\ \cdot \\ \sum_{i=1}^n v^i T_i^{\ n} \end{pmatrix},$$

which looks suspiciously like matrix multiplication. In fact, we can define the ***matrix of*** T ***in the basis*** $\mathcal{B}$, denoted $[T]_{\mathcal{B}}$, by the matrix equation

$$\boxed{[T(v)]_{\mathcal{B}} = [T]_{\mathcal{B}}[v]_{\mathcal{B}}},$$

where the product on the right-hand side is given by the usual matrix multiplication. Comparison of the last two equations then shows that

$$[T]_{\mathcal{B}} = \begin{pmatrix} T_1^{\ 1} & T_2^{\ 1} & \dots & T_n^{\ 1} \\ T_1^{\ 2} & T_2^{\ 2} & \dots & T_n^{\ 2} \\ \cdot & \cdot & \cdot & \cdot \\ \cdot & \cdot & \cdot & \cdot \\ \cdot & \cdot & \cdot & \cdot \\ T_1^{\ n} & T_2^{\ n} & \dots & T_n^{\ n} \end{pmatrix}. \tag{2.22}$$

Thus, we really can use the components of T to represent it as a matrix, and once we do so the action of T becomes just matrix multiplication by $[T]_{\mathcal{B}}$! Furthermore, if we have two linear operators A and B and we define their ***product*** (or ***composition***) AB as the linear operator

$$(AB)(v) \equiv A(B(v)),$$

you can then show that $[AB] = [A][B]$. Thus, composition of operators becomes matrix multiplication of the corresponding matrices.

---

[13]Nomenclature to be justified in the next chapter.

Box 2.4 *On Matrix Indices*
One caveat about (2.22): the left (lower) index of $T_i{}^j$ labels the column, not the row, in contrast to the usual convention (which we will also employ in this text, mostly in Part II). It's also helpful to note that the ith column of (2.22) is just the column vector $[T(e_i)]$.

Exercise 2.10. For two linear operators A and B on a vector space V, show that $[AB] = [A][B]$ in any basis.

Example 2.16. $L_z, \mathcal{H}_l(\mathbb{R}^3)$ *and Spherical Harmonics*

Recall that $\mathcal{H}_1(\mathbb{R}^3)$ is the set of all linear functions on $\mathbb{R}^3$ and that

$$\{rY^1_m\}_{-1\leq m\leq 1} = \left\{\frac{1}{\sqrt{2}}(x+iy), z, -\frac{1}{\sqrt{2}}(x-iy)\right\} \quad \text{and} \quad \{x, y, z\}$$

are both bases for this space.[14] Now consider the familiar angular momentum operator $L_z = -i(x\frac{\partial}{\partial y} - y\frac{\partial}{\partial x})$ on this space. You can check that

$$\begin{aligned}
\frac{1}{\sqrt{2}}L_z(x+iy) &= \frac{1}{\sqrt{2}}(x+iy) \\
L_z(z) &= 0 \\
-\frac{1}{\sqrt{2}}L_z(x-iy) &= \frac{1}{\sqrt{2}}(x-iy),
\end{aligned}$$

which implies by (2.21) that the components of L_z in this basis are

$$\begin{aligned}
(L_z)_1{}^1 &= 1 \\
(L_z)_1{}^2 = (L_z)_1{}^3 &= 0 \\
(L_z)_2{}^i &= 0\ \forall\, i \\
(L_z)_3{}^3 &= -1 \\
(L_z)_3{}^1 = (L_z)_3{}^2 &= 0.
\end{aligned}$$

Thus in the spherical harmonic basis,

$$[L_z]_{\{rY^1_m\}} = \begin{pmatrix} 1 & 0 & 0 \\ 0 & 0 & 0 \\ 0 & 0 & -1 \end{pmatrix}.$$

[14] We have again ignored the overall normalization of the spherical harmonics to avoid unnecessary clutter.

This of course just says that the wavefunctions $\frac{x+iy}{\sqrt{2}}, z$, and $-\frac{(x-iy)}{\sqrt{2}}$ have L_z eigenvalues of $1, 0$, and -1 respectively. Meanwhile,

$$L_z(x) = iy$$
$$L_z(y) = -ix$$
$$L_z(z) = 0$$

so in the cartesian basis,

$$[L_z]_{\{x,y,z\}} = \begin{pmatrix} 0 & -i & 0 \\ i & 0 & 0 \\ 0 & 0 & 0 \end{pmatrix}, \tag{2.23}$$

a very different looking matrix. Though the cartesian form of L_z is not usually used in physics texts, it has some very important mathematical properties, as we'll see in Part II of this book. For a preview of these properties, see Problem 2-4.

Exercise 2.11. Compute the matrices of $L_x = -i(y\frac{\partial}{\partial z} - z\frac{\partial}{\partial y})$ and $L_y = -i(z\frac{\partial}{\partial x} - x\frac{\partial}{\partial z})$ acting on $\mathcal{H}_1(\mathbb{R}^3)$ in both the cartesian and spherical harmonic bases.

Before concluding this section we should remark that there is much more one can say about linear operators, particularly concerning eigenvectors, eigenvalues, and diagonalization. Though these topics are relevant for physics, we will not need them in this text and good references for them abound, so we omit them. The interested reader can consult the first chapter of Sakurai [17] for a practical introduction, or Hoffman and Kunze [13] for a thorough discussion.

2.5 Dual Spaces

Another basic construction associated with a vector space, essential for understanding tensors and usually left out of the typical "mathematical methods for physicists" courses, is that of a dual vector. Roughly speaking, a dual vector is an object that eats a vector and spits out a number. This may not sound like a very natural operation, but it turns out that this notion is what underlies bras in quantum mechanics, as well as the raising and lowering of indices in relativity. We'll explore these applications in Sect. 2.7.

Now for the precise definitions. Given a vector space V with scalars C, a ***dual vector*** (or ***linear functional***) on V is a C-valued linear function f on V, where "linear" again means

$$f(cv + w) = cf(v) + f(w). \tag{2.24}$$

(Note that $f(v)$ and $f(w)$ are scalars, so that on the left-hand side of (2.24) the addition takes place in V, whereas on the right side it takes place in C.) The set of all dual vectors on V is called the ***dual space*** of V, and is denoted V^*. It's easily checked that the usual definitions of addition and scalar multiplication and the zero function turn V^* into a vector space over C.

The most basic examples of dual vectors are the following: let $\{e_i\}$ be a (not necessarily finite) basis for V, so that an arbitrary vector v can be written as $v = \sum_{i=1}^n v^i e_i$. Then for each i, we can define a dual vector e^i by

$$\boxed{e^i(v) \equiv v^i.} \tag{2.25}$$

This just says that e^i "picks off" the ith component of the vector v. Note that in order for this to make sense, a *basis has to be specified.*

A key property of any dual vector f is that it is entirely determined by its values on basis vectors. By linearity we have

$$\begin{aligned} f(v) &= f\left(\sum_{i=1}^n v^i e_i\right) \\ &= \sum_{i=1}^n v^i f(e_i) \\ &\equiv \sum_{i=1}^n v^i f_i, \end{aligned} \tag{2.26}$$

where in the last line we have defined

$$\boxed{f_i \equiv f(e_i)},$$

which we unsurprisingly refer to as the ***components*** of f in the basis $\{e_i\}$. To justify this nomenclature, notice that the e^i defined in (2.25) satisfy

$$e^i(e_j) = \delta^i_j. \tag{2.27}$$

If V is finite-dimensional with dimension n, it's then easy to check (by evaluating both sides on basis vectors) that we can write[15]

$$f = \sum_{i=1}^n f_i e^i$$

[15]If V is infinite-dimensional, then this may not work as the sum required may be infinite, and as mentioned before care must be taken in defining infinite linear combinations.

so that the f_i really are the components of f. Since f was arbitrary, this means that the e^i span V^*. In Exercise 2.12 below you will show that the e^i are actually linearly independent, so $\{e^i\}_{i=1\dots n}$ is actually a *basis* for V^*. Because of this and the relation (2.27) we sometimes say that the e^i are ***dual to*** the e_i. Note that we have shown that V and V^* always have the same dimension. We can use the dual basis $\{e^i\} \equiv \mathcal{B}^*$ to write f in components,

$$[f]_{\mathcal{B}^*} = \begin{pmatrix} f_1 \\ f_2 \\ \cdot \\ \cdot \\ \cdot \\ f_n \end{pmatrix}.$$

In terms of the row vector $[f]^T_{\mathcal{B}^*}$ we can write (2.26) as

$$f(v) = [f]^T_{\mathcal{B}^*}[v]_{\mathcal{B}} = [f]^T[v],$$

where in the last equality we dropped the subscripts indicating the bases. Again, we allow ourselves to do this whenever there is no ambiguity about which basis for V we're using, and in all such cases we assume that the basis being used for V^* is just the one dual to the basis for V.

Finally, since $e^i(v) = v^i$ we note that we can alternatively think of the ith component of a vector as the value of e^i on that vector. This duality of viewpoint will crop up repeatedly in the rest of the text.

Exercise 2.12. By carefully working with the definitions, show that the e^i defined in (2.25) and satisfying (2.27) are linearly independent.

Example 2.17. *Dual spaces of* $\mathbb{R}^n$, $\mathbb{C}^n$, $M_n(\mathbb{R})$ *and* $M_n(\mathbb{C})$

Consider the basis $\{e_i\}$ of $\mathbb{R}^n$ and $\mathbb{C}^n$, where e_i is the vector with a 1 in the ith place and 0's everywhere else; this is just the standard basis described in Example 2.8. Now consider the element f^j of V^* which eats a vector in $\mathbb{R}^n$ or $\mathbb{C}^n$ and spits out the jth component; clearly, $f^j(e_i) = \delta^j_i$ so the f^j are just the dual vectors e^j described above. Similarly, for $M_n(\mathbb{R})$ or $M_n(\mathbb{C})$ consider the dual vector f^{ij} defined by $f^{ij}(A) = A_{ij}$; these vectors are clearly dual to the E_{ij} and thus form the corresponding dual basis. While the f^{ij} may seem a little unnatural or artificial, you should note that there is one linear functional on $M_n(\mathbb{R})$ and $M_n(\mathbb{C})$ which is familiar: the ***trace*** functional, denoted Tr and defined by

$$\mathrm{Tr}(A) = \sum_{i=1}^{n} A_{ii}.$$

Can you express Tr as a linear combination of the f^{ij}?

Not Quite Example *Dual space of* $L^2([-a,a])$
We haven't yet properly treated $L^2([-a,a])$ so we clearly cannot yet properly treat its dual, but we would like to point out here that in infinite-dimensions, dual spaces get much more interesting. In finite-dimensions, we saw above that a basis $\{e_i\}$ for V induces a dual basis $\{e^i\}$ for V^*, so in a sense V^* "looks" very much like V. This is not true in infinite dimensions; in this case, we still have linear functionals dual to a given basis, but these may not span the dual space. Consider the case of $L^2([-a,a])$; you can check that $\{e^n\}_{n\in\mathbb{Z}}$ defined by

$$e^n(f(x)) \equiv \frac{1}{2a}\int_{-a}^{a} e^{-i\frac{n\pi x}{a}} f(x)dx$$

satisfy $e^n(e^{i\frac{m\pi x}{a}}) = \delta_n^m$ and are hence dual to $\{e^{i\frac{m\pi x}{a}}\}_{m\in\mathbb{Z}}$. In fact, these linear functionals just eat a function and spit out its nth Fourier coefficient. There are linear functionals, however, that can't be written as a linear combination of the e^i; one such linear functional is the ***Dirac delta functional*** δ, defined by

$$\delta(f(x)) \equiv f(0). \tag{2.28}$$

You are probably instead used to the Dirac delta *function*, which is a "function" $\delta(x)$ with the defining property that

$$\int \delta(x) f(x)dx = f(0) \quad \forall f \in L^2([-a,a]). \tag{2.29}$$

Note the similarity, in that both are used to evaluate arbitrary functions at 0. Later, in Sect. 2.7, we will clarify the nature of the Dirac delta "function" $\delta(x)$ and its relationship to the Dirac delta functional δ, as well as prove that δ can't be written as a linear combination of the e^i.

2.6 Non-degenerate Hermitian Forms

Non-degenerate Hermitian forms, of which the Euclidean dot product, Minkowski metric, and Hermitian scalar product of quantum mechanics are but a few examples, are very familiar to most physicists. We introduce them here not just to formalize their definition but also to make the fundamental but usually unacknowledged connection between these objects and dual spaces.

A ***non-degenerate Hermitian form*** on a vector space V is a C-valued function $(\cdot\,|\,\cdot)$ which assigns to an ordered pair of vectors $v, w \in V$ a scalar, denoted $(v|w)$, having the following properties:

1. $(v|w_1 + cw_2) = (v|w_1) + c(v|w_2)$ (linearity in the second argument)
2. $(v|w) = \overline{(w|v)}$ (***Hermiticity***; the bar denotes complex conjugation)

3. For each $v \neq 0 \in V$, there exists $w \in V$ such that $(v|w) \neq 0$ (non-degeneracy)

Note that conditions 1 and 2 imply that $(cv|w) = \bar{c}(v|w)$, so $(\cdot\,|\,\cdot)$ is ***conjugate-linear*** in the first argument. Also note that for a real vector space, condition 2 implies that $(\cdot\,|\,\cdot)$ is ***symmetric***, i.e. $(v|w) = (w|v)$[16]; in this case, $(\cdot\,|\,\cdot)$ is called a ***metric***. Condition 3 is not immediately intuitive but means that any two nonzero vectors $v, v' \in V$ can always be distinguished by $(\cdot\,|\,\cdot)$; see Box 2.5 for the precise statement of this.

If, in addition to the above three conditions, the Hermitian form obeys

4. $(v|v) > 0$ for all $v \in V, v \neq 0$ (positive-definiteness)

then we say that $(\cdot\,|\,\cdot)$ is an ***inner product***, and a vector space with such a Hermitian form is called an ***inner product space***. In this case we can think of $(v|v)$ as the "length squared" of the vector v, and the notation

$$||v|| \equiv \sqrt{(v|v)}$$

for the length (or ***norm***) of v is sometimes used. Note that condition 4 implies 3 (why?). Our reason for separating condition 4 from the rest of the definition will become clear when we consider the examples.

One very important use of non-degenerate Hermitian forms is to define preferred sets of bases known as ***orthornormal bases***. Such bases $\mathcal{B} = \{e_i\}$ by definition satisfy $(e_i|e_j) = \pm\delta_{ij}$ and are extremely useful for computation, and ubiquitous in physics for that reason. If $(\cdot\,|\,\cdot)$ is positive-definite (hence an inner product), then orthonormal basis vectors satisfy $(e_i|e_j) = \delta_{ij}$ and may be constructed out of arbitrary bases by the Gram–Schmidt process. If $(\cdot|\cdot)$ is not positive-definite, then orthonormal bases may still be constructed out of arbitrary bases, though the process is slightly more involved. See Hoffman and Kunze [13], Sections 8.2 and 10.2 for details.

Box 2.5 *The Meaning of Non-degeneracy*
The non-degeneracy condition given above is somewhat opaque. What does it mean, and what does it have to do with degeneracy? To answer this, suppose that the condition is violated, so that there exists $v^* \in V$ such that $(v^*|w) = 0\ \forall\, w \in V$. Then for any other $v \in V$, the vectors v and $v + v^*$ are indistinguishable by $(\cdot\,|\,\cdot)$, i.e.

$$(v + v^*|w) = (v|w) + (v^*|w) = (v|w) \quad \text{for all } w \in V.$$

It is in this sense that v and $v + v^*$ are *degenerate*. You will show in Exercise 2.14 below that the converse is also true, i.e. that if two vectors in

[16] In this case, $(\cdot\,|\,\cdot)$ is linear in the first argument as well as the second and would be referred to as ***bilinear***.

V are degenerate in this sense then the non-degeneracy condition is violated. Thus, the non-degeneracy condition is exactly what is required to ensure non-degeneracy!

Exercise 2.13. Let $(\cdot|\cdot)$ be an inner product. If a set of nonzero vectors $e_1, \ldots, e_k$ is ***orthogonal***, i.e. $(e_i|e_j) = 0$ when $i \neq j$, show that they are linearly independent. Note that an ***orthonormal*** set (i.e., $(e_i|e_j) = \pm\delta_{ij}$) is just an orthogonal set in which the vectors have unit length.

Exercise 2.14. Let v, v' be two nonzero vectors in V. Show that if $(v|w) = (v'|w) \quad \forall\, w \in W$, then condition 3 above is violated.

Example 2.18. *The dot product (or Euclidean metric) on* $\mathbb{R}^n$

Let $v = (v^1, \ldots, v^n),\ w = (w^1, \ldots, w^n) \in \mathbb{R}^n$. Define $(\cdot|\cdot)$ on $\mathbb{R}^n$ by

$$(v|w) \equiv \sum_{i=1}^{n} v^i w^i .$$

This is sometimes written as $v \cdot w$. You can check that $(\cdot|\cdot)$ is an inner product, and that the standard basis given in Example 2.8 is an orthonormal basis.

Example 2.19. *The Hermitian scalar product on* $\mathbb{C}^n$

Let $v = (v^1, \ldots, v^n),\ w = (w^1, \ldots, w^n) \in \mathbb{C}^n$. Define $(\cdot|\cdot)$ on $\mathbb{C}^n$ by

$$(v|w) \equiv \sum_{i=1}^{n} \bar{v}^i w^i . \tag{2.30}$$

Again, you can check that $(\cdot|\cdot)$ is an inner product, and that the standard basis given in Example 2.8 is an orthonormal basis. Such inner products on complex vector spaces are sometimes referred to as ***Hermitian scalar products*** and are present on every quantum-mechanical vector space. In this example we see the importance of condition 2, manifested in the conjugation of the v^i in (2.30); if that conjugation wasn't there, a vector like $v = (i, 0, \ldots, 0)$ would have $(v|v) = -1$ and $(\cdot|\cdot)$ wouldn't be an inner product.

Exercise 2.15. Let $A, B \in M_n(\mathbb{C})$. Define $(\cdot|\cdot)$ on $M_n(\mathbb{C})$ by

$$(A|B) = \frac{1}{2}\mathrm{Tr}(A^\dagger B). \tag{2.31}$$

Check that this is indeed an inner product by confirming properties 1 through 4 above (if you've been paying close attention, though, you'll see that it's only necessary to check properties 1, 2, and 4). Also check that the basis $\{I, \sigma_x, \sigma_y, \sigma_z\}$ for $H_2(\mathbb{C})$ is orthonormal with respect to this inner product.

Example 2.20. *The Minkowski Metric on 4-D Spacetime*

Consider two vectors[17] $v_i = (x_i, y_i, z_i, t_i) \in \mathbb{R}^4$, $i = 1, 2$. The Minkowski metric, denoted η, is defined to be[18]

$$\eta(v_1, v_2) \equiv x_1x_2 + y_1y_2 + z_1z_2 - t_1t_2. \tag{2.32}$$

η is clearly linear in both its arguments (i.e., it's bilinear) and symmetric, hence satisfies conditions 1 and 2, and you will check condition 3 in Exercise 2.16 below. Notice that for $v = (1, 0, 0, 1)$, $\eta(v, v) = 0$ so η is *not* positive-definite, hence not an inner product. This is why we separated condition 4, and considered the more general non-degenerate Hermitian forms instead of just inner products.

Exercise 2.16. Let $v = (x, y, z, t)$ be an arbitrary nonzero vector in $\mathbb{R}^4$. Show that η is non-degenerate by finding another vector w such that $\eta(v, w) \neq 0$.

We should point out here that the Minkowski metric can be written in components as a matrix, just as a linear operator can. Taking the standard basis $\mathcal{B} = \{e_i\}_{i=1,\ldots,4}$ in $\mathbb{R}^4$, we can define the ***components of*** η, denoted η_{ij}, as

$$\boxed{\eta_{ij} \equiv \eta(e_i, e_j).}$$

Then, just as was done for linear operators, you can check that if we define the ***matrix of*** η ***in the basis*** $\mathcal{B}$, denoted $[\eta]_{\mathcal{B}}$, as the matrix

$$[\eta]_{\mathcal{B}} = \begin{pmatrix} \eta_{11} & \eta_{21} & \eta_{31} & \eta_{41} \\ \eta_{12} & \eta_{22} & \eta_{32} & \eta_{42} \\ \eta_{13} & \eta_{23} & \eta_{33} & \eta_{43} \\ \eta_{14} & \eta_{24} & \eta_{34} & \eta_{44} \end{pmatrix} = \begin{pmatrix} 1 & 0 & 0 & 0 \\ 0 & 1 & 0 & 0 \\ 0 & 0 & 1 & 0 \\ 0 & 0 & 0 & -1 \end{pmatrix}, \tag{2.33}$$

then we can write

$$\eta(v_1, v_2) = [v_2]^T [\eta][v_1] = (x_2, y_2, z_2, t_2) \begin{pmatrix} 1 & 0 & 0 & 0 \\ 0 & 1 & 0 & 0 \\ 0 & 0 & 1 & 0 \\ 0 & 0 & 0 & -1 \end{pmatrix} \begin{pmatrix} x_1 \\ y_1 \\ z_1 \\ t_1 \end{pmatrix} \tag{2.34}$$

as some readers may be used to from computations in relativity. Note that the symmetry of η implies that $[\eta]_{\mathcal{B}}$ is a symmetric matrix for any basis $\mathcal{B}$.

---

[17]These are often called "events" in the physics literature.

[18]We are, of course, arbitrarily choosing the $+ + + -$ signature; we could equally well choose $- - - +$.

Example 2.21. *The Hermitian scalar product on $L^2([-a,a])$*

For $f, g \in L^2([-a,a])$, define

$$(f|g) \equiv \frac{1}{2a}\int_{-a}^{a} \bar{f} g\, dx. \tag{2.35}$$

You can easily check that this defines an inner product on $L^2([-a,a])$, and that $\{e^{i\frac{n\pi x}{a}}\}_{n\in\mathbb{Z}}$ is an orthonormal set. What's more, this inner product turns $L^2([-a,a])$ into a ***Hilbert Space***, which is an inner product space that is ***complete***. The notion of completeness is a technical one, so we will not give its precise definition, but in the case of $L^2([-a,a])$ one can think of it as meaning roughly that a limit of a sequence of square-integrable functions is again square-integrable. Making this precise and proving it for $L^2([-a,a])$ is the subject of real analysis textbooks and far outside the scope of this text,[19] so we'll content ourselves here with just mentioning completeness and noting that it is responsible for many of the nice features of Hilbert spaces, in particular the generalized notion of a basis which we now describe.

Given a Hilbert space $\mathcal{H}$ and an orthonormal (and possibly infinite) set $\{e_i\} \subset \mathcal{H}$, the set $\{e_i\}$ is said to be an ***orthonormal basis for*** $\mathcal{H}$ if

$$(e_i|f) = 0\ \forall i \implies f = 0. \tag{2.36}$$

You can check (see Exercise 2.17 below) that in the finite-dimensional case this definition is equivalent to our previous definition of an orthonormal basis. In the infinite-dimensional case, however, this definition differs substantially from the old one in that we no longer require $\text{Span}\{e_i\} = \mathcal{H}$ (recall that spans only include *finite* linear combinations). Does this mean, though, that we now allow *arbitrary* infinite combinations of the basis vectors? If not, which ones are allowed? For $L^2([-a,a])$, for which $\{e^{i\frac{n\pi x}{a}}\}_{n\in\mathbb{Z}}$ is an orthonormal basis, we mentioned in Example 2.14 that any $f \in L^2([-a,a])$ can be written as

$$f = \sum_{n=-\infty}^{\infty} c_n e^{i\frac{n\pi x}{a}}, \tag{2.37}$$

where

$$\frac{1}{2a}\int_{-a}^{a} |f|^2 dx = \sum_{n=-\infty}^{\infty} |c_n|^2 < \infty. \tag{2.38}$$

(The first equality in (2.38) should be familiar from quantum mechanics and follows from Exercise 2.18 below.) The converse to this is also true, and this is where the

[19] See Rudin [16], for instance, for this and for proofs of all the claims made in this example.

completeness of $L^2([-a,a])$ is essential: if a set of numbers c_n satisfy (2.38), then the series

$$g(x) \equiv \sum_{n=-\infty}^{\infty} c_n e^{i\frac{n\pi x}{a}} \tag{2.39}$$

converges, yielding a square-integrable function g. So $L^2([-a,a])$ is the set of all expressions of the form (2.37), *subject to the condition* (2.38). Now we know how to think about infinite-dimensional Hilbert spaces and their bases: a basis for a Hilbert space is an infinite set whose infinite linear combinations, *together with some suitable convergence condition*, form the entire vector space.

Exercise 2.17. Show that the definition (2.36) of a Hilbert space basis is equivalent to our original definition of an (orthonormal) basis for a *finite*-dimensional inner product space V.

Exercise 2.18. Show that for $f = \sum_{n=-\infty}^{\infty} c_n e^{i\frac{n\pi x}{a}}$ and $g = \sum_{m=-\infty}^{\infty} d_m e^{i\frac{m\pi x}{a}} \in L^2([-a,a])$, their inner product can be written as

$$(f|g) = \sum_{n=-\infty}^{\infty} \bar{c}_n d_n. \tag{2.40}$$

Thus $(\cdot|\cdot)$ on $L^2([-a,a])$ can be viewed as the infinite-dimensional version of the standard Hermitian scalar product on $\mathbb{C}^n$.

Example 2.22. *Various inner products on the space of real polynomials* $P(\mathbb{R})$

All of the examples we've seen of non-degenerate Hermitian forms, with the exception of the Minkowski metric of Example 2.20, are actually positive definite, hence inner products. Furthermore, these inner products probably seem somewhat "natural," in the sense that it's hard to imagine what other kind of inner product one would want to define on, say, $\mathbb{R}^n$ or $\mathbb{C}^n$. This might suggest that inner products are an inherent part of the vector spaces they're defined on, as opposed to additional structure that we impose. After all, when does one come across a single vector space that has multiple different, useful inner products defined on it? In this example we will meet one such vector space, and find that we have met the different inner products on it through our study of differential equations in physics.

The vector space in question is the space $P(\mathbb{R})$ of polynomials in one real variable x, with real coefficients. $P(\mathbb{R})$ is just the set of all functions of the form

$$f(x) = c_0 + c_1 x + c_2 x^2 + \cdots + c_n x^n,$$

where $c_i \in \mathbb{R}\ \forall\, i$ and n is arbitrary. It is straightforward to verify that with the usual addition and scalar multiplication of polynomials, this set is in fact a vector space. Since the degree of a polynomial $f \in P(\mathbb{R})$ is arbitrary, no finite basis exists and this space is infinite-dimensional (for more detail on this, see Exercise 2.19 below).

An obvious basis for this space would be $\mathcal{B} = \{1, x, x^2, x^3, \ldots\}$, but this doesn't necessarily turn out to be the most useful choice. Instead, it's more convenient to choose an orthonormal basis. This leads immediately to a conundrum, however; what inner product do we use to define orthonormality? It turns out that a very useful *family* of inner products is given by

$$(f|g) \equiv \int_a^b f(x)g(x)W(x)\,dx \qquad f, g \in P(\mathbb{R}),$$

where $W(x)$ is a nonnegative ***weight function***. Each different choice of integration range $[a, b]$ and weight function $W(x)$ gives a different inner product, and will yield different orthonormal bases (one can obtain the bases by just applying the Gram–Schmidt process to the basis $\mathcal{B} = \{1, x, x^2, x^3, \ldots\}$). Amazingly, these orthonormal bases turn out to be the various orthogonal polynomials (Legendre, Hermite, Laguerre, etc.) one meets in studying the various differential equations that arise in electrostatics and quantum mechanics!

As an example, let $[a, b] = [-1, 1]$ and $W(x) = 1$. Then in Exercise 2.20 below you will show that applying Gram–Schmidt to the set $S = \{1, x, x^2, x^3\} \subset \mathcal{B}$ yields (up to normalization) the first four ***Legendre polynomials***

$$\begin{aligned}
P_0(x) &= 1\\
P_1(x) &= x\\
P_2(x) &= \frac{1}{2}(3x^2 - 1)\\
P_3(x) &= \frac{1}{2}(5x^3 - 3x).
\end{aligned}$$

Recall that the Legendre polynomials show up in the solutions to the differential equation (2.46), where we make the identification $x = \cos\theta$. Since $-1 \le \cos\theta \le 1$, this explains the range of integration $[a, b] = [-1, 1]$.

One can obtain the other familiar orthogonal polynomials in this way as well. For instance, let $[a, b] = (-\infty, \infty)$ and $W(x) = e^{-x^2}$. Again applying Gram Schmidt yields the first four ***Hermite Polynomials***

$$\begin{aligned}
H_0(x) &= 1\\
H_1(x) &= 2x\\
H_2(x) &= 4x^2 - 2\\
H_3(x) &= 8x^3 - 12x.
\end{aligned}$$

These polynomials arise in the solution to the Schrödinger equation for a one-dimensional harmonic oscillator. In fact, you may have noticed that $W(x) = e^{-x^2}$ is just a Gaussian, which is the form of the ground-state wavefunction for this system.

Note also that the range of integration $[a, b] = (-\infty, \infty)$ corresponds to the range of the position variable, as expected.

Further examples are given in Problem 2-9. Remember, though, that the point here is that a single vector space may support many different inner products, all of which may be of physical relevance!

Exercise 2.19. Verify that $P(\mathbb{R})$ is a (real) vector space. Then show that $P(\mathbb{R})$ is infinite-dimensional by showing that, for any finite set $S \subset P(\mathbb{R})$, there is a polynomial that is not in Span S.

Exercise 2.20. Verify that applying the Gram–Schmidt process to $S = \{1, x, x^2, x^3\}$ with the inner product

$$(f|g) \equiv \int_{-1}^{1} f(x)g(x)\,dx \quad f, g \in P(\mathbb{R})$$

yields, up to normalization, the first four Legendre polynomials, as claimed above. Do the same for the Hermite polynomials, using $[a, b] = (-\infty, \infty)$ and $W(x) = e^{-x^2}$.

2.7 Non-degenerate Hermitian Forms and Dual Spaces

We are now ready to explore the connection between dual vectors and non-degenerate Hermitian forms. This will allow us to put bras and kets, covariant and contravariant indices, and the Dirac delta function all into their proper context. After that we'll be ready to discuss tensors in earnest in Chap. 3.

Given a non-degenerate Hermitian form $(\cdot|\cdot)$ on a finite-dimensional vector space V, we can associate with any $v \in V$ a dual vector $\tilde{v} \in V^*$ defined by

$$\boxed{\tilde{v}(w) \equiv (v|w).} \tag{2.41}$$

This defines a very important map

$$\boxed{\begin{aligned} L : V &\to V^* \\ v &\mapsto \tilde{v}, \end{aligned}}$$

which will crop up repeatedly in this chapter and the next. We'll sometimes write $\tilde{v}$ as $L(v)$ or $(v|\cdot)$, and refer to it as the ***metric dual*** of v.[20]

Now, L is conjugate-linear since for $v = cx + z$, where $v, z, x \in V$,

$$L(v) = (v|\cdot) = (cx + z|\cdot) = \bar{c}(x|\cdot) + (z|\cdot) = \bar{c}L(x) + L(z),$$

[20]The $\cdot$ in the notation $(v|\cdot)$ signifies the slot into which a vector w is to be inserted, yielding the number $(v|w)$.

where in the third equality we used the Hermiticity of our non-degenerate Hermitian form. A word of warning here: as a map from $V \to V^*$, L is conjugate linear, but any $L(v)$ in the range of L is a dual vector, hence a *fully linear* map from $V \to C$.

In Exercise 2.21 below you will show that the non-degeneracy of $(\cdot\,|\,\cdot)$ implies that L is one-to-one and onto, so L is an invertible map from V to V^*. This allows us to identify V with V^*, a fact we will discuss further below.

Exercise 2.21. Use the non-degeneracy of $(\cdot\,|\,\cdot)$ to show that L is one-to-one, i.e. that $L(v) = L(w) \implies v = w$. Combine this with the argument used in Exercise 2.9 to show that L is onto as well.

Box 2.6 e^i *vs.* $L(e_i)$

Suppose we are given basis vectors $\{e_i\}_{i=1\ldots n}$ for V. We then have *two* sets of dual vectors corresponding to these basis vectors e_i : the dual basis vectors e^i defined by $e^i(e_j) = \delta^i_{\ j}$, and the metric duals $L(e_i)$. It's important to understand that, for a given i, **the dual vector e^i is not necessarily the same as the metric dual** $L(e_i)$; in particular, the dual basis vector e^i is defined relative to the whole *basis* $\{e_i\}_{i=1,\ldots,n}$, whereas the metric dual $L(e_i)$ only depends on what e_i is, and doesn't care if we change the other basis vectors. Furthermore, $L(e_i)$ depends on your non-degenerate Hermitian form (that's why we call it a *metric* dual), whereas e^i does not. In fact, we introduced the dual basis vectors e^i in Sect. 2.5, before we even knew what non-degenerate Hermitian forms were!

You may wonder if there are special circumstances when it *is* true that $e^i = L(e_i)$; this is Exercise 2.22.

Exercise 2.22. Given a basis $\{e_i\}_{i=1\ldots n}$, under what circumstances do we have $e^i = \tilde{e}_i$ for all i?

Let's now proceed to some examples, where you'll see that you're already familiar with L from a couple of different contexts.

Example 2.23. *Bras and kets in quantum mechanics*

Let $\mathcal{H}$ be a quantum-mechanical Hilbert space with inner product $(\cdot\,|\,\cdot)$. In Dirac notation, a vector $\psi \in \mathcal{H}$ is written as a ket $|\psi\rangle$ and the inner product $(\psi|\phi)$ is written $\langle\psi|\phi\rangle$. What about bras, written as $\langle\psi|$? What, exactly, are they? Most quantum mechanics texts gloss over their definition, just telling us that they are in 1–1 correspondence with kets and can be combined with kets as $\langle\psi|\phi\rangle$ to get a scalar. We are also told that the correspondence between bras and kets is conjugate-linear, i.e. that the bra corresponding to $c|\psi\rangle$ is $\bar{c}\langle\psi|$. From what we have seen in this section, then, we can conclude the following:

Bras are nothing but dual vectors.

These dual vectors are labeled in the same way as regular vectors, because the map L allows us to identify the two. In short, $\langle\psi|$ is really just $L(\psi)$, or equivalently $(\psi|\cdot)$.

Example 2.24. *Raising and lowering indices in relativity*

Consider $\mathbb{R}^4$ with the Minkowski metric, let $\mathcal{B} = \{e_\mu\}_{\mu=1,\ldots,4}$ and $\mathcal{B}^* = \{e^\mu\}_{\mu=1,\ldots,4}$ be the standard basis and dual basis for $\mathbb{R}^4$ (we use a Greek index to conform with standard physics notation[21]), and let $v = \sum_{\mu=1}^4 v^\mu e_\mu \in \mathbb{R}^4$. What are the components of the dual vector $\tilde{v}$ in terms of the v^μ? Well, as we saw in Sect. 2.5, the components of a dual vector are just given by evaluation on the basis vectors, so

$$\tilde{v}_\mu = \tilde{v}(e_\mu) = (v|e_\mu) = \sum_\nu v^\nu (e_\nu|e_\mu) = \sum_\nu v^\nu \eta_{\nu\mu}. \tag{2.42}$$

In matrices, this reads

$$[\tilde{v}]_{\mathcal{B}*} = [\eta]_{\mathcal{B}}[v]_{\mathcal{B}}$$

so matrix multiplication of a vector by the metric matrix gives the corresponding dual vector in the dual basis. Thus, the map L is implemented in coordinates by $[\eta]$. Now, we mentioned above that L is invertible; what does L^{-1} look like in coordinates? Well, by the above, L^{-1} should be given by matrix multiplication by $[\eta]^{-1}$, the matrix inverse to $[\eta]$. Denoting the components of this matrix by $\eta^{\mu\nu}$ (so that $\eta^{\tau\mu}\eta_{\nu\tau} = \delta^\mu_\nu$) and writing $\tilde{f} \equiv L^{-1}(f)$ where f is a dual vector, we have

$$[\tilde{f}]_{\mathcal{B}} = [\eta]^{-1}{}_{\mathcal{B}*}[f]_{\mathcal{B}*}$$

or in components

$$\tilde{f}^\mu = \sum_\nu \eta^{\nu\mu} f_\nu. \tag{2.43}$$

The expressions in (2.42) and (2.43) are probably familiar to you. In physics one usually works with components of vectors, and in the literature on relativity the numbers v^μ are called the ***contravariant components*** of v and the numbers $v_\mu \equiv \sum_\nu v^\nu \eta_{\nu\mu}$ of (2.42) are referred to as the ***covariant components*** of v. We see now that the contravariant components of a vector are just its usual components, **while its covariant components are actually the components of the associated dual vector** $\tilde{v}$. For a dual vector f, the situation is reversed—the covariant components f_μ are its actual components, and the contravariant components are the components of $\tilde{f}$. Since L allows us to turn vectors into dual vectors and vice-versa, we usually don't bother trying to figure out whether something is "really" a vector or a dual vector; it can be either, depending on which components we use.

The above discussion shows that the familiar process of "raising" and "lowering" indices is just the application of the map L (and its inverse) in components. For an interpretation of $[\eta]^{-1}$ as the matrix of a metric on $\mathbb{R}^{4*}$, see Problem 2-8.

---

[21] As long as we're talking about "standard" physics notation, you should also be aware that in many texts the indices run from 0 to 3 instead of 1 to 4, and in that case the 0th coordinate corresponds to time.

Exercise 2.23. Consider $\mathbb{R}^3$ with the Euclidean metric. Show that the covariant and contravariant components of a vector in an orthonormal basis are identical. This explains why we never bother with this terminology, nor the concept of dual spaces, in basic physics where $\mathbb{R}^3$ is the relevant vector space. Is the same true for $\mathbb{R}^4$ with the Minkowski metric?

Example 2.25. *$L^2([-a,a])$ and its dual*

In our above discussion of the map L we stipulated that V should be finite-dimensional. Why? If you examine the discussion closely, you'll see that the only place where we use the finite-dimensionality of V is in showing that L is onto. Does this mean that L is not necessarily onto in infinite-dimensions? Consider the Dirac delta functional $\delta \in L^2([-a,a])^*$ defined in (2.28). Is there a function $\delta(x) \in L^2([-a,a])$ such that $L(\delta(x)) = \delta$? If so, then we'd have

$$g(0) = \delta(g) = (\delta(x)|g) = \int_{-a}^{a} \delta(x)g(x)dx, \tag{2.44}$$

and then $\delta(x)$ would be nothing but the Dirac delta function defined in (2.29). Does such a square-integrable function exist? If we write g as

$$g(x) = \sum_{n=-\infty}^{\infty} d_n e^{i\frac{n\pi x}{a}},$$

then simply evaluating this at $x = 0$ gives

$$g(0) = \sum_{n=-\infty}^{\infty} d_n \stackrel{?}{=} (\delta(x)|g). \tag{2.45}$$

Comparing this with (2.40) tells us that the function $\delta(x)$ must have Fourier coefficients $c_n = 1$ for all n. Such c_n, however, do not satisfy (2.38), and hence $\delta(x)$ cannot be a square-integrable function. Thus, **there is no Dirac delta function, only the Dirac delta *functional*** δ. This also means that L is not onto in the case of $L^2([-a,a])$. It is so convenient, however, to associate vectors with dual vectors that we pretend that the delta function $\delta(x)$ exists and can be manipulated (say, by differentiation) as an ordinary function. If you look closely, however, you'll find that $\delta(x)$ only crops up in integral expressions, and can always be rewritten in terms of the delta functional. We'll see an example of this in Problem 2-6.

Epilogue: Tiers of Structure in Linear Algebra

In this chapter, as is often done in mathematics, we started with some basic notions (vectors and vector spaces) and gradually built complexity, culminating in the identification of a vector space V and its dual V^* via a non-degenerate Hermitian form on V. Let us step back at this point and survey what we've done, so that we understand the tiers of structure at play.

Before we could define an abstract vector space, we needed a candidate set V. Initially, this set, whatever it was (a set of n-tuples of numbers, or matrices, or functions, etc.), had no structure at all; it was just a set with elements. This is the first level of our hierarchy. To construct a vector space, we needed to combine this set with some notion of addition (really, just some map $+ : V \times V \to V$) along with some notion of scalar multiplication (really, just some map from $C \times V \to V$) that satisfied the axioms of Sect. 2.1. These notions are *additional structure* that one adds to V, and one calls the result a "vector space" only if those specific notions are compatible with the vector space axioms. Vector spaces are the second level of our hierarchy; see Table 2.1.

That last paragraph might seem pedantic. If $V = \mathbb{R}^n$, for instance, do we really consider the notion of addition of vectors to be extra structure? What else could addition be, in this case? Is it really useful or necessary to separate the set V from the obvious addition operation it carries? These questions may be reasonable for $\mathbb{R}^n$, but it's important to note that there are plenty of sets for which no useful or intuitive notion of addition exists. Consider the set

$$S^2 = \{(x, y, z) \in \mathbb{R}^3 \mid x^2 + y^2 + z^2 = 1\},$$

which is, of course, nothing but the 2-D surface of a sphere in $\mathbb{R}^3$ (see Fig. 2.2). This is a perfectly well-defined and intuitive set, but I know of no meaningful way to perform addition on this set. If you wanted to add the North Pole $N = (0, 0, 1)$ to the South Pole $S = (0, 0, -1)$, what should the resulting point of S^2 be? Note that we cannot simply add N to S as vectors in $\mathbb{R}^3$ because the resulting point, the origin O, is not in S^2.

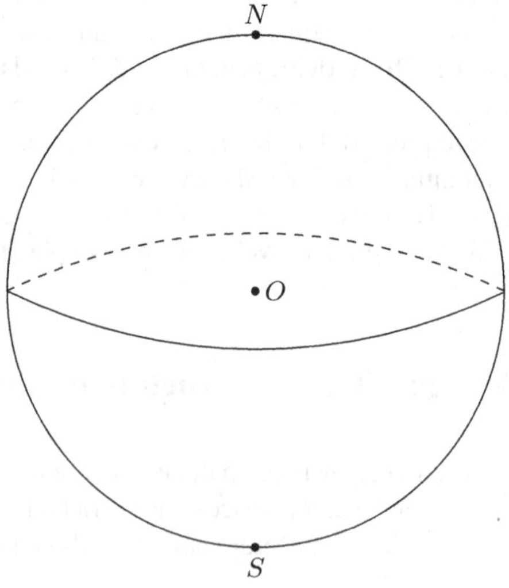

Fig. 2.2 The 2-sphere S^2 with North Pole $N = (0, 0, 1)$, South Pole $S = (0, 0, -1)$, and origin $O = (0, 0, 0)$. Note that $O \notin S^2$

Thus not all sets are cut out to be vector spaces, and even though the operations of addition and scalar multiplication may seem trivial or obvious in the examples of this chapter, they really do confer a significant amount of non-trivial structure on a space. This structure is manifest in all the constructions and definitions one then proceeds to make: linear independence, span, basis sets, linear operators, dual spaces, etc.

Note that none of these notions requires an inner product[22]; the former exist completely independently of the latter. With an inner product, though, one may then (and only then!) speak of orthogonality, length of vectors, and the like. Thus, inner product spaces comprise the third level of our hierarchy, illustrated again in Table 2.1. To be sure, once one introduces this third level of structure, there is then interplay between the levels (as exemplified in Exercise 2.11, which says that orthogonal vectors are linearly independent), but you don't *need* an inner product to know what linear independence means. Furthermore, level three of the hierarchy is rich enough that for a *given* set V with a *given* vector space structure, there can be multiple meaningful notions of an inner product. A good example of this

Table 2.1 Levels of structure in linear algebra. One starts with a set, whose only attendant notion is membership in the set. Including operations of addition and scalar multiplication compatible with the axioms turns a set into a vector space, and then one can speak of linear independence, bases, etc. Endowing the vector space with an inner product then yields notions of orthogonality, length, angles, etc.

Level 1	Set	(Membership)
	↓ Addition, Scalar multiplication	
Level 2	Vector space	(Linear independence, basis, span, etc.)
	↓ Inner product $(\cdot\|\cdot)$	
Level 3	Inner Product Space	(Orthogonality, length, unit vector, etc.)

[22]We could use non-degenerate Hermitian forms here rather than inner products to make a similar point, but will stick with inner products for definiteness.

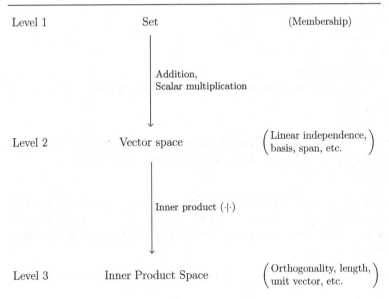

is the vector space $P(\mathbb{R})$ of polynomials of one real variable, which as we saw in Example 2.22 and Problem 2-9 have many different inner products, all of which have direct applications to quantum mechanics!

This means that one cannot ask in general whether the polynomials $2x$ and $4x^2 - 2$ are orthogonal; the question is only meaningful once an inner product on $P(\mathbb{R})$ is specified. Once this is done, however, one can then speak of orthogonality, length, and the angle between vectors—notions which are absent for a vector space without inner product, as well as for a mere set.

Chapter 2 Problems

Note: Problems marked with an "*" tend to be longer, and/or more difficult, and/or more geared towards the completion of proofs and the tying up of loose ends. Though these problems are still worthwhile, they can be skipped on a first reading.

2-1. For each notion below, determine whether it is defined on all vector spaces, or just those with non-degenerate Hermitian forms, or just those with an inner product (i.e., inner product spaces):

a) Unit vector $\hat{\mathbf{x}}$ for any $\mathbf{x} \in V$
b) Basis
c) Linear independence
d) Length or norm
e) Span
f) Orthogonality
g) Linear operator
h) Angle between two vectors v and w, as defined by $\cos\theta \equiv \frac{(v|w)}{\|v\|\,\|w\|}$
i) Dual space
j) Metric dual

2-2. Prove that $L^2([-a,a])$ is closed under addition. You'll need the triangle inequality, as well as the following inequality, valid for all $\lambda \in \mathbb{R}$: $0 \le \int_{-a}^{a}(|f| + \lambda|g|)^2 dx$.

2-3. (*) In this problem we show that $\{r^l Y^l_m\}$ is a basis for $\mathcal{H}_l(\mathbb{R}^3)$, which implies that $\{Y^l_m\}$ is a basis for $\tilde{\mathcal{H}}_l$. We'll gloss over a few subtleties here; for a totally rigorous discussion see Sternberg [19] or our discussion in Chap. 4.

a) Let $f \in \mathcal{H}_l(\mathbb{R}^3)$, and write f as $f = r^l Y(\theta,\phi)$. Then we know that

$$\Delta_{S^2} Y = -l(l+1)Y. \tag{2.46}$$

If you have never done so, use the expression (2.6) for Δ_{S^2} and the expressions (2.17) for the angular momentum operators to show that

$$\Delta_{S^2} = L_x^2 + L_y^2 + L_z^2 \equiv \mathbf{L}^2$$

so that (2.46) says that Y is an eigenfunction of $\mathbf{L}^2$, as expected. The theory of angular momentum[23] then tells us that $\mathcal{H}_l(\mathbb{R}^3)$ has dimension $2l + 1$.

b) Exhibit a basis for $\mathcal{H}_l(\mathbb{R}^3)$ by considering the function $f_0^l \equiv (x + iy)^l$ and showing that

$$L_z(f_0^l) = lf_0^l, \quad L_+(f_0^l) \equiv (L_x + iL_y)(f_0^l) = 0.$$

The theory of angular momentum then tells us that $(L_-)^k f_0^l \equiv f_k^l$ satisfies $L_z(f_k^l) = (l - k) f_k^l$ and that $\{f_k^l\}_{0 \le k \le 2l}$ is a basis for $\mathcal{H}_l(\mathbb{R}^3)$.

c) Writing $f_k^l = r^l Y_{l-k}^l$ we see that Y_m^l satisfies $\mathbf{L}^2 Y_m^l = -l(l+1) Y_m^l$ and $L_z Y_m^l = m Y_m^l$ as expected. Now use this definition of Y_m^l to compute all the spherical harmonics for $l = 1, 2$ and show that this agrees, up to normalization, with the spherical harmonics as tabulated in any quantum mechanics textbook. If you read Example 2.6 and did Exercise 2.2, then all you have to do is compute f_k^1, $0 \le k \le 2$ and f_k^2, $0 \le k \le 4$ and show that these functions agree with the ones given there.

2-4. In discussions of quantum mechanics you may have heard the phrase "angular momentum generates rotations." What this means is that if one takes a component of the angular momentum such as L_z and exponentiates it, i.e. if one considers the operator

$$\begin{aligned}\exp(-i\phi L_z) &\equiv \sum_{n=0}^{\infty} \frac{1}{n!}(-i\phi L_z)^n \\ &= I - i\phi L_z + \frac{1}{2!}(-i\phi L_z)^2 + \frac{1}{3!}(-i\phi L_z)^3 + \dots\end{aligned}$$

(the usual power series expansion for e^x) then one gets the operator which represents a rotation about the z axis by an angle ϕ. Confirm this in one instance by explicitly summing the power series for the operator $[L_z]_{\{x,y,z\}}$ of Example 2.16 to get

$$\exp\left(-i[L_z]_{\{x,y,z\}}\right) = \begin{pmatrix} \cos\phi & -\sin\phi & 0 \\ \sin\phi & \cos\phi & 0 \\ 0 & 0 & 1 \end{pmatrix},$$

the usual matrix for a rotation about the z-axis.

[23] See Sakurai [17] or Gasiorowicz [7] or our discussion in Chap. 4.

2-5. Let V be finite-dimensional, and consider the "double dual" space $(V^*)^*$. Define a map

$$J : V \to (V^*)^*$$
$$v \mapsto J_v,$$

where J_v acts on $f \in V^*$ by

$$J_v(f) \equiv f(v).$$

Show that J is linear, one-to-one, and onto. This means that we can identify $(V^*)^*$ with V itself, and so the process of taking duals repeats itself after two iterations. Note that no non-degenerate Hermitian form was involved in the definition of J!

2-6. Consider a linear operator A on a vector space V. We can define a linear operator on V^* called the ***transpose*** of A and denoted by A^T as follows:

$$(A^T(f))(v) \equiv f(Av) \qquad \text{where } v \in V,\ f \in V^*.$$

a) If $\mathcal{B}$ is a basis for V and $\mathcal{B}^*$ the corresponding dual basis, show that

$$[A^T]_{\mathcal{B}^*} = [A]^T_{\mathcal{B}}.$$

Thus the transpose of a matrix really has meaning; it's the matrix representation of the transpose of the linear operator represented by the original matrix!

b) Consider the linear operator $\frac{d}{dx}$ on $L^2([-a,a])$, as well as the Dirac delta functional $\delta \in L^2([-a,a])^*$ defined in (2.28). Show that

$$\left(\frac{d}{dx}^T \delta\right)(f) = \frac{df}{dx}(0) \quad \forall\, f \in L^2([-a,a]).$$

This is the sense in which one may "differentiate the delta function."

2-7. a) Let A be a linear operator on a finite-dimensional, real or complex vector space V with inner product $(\cdot\,|\,\cdot)$. Using the transpose A^T from the previous problem, as well as the map $L : V \to V^*$ defined in Sect. 2.7, we can construct a *new* linear operator $A^\dagger$; this is known as the ***Hermitian adjoint*** of A, and is defined as

$$A^\dagger \equiv L^{-1} \circ A^T \circ L : V \to V. \tag{2.47}$$

Show that $A^\dagger$ satisfies

$$(A^\dagger v|w) = (v|Aw), \tag{2.48}$$

which is the equation that is usually taken as the definition of the adjoint operator. Use this to show that $(A^\dagger)^\dagger = A$, and use that to show that

$$(Av|w) = (v|A^\dagger w). \tag{2.49}$$

The advantage of our definition (2.47) is that it gives an interpretation to $A^\dagger$; it's the operator you get by transplanting A^T, which originally acts on V^*, over to V via L. Note that $A^\dagger$ is defined using the inner product $(\cdot\,|\,\cdot)$, and so combines information from both A and $(\cdot\,|\,\cdot)$.

b) Show that in an orthonormal basis $\{e_i\}_{i=1\ldots n}$, $[A^\dagger] = [A]^\dagger$, where the dagger outside the brackets denotes the usual conjugate transpose of a matrix (if V is a real vector space, then the dagger outside the brackets will reduce to just the transpose). You may want to prove and use the fact $A_j{}^i = e^i(Ae_j)$.

c) If A satisfies $A = A^\dagger$, A is then said to be ***self-adjoint*** or ***Hermitian***. Since $A^\dagger$ is defined with respect to the inner product $(\cdot\,|\,\cdot)$, self-adjointness indicates a certain compatibility between A and $(\cdot\,|\,\cdot)$. Show that even when V is a complex vector space, any eigenvalue of A must be real.

d) In part b) you showed that *in an orthonormal basis* the matrix of a Hermitian operator is a Hermitian matrix. Is this necessarily true in a non-orthonormal basis? (*Hint: If you think this is true, you should prove it. If you think it isn't, you should find a counterexample. A simple candidate counterexample would be the matrix of the L_z operator of Example 2.16 in a non-orthonormal basis such as* $\{(1,0,0), (1,1,0), (0,0,1)\}$.)

2-8. Let g be a non-degenerate bilinear form on a vector space V (we have in mind the Euclidean metric on $\mathbb{R}^3$ or the Minkowski metric on $\mathbb{R}^4$). Pick an arbitrary (not necessarily orthonormal) basis, let $[g]^{-1}$ be the matrix inverse of $[g]$ in this basis, and write $g^{\mu\nu}$ for the components of $[g]^{-1}$. Also let $f, h \in V^*$. Define a non-degenerate bilinear form $\tilde{g}$ on V^* by

$$\tilde{g}(f, h) \equiv g(\tilde{f}, \tilde{h}),$$

where $\tilde{f} = L^{-1}(f)$ as in Example 2.24. Show that

$$\tilde{g}^{\mu\nu} \equiv \tilde{g}(e^\mu, e^\nu) = g^{\mu\nu}$$

so that $[g]^{-1}$ is truly a matrix representation of a non-degenerate bilinear form on V^*.

2-9. This problem builds on Example 2.22 and further explores different bases for the vector space $P(\mathbb{R})$, the polynomials in one variable x with real coefficients.

a) Compute the matrix corresponding to the operator $\frac{d}{dx} \in \mathcal{L}(P(\mathbb{R}))$ with respect to the basis $\mathcal{B} = \{1, x, x^2, x^3, \ldots\}$.
b) Consider the inner product

$$(f|g) \equiv \int_0^\infty f(x)g(x)e^{-x}\,dx$$

on $P(\mathbb{R})$. Apply the Gram–Schmidt process to the set $S = \{1, x, x^2, x^3\} \subset \mathcal{B}$ to get (up to normalization) the first four ***Laguerre Polynomials***

$$\begin{aligned} L_0(x) &= 1 \\ L_1(x) &= -x + 1 \\ L_2(x) &= \frac{1}{2}(x^2 - 4x + 2) \\ L_3(x) &= \frac{1}{6}(-x^3 + 9x^2 - 18x + 6). \end{aligned}$$

These polynomials arise as solutions to the radial part of the Schrödinger equation for the Hydrogen atom. In this case x is interpreted as a radial variable, hence the range of integration $(0, \infty)$.

Chapter 3
Tensors

Now that we're familiar with vector spaces we can finally approach the main subject of Part I, tensors. We'll give the modern component-free definition, from which will follow the usual transformation laws that used to *be* the definition. We'll then introduce the tensor product and apply it liberally in both classical and quantum physics, before specializing to its symmetric and antisymmetric variants.

As in the previous chapter, the mathematics we'll learn will unify and hopefully illuminate many disparate and enigmatic topics. These include more mathematical objects such as the cross product, determinant, and pseudovectors, as well as physical constructions such as entanglement, addition of angular momenta, and multipole moments.

Also, in this chapter we'll treat mostly finite-dimensional vector spaces and ignore the complications and technicalities that arise in the infinite-dimensional case. In the few examples where we apply our results to infinite-dimensional spaces, you should rest assured that these applications are legitimate (if not explicitly justified), and that rigorous justification can be found in the literature.

Box 3.1 *Einstein Summation Convention*

From here on out we will employ the ***Einstein summation convention***, which is that whenever an index is repeated in an expression, once as a superscript and once as a subscript, then summation over that index is implied. Thus an expression like $v = \sum_{i=1}^{n} v^i e_i$ becomes $v = v^i e_i$, and $Te_j = \sum_{i=1}^{n} T_j{}^i e_i$ (where T is a linear operator) becomes

$$Te_j = T_j{}^i e_i. \tag{3.1}$$

Any index that is summed over, either implicitly (via the Einstein convention) or explicitly (via a summation sign) is referred to as a ***dummy index***. In contrast, any index which is not summed over is referred to as a ***free index***. In both examples above i is a dummy index, and in (3.1) j is a free index.

N. Jeevanjee, *An Introduction to Tensors and Group Theory for Physicists*,
DOI 10.1007/978-3-319-14794-9_3

When writing equations that hold in arbitrary bases or coordinates, *free indices must always appear exactly once on either side, in the same position (upper or lower)*. This is very different from how dummy indices appear in such equations, where (since they are indices of summation) it's perfectly legal for them to appear on one side only. This differing behavior of free and dummy indices is evident in (3.1).

We'll comment further on the Einstein convention in Sect. 3.2.

3.1 Definition and Examples

Recall that in Chap. 1 we heuristically defined a rank r tensor as a multilinear function that eats r vectors and produces a number. We now reiterate this definition, but generalize it so that tensors can eat r vectors as well as s dual vectors. We thus define a ***tensor of type*** (r, s) on a vector space V as a C-valued function T on

$$\underbrace{V \times \cdots \times V}_{r \text{ times}} \times \underbrace{V^* \times \cdots \times V^*}_{s \text{ times}}$$

which is linear in each argument, i.e.

$$\begin{aligned} &T(v_1 + cw, v_2, \ldots, v_r, f_1, \ldots, f_s) \\ &\quad = T(v_1, \ldots, v_r, f_1, \ldots, f_s) \\ &\qquad + cT(w, v_2, \ldots, f_1, \ldots, f_s) \end{aligned}$$

and similarly for all the other arguments. This property is called ***multilinearity***. Note that dual vectors are $(1, 0)$ tensors, and that vectors can be viewed as $(0, 1)$ tensors as follows:

$$v(f) \equiv f(v) \quad \text{where} \quad v \in V, f \in V^*. \tag{3.2}$$

Similarly, linear operators can be viewed as $(1, 1)$ tensors as

$$A(v, f) \equiv f(Av). \tag{3.3}$$

We take $(0, 0)$ tensors to be scalars, as a matter of convention. You will show in Exercise 3.1 below that the set of all tensors of type (r, s) on a vector space V, denoted $\mathcal{T}_s^r(V)$ or just $\mathcal{T}_s^r$, form a vector space. This should not come as much of a surprise since we already know that vectors, dual vectors, and linear operators all form vector spaces. Also, just as linearity implies that dual vectors and linear operators are determined by their values on the basis vectors, multilinearity implies the same thing for general tensors. To see this, let $\{e_i\}_{i=1\dots n}$ be a basis for V and $\{e^i\}_{i=1\dots n}$ the corresponding dual basis. Then, denoting the ith component of the vector v_p as v_p^i and the jth component of the dual vector f_q as f_{qj}, repeated application of multilinearity gives (see Box 3.2 for help with the notation)

$$\begin{aligned} T(v_1,\dots,v_r,f_1,\dots,f_s) &= v_1^{i_1}\dots v_r^{i_r} f_{1j_1}\dots f_{sj_s} T(e_{i_1},\dots,e_{i_r},e^{j_1},\dots,e^{j_s}) \\ &\equiv v_1^{i_1}\dots v_r^{i_r} f_{1j_1}\dots f_{sj_s} T_{i_1,\dots,i_r}{}^{j_1\dots j_s}, \end{aligned} \tag{3.4}$$

where, as before, the numbers

$$\boxed{T_{i_1\dots i_r}{}^{j_1\dots j_s} \equiv T(e_{i_1},\dots,e_{i_r},e^{j_1},\dots,e^{j_s})} \tag{3.5}$$

are referred to as the ***components of*** T in the basis $\{e_i\}_{i=1\dots n}$. You should check that this definition of the components of a tensor, when applied to vectors, dual vectors, and linear operators, agrees with the definitions given earlier. Also note that (3.5) gives us a concrete way to think about the components of tensors: they are the *values of the tensor on the basis vectors.*

Exercise 3.1. By choosing suitable definitions of addition and scalar multiplication, show that $\mathcal{T}_s^r(V)$ is a vector space.

Box 3.2 *Making Sense of Tensor Indices*
The profusion of subscripts, superscripts, and indices in equations like (3.4) can be quite intimidating at first. Normally, indices like i don't have subscripts on them. For instance, if we wrote (3.4) for a (2,0) tensor we'd just have

$$T(v,w) = v^i w^j T_{ij}$$

which is a fairly benign looking equation. When treating a general tensor, however, we must accommodate *arbitrary* numbers r, s of vector and dual vector arguments. If we have r vectors $v_1, v_2, \dots, v_r$, then we'll need r dummy indices to label their components, and since r is indefinite it's impractical to choose our indices as different Roman letters i, j, k, etc. Instead, we adopt index subscripts and take $i_1, i_2, \dots, i_r$ as the indices for $v_1, v_2, \dots, v_r$. Understanding this switch in notation should help in untangling the indices in equations such as (3.4).

If we have a non-degenerate bilinear form on V, then we may change the type of T by precomposing with the map L or L^{-1}. If T is of type (1,1) with components $T_i{}^j$, for instance, then we may turn it into a tensor $\tilde{T}$ of type (2,0) by defining $\tilde{T}(v,w) = T(v,L(w))$. This corresponds to lowering the second index, and we write the components of $\tilde{T}$ as T_{ij}, omitting the tilde since the fact that we lowered the second index implies that we precomposed with L.[1] This is in accord with the conventions in relativity, where given a vector $v \in \mathbb{R}^4$ we write v_μ for the components of $\tilde{v}$ when we should really write $\tilde{v}_\mu$. From this point on, if we have a non-degenerate bilinear form on a vector space, then we permit ourselves to raise and lower indices at will and without comment. In such a situation we often don't discuss the type of a tensor, speaking instead of its ***rank***, equal to $r + s$, which obviously doesn't change as we raise and lower indices.

Example 3.1. *Linear operators in quantum mechanics*

Thinking about linear operators as $(1, 1)$ tensors may seem a bit strange, but in fact this is what one does in quantum mechanics all the time! Given an operator H on a quantum-mechanical Hilbert space spanned by orthonormal vectors $\{e_i\}$ (which in Dirac notation we would write as $\{|i\rangle\}$), we usually write $H|i\rangle$ for $H(e_i)$, $\langle j|i\rangle$ for $\tilde{e}_j(e_i) = (e_j|e_i)$, and $\langle j|H|i\rangle$ for $(e_j|He_i)$. Thus, (3.3) would tell us that (using orthonormal basis vectors instead of arbitrary vectors)

$$\begin{aligned} H_i{}^j &= H(e_i, e^j) \\ &= e^j(He_i) \\ &= \langle j|H|i\rangle, \end{aligned}$$

where we converted to Dirac notation in the last equality to obtain the familiar quantum-mechanical expression for the components of a linear operator. These components are often referred to as ***matrix elements***, since when we write operators as matrices the elements of the matrices are just the components arranged in a particular fashion, as in (2.22).

Example 3.2. *The Levi–Civita Tensor*

Note: If you read Chap. 1, then the material in the next two examples will be familiar.

Consider the $(3, 0)$ ***Levi–Civita tensor*** ϵ on $\mathbb{R}^3$ defined by

$$\epsilon(u, v, w) \equiv (u \times v) \cdot w, \quad u, v, w \in \mathbb{R}^3. \tag{3.6}$$

You will check below that ϵ really is multilinear, hence a tensor. It is well known from vector calculus that $(u \times v) \cdot w$ is the (oriented) volume of a parallelepiped spanned by u, v, and w (see Fig. 3.1 below), so one can think of the Levi–Civita tensor as a kind of "volume operator" which eats three vectors and spits out the volume that they span.

What about the components of the Levi–Civita tensor? If $\{e_1, e_2, e_3\}$ is the standard basis for $\mathbb{R}^3$, then (3.6) yields

[1]The desire to raise and lower indices at will is one reason why we *offset* tensor indices and write $T(e_i, e^j)$ as $T_i{}^j$, rather than T_i^j. Raising and lowering indices on the latter would be ambiguous!

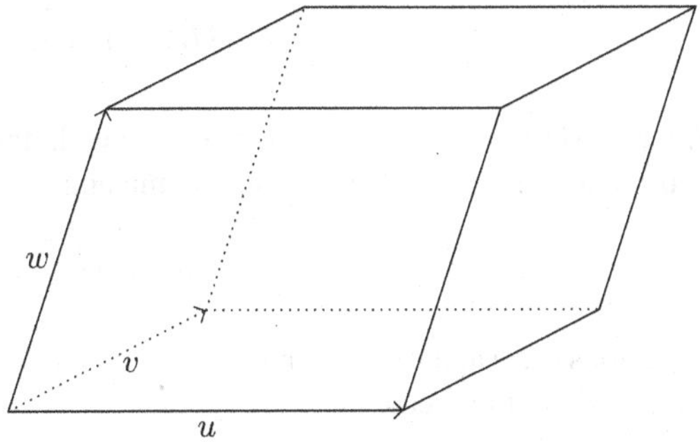

Fig. 3.1 The parallelepiped spanned by u, v, and w

$$\begin{aligned}\epsilon_{ijk} &= \epsilon(e_i, e_j, e_k) \\ &= (e_i \times e_j) \cdot e_k \\ &= \bar{\epsilon}_{ijk},\end{aligned}$$

where $\bar{\epsilon}_{ijk}$ is the usual Levi–Civita symbol. Thus the Levi–Civita symbol represents the components of an actual tensor, the Levi–Civita tensor! Furthermore, keeping in mind the interpretation of the Levi–Civita tensor as a volume operator, as well as the fact that the components ϵ_{ijk} are just the values of the tensor on the basis vectors, then we find that the usual definition of the Levi–Civita symbol,

$$\bar{\epsilon}_{ijk} = \begin{cases} +1 \text{ if } \{ijk\} = \{1,2,3\},\ \{2,3,1\},\ \text{or } \{3,1,2\} \\ -1 \text{ if } \{ijk\} = \{3,2,1\},\ \{1,3,2\},\ \text{or } \{2,1,3\} \\ 0 \quad \text{otherwise,} \end{cases}$$

is just telling us, for instance, that a parallelepiped spanned by $\{e_1, e_2, e_3\}$ has oriented volume +1!

Exercise 3.2. Verify that the Levi–Civita tensor as defined by (3.6) really is multilinear.

Exercise 3.3. Using the Euclidean metric on $\mathbb{R}^3$ we can raise the last index on ϵ to get a (2,1) tensor $\tilde{\epsilon}$, where

$$\tilde{\epsilon}(v, w, f) \equiv \epsilon(v, w, L^{-1}(f)) \quad \forall\, v, w \in V,\ f \in V^*.$$

Now, just as we can interpret a (1,1) tensor as a map from $V \to V$, we can interpret the (2,1) tensor $\tilde{\epsilon}$ as a map $\alpha_\epsilon : V \times V \to V$ via the equation $\tilde{\epsilon}(v, w, f) \equiv f(\alpha_\epsilon(v, w))$. Compute α_ϵ. It should look familiar!

Example 3.3. *The Moment of Inertia Tensor*

The moment of inertia tensor, denoted $\mathcal{I}$, is the symmetric (2,0) tensor on $\mathbb{R}^3$ which, when evaluated on the angular velocity vector, yields the kinetic energy of a rigid body, i.e.

$$\frac{1}{2}\mathcal{I}(\omega, \omega) = KE. \tag{3.7}$$

Alternatively we can raise an index on $\mathcal{I}$ and define it to be the linear operator which eats the angular velocity and spits out the angular momentum, i.e.

$$\mathbf{L} = \mathcal{I}\omega. \tag{3.8}$$

Equations (3.7) and (3.8) are most often seen in components (referred to a cartesian basis), where they read

$$\begin{aligned} KE &= \tfrac{1}{2}[\omega]^T[\mathcal{I}][\omega] \\ [\mathbf{L}] &= [\mathcal{I}][\omega]. \end{aligned}$$

Note that since we raise and lower indices with an inner product and usually use orthornormal bases, the components of $\mathcal{I}$ when viewed as a (2,0) tensor and when viewed as a (1,1) tensor are the same, cf. Exercise 2.23.

Example 3.4. *Multipole moments*

It is a standard result from electrostatics that the scalar potential $\Phi(\mathbf{r})$ of a charge distribution $\rho(\mathbf{r}')$ localized around the origin in $\mathbb{R}^3$ can be expanded in a Taylor series as[2]

$$\Phi(\mathbf{r}) = \frac{1}{4\pi}\left[\frac{Q_0}{r} + \frac{Q_1(\mathbf{r})}{r^3} + \frac{1}{2!}\frac{Q_2(\mathbf{r},\mathbf{r})}{r^5} + \frac{1}{3!}\frac{Q_3(\mathbf{r},\mathbf{r},\mathbf{r})}{r^7}\ldots\right],$$

where the Q_i are ith rank tensors known as the ***multipole moments*** of the charge distribution $\rho(\mathbf{r}')$. The first few multipole moments are familiar to most physicists: the first, Q_0, is just the total charge or ***monopole moment*** of the charge distribution and is given by

$$Q_0 = \int \rho(\mathbf{r}')\, d^3r'.$$

The second, Q_1, is a dual vector known as the ***dipole moment*** (often denoted as $\mathbf{p}$), which has components

$$p_i = \int x_i'\, \rho(\mathbf{r}')\, d^3r'.$$

The third multipole moment, Q_2, is known as the ***quadrupole moment*** and has components given by

---

[2]Here and below we set all physical constants such as c and ϵ_0 equal to 1.

$$Q_{ij} = \int (3x'_i x'_j - r'^2 \delta_{ij})\, d^3 r'.$$

Notice that the Q_{ij} are symmetric in i and j, and that $\sum_i Q_{ii} = 0$. Analogous properties hold for the higher order multipole moments as well (i.e., the ***octopole moment*** Q_3 has components Q_{ijk} which are totally symmetric and which satisfy $\sum_i Q_{iij} = \sum_i Q_{iji} = \sum_i Q_{jii} = 0$). We will explain these curious features of the Q_l at the end of this chapter.

Example 3.5. *Metric Tensors*

We met the Euclidean metric on $\mathbb{R}^n$ in Example 2.18 and the Minkowski metric on $\mathbb{R}^4$ in Example 2.20, and it's easy to verify that both are (2,0) tensors (why isn't the Hermitian scalar product of Example 2.19 included?). We also have the inverse metrics, defined in Problem 2-8, and you can verify that these are (0,2) tensors.

Exercise 3.4. Show that for a metric g on V,

$$g_i{}^j = \delta_i{}^j,$$

so the $(1,1)$ tensor associated with g (via $g!$) is just the identity operator. You will need the components g^{ij} of the inverse metric, defined in Problem 2-8.

3.2 Change of Basis

Now we are in a position to derive the usual transformation laws that historically were taken as the *definition* of a tensor. Suppose we have a vector space V and two bases for V, $\mathcal{B} = \{e_i\}_{i=1\dots n}$ and $\mathcal{B}' = \{e_{i'}\}_{i=1\dots n}$. Since $\mathcal{B}$ is a basis, each of the $e_{i'}$ can be expressed as

$$e_{i'} = A^j_{i'}\, e_j \tag{3.9}$$

for some numbers $A^j_{i'}$. Likewise, there exist numbers $A^{j'}_i$ (note that here the upper index is primed) such that

$$e_i = A^{j'}_i\, e_{j'}.$$

We then have

$$e_i = A^{j'}_i\, e_{j'} = A^{j'}_i\, A^k_{j'} e_k \tag{3.10}$$

and can then conclude that

$$A_i^{j'} A_{j'}^k = \delta_i^k. \tag{3.11}$$

Considering (3.10) with the primed and unprimed indices switched also yields

$$A_{i'}^j A_j^{k'} = \delta_{i'}^{k'}, \tag{3.12}$$

so, in a way, $A_i^{j'}$ and $A_{i'}^j$ are inverses of each other. Notice that $A_{i'}^j$ and $A_i^{j'}$ are *not* to be interpreted as the components of tensors, as their indices refer to different bases.[3] How do the corresponding dual bases transform? Let $\{e^i\}_{i=1\ldots n}$ and $\{e^{i'}\}_{i=1\ldots n}$ be the bases dual to $\mathcal{B}$ and $\mathcal{B}'$. Then the components of $e^{i'}$ with respect to $\{e^i\}_{i=1\ldots n}$ are

$$e^{i'}(e_j) = e^{i'}(A_j^{k'} e_{k'}) = A_j^{k'} \delta_{k'}^{i'} = A_j^{i'}, \tag{3.13}$$

i.e.

$$e^{i'} = A_j^{i'} e^j. \tag{3.14}$$

Likewise,

$$e^i = A_{j'}^i e^{j'}. \tag{3.15}$$

Notice how well the Einstein summation convention and our convention for priming indices work together in the transformation laws.[4]

Now we are ready to see how the components of tensors transform. Before proceeding to the general case, let's warm up by considering a (1,1) tensor with components $T_i{}^j$. Its components in the primed basis are given by

$$\begin{aligned} T_{i'}{}^{j'} &= T(e_{i'}, e^{j'}) && \text{by (3.5)} \\ &= T(A_{i'}^k e_k, A_l^{j'} e^l) && \text{by (3.9) and (3.14)} \\ &= A_{i'}^k A_l^{j'} T(e_k, e^l) && \text{by multilinearity} \\ &= A_{i'}^k A_l^{j'} T_k{}^l. && \end{aligned} \tag{3.16}$$

[3] This is also why we wrote the upper index directly above the lower index, rather than with a horizontal offset as is customary for tensors. For more about these numbers and a possible interpretation, see the beginning of the next section.

[4] This convention, in which we prime the *indices* rather than the objects themselves, is sometimes known as the ***Schouten convention***; for more on this, see Battaglia and George [3].

Equation (3.16) may look familiar to you. Note that we have *derived* it, rather than taking it as the definition of a (1,1) tensor. Note also how there are two factors of the As, one for each argument of T. Thus the rank of the tensor determines the number of factors of A that occur in the transformation law. Also notice that the free indices i' and j' appear in the same position (lower and upper, respectively) on both sides of (3.16), in accordance with the discussion in Box 3.1.

We can now generalize (3.16) to an arbitrary (r, s) tensor T. We'll proceed exactly as above, except that we'll need r covariant indices $i_1, \ldots, i_r$ and s contravariant indices $j_1, \ldots, j_s$ as discussed in Box 3.2. We have

$$\begin{aligned} T_{i'_1,\ldots,i'_r}{}^{j'_1\ldots j'_s} &= T(e_{i'_1}, \ldots, e_{i'_r}, e^{j'_1}, \ldots, e^{j'_s}) \\ &= T(A^{k_1}_{i'_1} e_{k_1}, \ldots, A^{k_r}_{i'_r} e_{k_r}, A^{j'_1}_{l_1} e^{l_1}, \ldots, A^{j'_s}_{l_s} e^{l_s}) \\ &= A^{k_1}_{i'_1} \ldots A^{k_r}_{i'_r} A^{j'_1}_{l_1} \ldots A^{j'_s}_{l_s} T(e_{k_1}, \ldots, e_{k_r}, e^{l_1}, \ldots, e^{l_s}) \\ &= \boxed{A^{k_1}_{i'_1} \ldots A^{k_r}_{i'_r} A^{j'_1}_{l_1} \ldots A^{j'_s}_{l_s} T_{k_1\ldots k_r}{}^{l_1\ldots l_s}.} \end{aligned} \tag{3.17}$$

Equation (3.17) is the standard general tensor transformation law, which as remarked above is taken as the *definition* of a tensor in much of the physics literature; here, we have *derived* it as a consequence of our definition of a tensor as a multilinear function on V and V^*. The two are equivalent, however, as you will check in Exercise 3.5 below.

With the general transformation law in hand, we'll now look at specific types of tensors and derive their *matrix* transformation laws; to this end, it will be useful to introduce the matrices

$$\boxed{A = \begin{pmatrix} A^{1'}_1 & A^{1'}_2 & \ldots & A^{1'}_n \\ A^{2'}_1 & A^{2'}_2 & \ldots & A^{2'}_n \\ \vdots & \vdots & \vdots & \vdots \\ A^{n'}_1 & A^{n'}_{2'} & \ldots & A^{n'}_n \end{pmatrix}, \quad A^{-1} = \begin{pmatrix} A^1_{1'} & A^1_{2'} & \ldots & A^1_{n'} \\ A^2_{1'} & A^2_{2'} & \ldots & A^2_{n'} \\ \vdots & \vdots & \vdots & \vdots \\ A^n_{1'} & A^n_{2'} & \ldots & A^n_{n'} \end{pmatrix}.} \tag{3.18}$$

By virtue of (3.11) and (3.12), these matrices satisfy

$$AA^{-1} = A^{-1}A = I \tag{3.19}$$

as our notation suggests.

Exercise 3.5. Consider a function which assigns to a basis $\{e_i\}_{i=1\ldots n}$ a set of numbers $\{T_{k_1\ldots k_r}{}^{l_1\ldots l_s}\}$ which transform according to (3.17) under a change of basis. Use this assignment to define a multilinear function T of type (r, s) on V, and be sure to check that your definition is basis independent (i.e., that the value of T does not depend on which basis $\{e_i\}_{i=1\ldots n}$ you choose).

Fig. 3.2 The standard basis $\mathcal{B}$ and a new one gray $\mathcal{B}'$ obtained by rotation through an angle ϕ

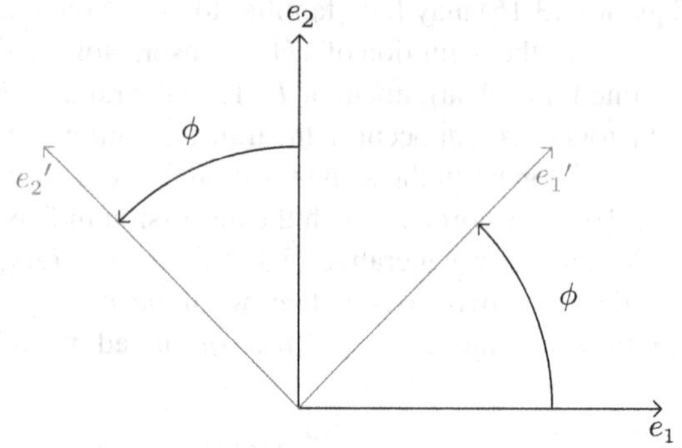

Example 3.6. *Change of basis matrix for a 2-D rotation*

As a simple illustration of the formalism, consider the standard basis $\mathcal{B}$ in $\mathbb{R}^2$ and another basis $\mathcal{B}'$ obtained by rotating $\mathcal{B}$ by an angle θ. This is illustrated in Fig. 3.2.

By inspection we have

$$\begin{aligned} e_{1'} &= \cos\theta\, e_1 + \sin\theta\, e_2 \\ e_{2'} &= -\sin\theta\, e_1 + \cos\theta\, e_2 \end{aligned} \tag{3.20}$$

and so by (3.9) we have

$$\begin{aligned} A^1_{1'} &= \cos\theta \qquad & A^1_{2'} &= -\sin\theta \\ A^2_{1'} &= \sin\theta \qquad & A^2_{2'} &= \cos\theta. \end{aligned}$$

Equation (3.18) then tells us that

$$A^{-1} = \begin{pmatrix} \cos\theta & -\sin\theta \\ \sin\theta & \cos\theta \end{pmatrix}.$$

The numbers $A^{j'}_i$ and the corresponding matrix A can be computed by either inverting A^{-1} or equivalently by inverting the system (3.20) and proceeding as above. □

Example 3.7. *Vectors and Dual Vectors*

Given a vector v [considered as a (0, 1) tensor as per (3.2)], Eq. (3.17) tells us that its components transform as

$$v^{i'} = A^{i'}_j v^j \tag{3.21}$$

while the components of a dual vector f transform as

$$f_{i'} = A^{j}_{i'} f_j. \tag{3.22}$$

Notice that the components of v transform with the $A^{i'}_{j}$ whereas the basis vectors transform with the $A^{j}_{i'}$, so the components of a vector obey the law opposite ('contra') to the basis vectors. This is the origin of the term "contravariant." Note also that the components of a dual vector transform in the *same* way as the basis vectors, hence the term "covariant".[5] It makes sense that the basis vectors and the components of a vector should transform oppositely; v exists independently of any basis for V and shouldn't change under a change of basis, so if the e_i change one way, the v^i should change oppositely. Similar remarks apply to dual vectors.

Box 3.3 *More on the Einstein Summation Convention*
We can now explain a little bit more about the Einstein summation convention. We knew ahead of time that the components of dual vectors would transform like basis vectors, so we gave them both lower indices. We also knew that the components of vectors would transform like dual basis vectors, so we gave them both upper indices. Since the two transformation laws are opposite, we know (see below) that a summation over an upper index and lower index will yield an object that does not transform at all, so **the summation represents an object or a process that doesn't depend upon a choice of basis**. For instance, the expression $v^i e_i$ represents the vector v which is defined without reference to any basis, and the expression $f_i v^i$ is just $f(v)$, the action of the functional f on the vector v, which is also defined without reference to any basis. Processes such as these are so important and ubiquitous that it becomes very convenient to omit the summation sign for repeated upper and lower indices, and we thus have the summation convention.

What if we have two repeated upper indices or two repeated lower indices? In these cases we sometimes require summation and sometimes not, so we choose to indicate summation explicitly. An example where summation is required is (3.26) below; an example where summation is not desired is given by Euler's equations of rigid body motion,[6] which are sometimes written collectively as

$$\tau_i = \lambda_i \dot{\omega}_i + \sum_{j,k} \epsilon_{ijk} \lambda_k \omega_j \omega_k, \tag{3.23}$$

where $\boldsymbol{\tau}$ is the torque, λ_i are the eigenvalues of the moment of inertia tensor (the so-called ***principal moments of inertia***), and $\boldsymbol{\omega}$ is the angular velocity vector.

[5] For a more detailed and complete discussion of covariance and contravariance, see Fleisch [5].

[6] See [8] for details.

In the first term on the right-hand side of (3.23) summation over i is not desired since i is actually a free index, as evidenced by the left-hand side.

In both (3.23) and (3.26) the repeated upper or lower indices arise from the use of a particular basis; in (3.26), it's assumed that the basis is orthonormal, and in (3.23) it's assumed that the basis is orthonormal *and* that the axes are eigenvectors of the moment of inertia tensor (the so-called ***principal axes***). Such assumptions are usually the case when upper or lower indices are repeated, and one should then note that the resulting equations do not hold in arbitrary bases. Furthermore, if the repeated indices *are* summed over, then that summation represents an invariant process only when the accompanying assumption is satisfied.

Returning to our discussion of how components of vectors and dual vectors transform, we can write (3.21) and (3.22) in terms of matrices as

$$[v]_{\mathcal{B}'} = A[v]_{\mathcal{B}} \tag{3.24a}$$

$$[f]_{\mathcal{B}'} = A^{-1^T}[f]_{\mathcal{B}}, \tag{3.24b}$$

where the superscript T again denotes the transpose of a matrix. From Box 3.3, we know that $f(v)$ is basis-independent, but we also know that $f(v) = [f]^T_{\mathcal{B}}[v]_{\mathcal{B}}$. This last equation then must be true in *any* basis, and we can in fact prove this using (3.24): in a new basis $\mathcal{B}'$, we have

$$\begin{aligned} [f]^T_{\mathcal{B}'}[v]_{\mathcal{B}'} &= \left(A^{-1^T}[f]_{\mathcal{B}}\right)^T A[v]_{\mathcal{B}} \\ &= [f]^T_{\mathcal{B}}A^{-1}A[v]_{\mathcal{B}} \\ &= [f]^T_{\mathcal{B}}[v]_{\mathcal{B}}. \end{aligned} \tag{3.25}$$

This makes concrete our claim above that $[f]$ transforms "oppositely" to $[v]$, so that the basis-independent object $f(v)$ really *is* invariant under a change of basis.

Before moving on to our next example we should point out a minor puzzle: you showed in Exercise 2.23 that if we have an inner product $(\cdot\,|\,\cdot)$ on a real vector space V and an orthornormal basis $\{e_i\}_{i=1\ldots n}$ then the components of vectors and their corresponding dual vectors are identical, which is why we were able to ignore the distinction between them for so long. Equation (3.24) seems to contradict this, however, since it looks like the components of dual vectors transform very differently from the components of vectors. How do we explain this? Well, if we change from one *orthonormal* basis to another, we have

$$\delta_{i'j'} = (e_{i'}|e_{j'}) = A^k_{i'}A^l_{j'}(e_k|e_l) = \sum_{k=1}^{n} A^k_{i'}A^k_{j'} \tag{3.26}$$

which in matrices reads

$$A^{-1^T} A^{-1} = I$$

so we must have

$$A^{-1^T} = A$$

$$\iff \boxed{A^{-1} = A^T.}$$

Such matrices are known as ***orthogonal*** matrices, and we see here that a transformation from one orthornormal basis to another is always implemented by an orthogonal matrix. [7] For such matrices (3.24a) and (3.24b) are identical, resolving our contradiction.

Incidentally, for a *complex* inner product space you will show that orthonormal basis changes are implemented by matrices satisfying $A^{-1} = A^\dagger$. Such matrices are known as ***unitary*** matrices and should be familiar from quantum mechanics.

Exercise 3.6. Show that for any invertible matrix A, $(A^{-1})^T = (A^T)^{-1}$, justifying the sloppiness of our notation above.

Exercise 3.7. Show that for a *complex* inner product space V, the matrix A implementing an orthonormal change of basis satisfies $A^{-1} = A^\dagger$.

Example 3.8. *Linear Operators*

We already noted that linear operators can be viewed as (1,1) tensors as per (3.3). Equation (3.17) then tells us that, for a linear operator T on V,

$$T_{i'}{}^{j'} = A^k_{i'} A^{j'}_l T_k{}^l$$

which in matrix form reads

$$[T]_{\mathcal{B}'} = A[T]_{\mathcal{B}} A^{-1} \tag{3.27}$$

which is the familiar ***similarity transformation*** of matrices. It just says that to compute the action of T in the primed basis, you use A^{-1} to convert your column vector $[v]_{\mathcal{B}'}$ from the new basis back to the old; then, you operate with T via $[T]_{\mathcal{B}}$; then, you convert *back* to the new basis using A.

Incidentally, the similarity transformation (3.27) allows us to extend the trace functional of Example 2.17 from $n \times n$ matrices to linear operators as follows: Given $T \in \mathcal{L}(V)$ and a basis $\mathcal{B}$ for V, define the trace of T as

$$\mathrm{Tr}(T) \equiv \mathrm{Tr}([T]_{\mathcal{B}}).$$

You can then use (3.27) to show (see Exercise 3.10) that $\mathrm{Tr}(T)$ does not depend on the choice of basis $\mathcal{B}$.

[7] See Problem 3-1 for more on orthogonal matrices, as well as Chap. 4.

Exercise 3.8. Show that for $v \in V, f \in V^*, T \in \mathcal{L}(V)$, $f(Tv) = [f]^T[T][v]$ is invariant under a change of basis. Use the matrix transformation laws as we did in (3.25).

Exercise 3.9. Let $\mathcal{B} = \{x, y, z\}$, $\mathcal{B}' = \left\{\frac{1}{\sqrt{2}}(x+iy), z, -\frac{1}{\sqrt{2}}(x-iy)\right\}$ be bases for $\mathcal{H}_1(\mathbb{R}^3)$, and consider the operator L_z for which matrix expressions were found with respect to both bases in Example 2.16. Find the numbers $A^{i'}_{\ j}$ and $A^{j}_{\ i'}$ and use these, along with (3.27), to obtain $[L_z]_{\mathcal{B}'}$ from $[L_z]_{\mathcal{B}}$.

Exercise 3.10. Show that (3.27) implies that $\mathrm{Tr}([T]_{\mathcal{B}})$ does not depend on the choice of basis $\mathcal{B}$, so that $\mathrm{Tr}(T)$ is well defined.

Example 3.9. $(2,0)$ *Tensors*

(2,0) tensors g, which include important examples such as the Minkowski metric and the Euclidean metric, transform as follows according to (3.17):

$$g_{i'j'} = A^k_{\ i'} A^l_{\ j'} g_{kl}$$

or in matrix form

$$[g]_{\mathcal{B}'} = A^{-1^T} [g]_{\mathcal{B}} \, A^{-1}. \tag{3.28}$$

Notice that if g is an inner product and $\mathcal{B}$ and $\mathcal{B}'$ are orthonormal bases then $[g]_{\mathcal{B}'} = [g]_{\mathcal{B}} = I$ and (3.28) becomes

$$I = A^{-1^T} A^{-1},$$

again telling us that A must be orthogonal. Also note that if A is orthogonal, (3.28) is identical to (3.27), so we don't have to distinguish between (2,0) tensors and linear operators (as most of us haven't in the past!). In the case of the Minkowski metric η we aren't dealing with an inner product but we do have orthonormal bases, with respect to which[8] η takes the form

$$[\eta] = \begin{pmatrix} 1 & 0 & 0 & 0 \\ 0 & 1 & 0 & 0 \\ 0 & 0 & 1 & 0 \\ 0 & 0 & 0 & -1 \end{pmatrix}$$

so if we are changing from one orthonormal basis to another we have

[8] We assume here that the basis vector e_t satisfying $\eta(e_t, e_t) = -1$ is the fourth vector in the basis, which isn't necessary but is somewhat conventional in physics.

$$\begin{pmatrix} 1 & 0 & 0 & 0 \\ 0 & 1 & 0 & 0 \\ 0 & 0 & 1 & 0 \\ 0 & 0 & 0 & -1 \end{pmatrix} = A^{-1T} \begin{pmatrix} 1 & 0 & 0 & 0 \\ 0 & 1 & 0 & 0 \\ 0 & 0 & 1 & 0 \\ 0 & 0 & 0 & -1 \end{pmatrix} A^{-1}$$

or equivalently

$$\begin{pmatrix} 1 & 0 & 0 & 0 \\ 0 & 1 & 0 & 0 \\ 0 & 0 & 1 & 0 \\ 0 & 0 & 0 & -1 \end{pmatrix} = A^{T} \begin{pmatrix} 1 & 0 & 0 & 0 \\ 0 & 1 & 0 & 0 \\ 0 & 0 & 1 & 0 \\ 0 & 0 & 0 & -1 \end{pmatrix} A. \tag{3.29}$$

Matrices A satisfying (3.29) are known as ***Lorentz Transformations***. Notice that these matrices are not quite orthogonal, so the components of vectors will transform slightly differently than those of dual vectors under these transformations. This is in contrast to the case of $\mathbb{R}^n$ with a positive-definite metric, where if we go from one orthonormal basis to another then the components of vectors and dual vectors transform identically, as you showed in Exercise 2.23. □

Exercise 3.11. As in previous exercises, show using the matrix transformation laws that $g(v, w) = [w]^T[g][v]$ is invariant under a change of basis.

3.3 Active and Passive Transformations

Before we move on to the tensor product, we have a little unfinished business to conclude. In the last section when we said that the $A_i^{j'}$ were not the components of a tensor, we were lying a little; there *is* a related tensor lurking around, namely the linear operator U that takes the new basis vectors into the old, i.e. $U(e_{i'}) = e_i \;\forall\, i$ (the action of U on an arbitrary vector is then given by expanding that vector in the basis $\mathcal{B}'$ and using linearity). What are the components of this tensor? Well, in the old basis $\mathcal{B}$ we have

$$U_i{}^j = U(e_i, e^j) = e^j(Ue_i) = e^j(U(A_i^{k'} e_{k'})) = A_i^{k'} e^j(U(e_{k'})) = A_i^{k'} e^j(e_k) = A_i^{j'} \tag{3.30}$$

so the $A_i^{j'}$ actually are the components of a tensor[9]! Why did we lie, then? Well, the approach we have been taking so far is to try and think about things in a basis-independent way, and although U is a well-defined linear operator, its definition depends entirely on the two bases we've chosen, so we may as well work directly

[9]If the sleight-of-hand with the primed and unprimed indices in the last couple steps of (3.30) bothers you, puzzle it out and see if you can understand it. It may help to note that the prime on an index doesn't change its numerical value; it's just a reminder that it refers to the primed basis.

Table 3.1 Summary of active vs. passive transformations

$[e_i]_{\mathcal{B}} = A[e_{i'}]_{\mathcal{B}}$	Active
$[v]_{\mathcal{B}'} = A[v]_{\mathcal{B}}$	Passive

with the numbers that relate the bases. Also, using one primed index and one unprimed index makes it easy to remember transformation laws like (3.14) and (3.15), but is not consistent with our notation for the components of tensors.

If we write out the components of U as a matrix, you should verify that

$$[e_i]_{\mathcal{B}} = [U]_{\mathcal{B}}[e_{i'}]_{\mathcal{B}} = A[e_{i'}]_{\mathcal{B}} \tag{3.31}$$

which should be compared to (3.24a), which reads $[v]_{\mathcal{B}'} = A[v]_{\mathcal{B}}$. Equation (3.31) is called an ***active*** transformation, since we use the matrix A to change one vector into another, namely $e_{i'}$ into e_i. Note that in (3.31) all vectors are expressed in the same basis. Equation (3.24a), on the other hand, is called a ***passive*** transformation, since we use the matrix A not to change the vector v but rather to change the basis which v is referred to, hence changing its components. All this is summarized in Table 3.1.

The notation in most physics texts is not as explicit as ours; one usually sees matrix equations like

$$\mathbf{r}' = A\mathbf{r} \tag{3.32}$$

for both passive and active transformations, and one must rely on context to figure out how the equation is to be interpreted. In the active case, one considers the coordinate system fixed and interprets the matrix A as taking the physical vector $\mathbf{r}$ into a *new* vector $\mathbf{r}'$, where the components of both are expressed in the *same* coordinate system, just as in (3.31). In the passive case, the physical vector $\mathbf{r}$ doesn't change but the basis does, so one interprets the matrix A as taking the components of $\mathbf{r}$ in the old coordinate system and giving back the components of the *same* vector $\mathbf{r}$ in the new (primed) coordinate system, just as in (3.24a). All this is illustrated in Fig. 3.3.

Before we get to some examples, note that in the passive transformation (3.24a) the matrix A takes the old components to the new components, whereas in the active transformation (3.31) A takes the *new* basis vectors to the old ones. Thus when A is interpreted actively it corresponds to the *opposite* transformation as in the passive case. This dovetails with the fact that components and basis vectors transform oppositely, as discussed under (3.22).

Example 3.10. *Active and passive orthogonal transformations in two dimensions*

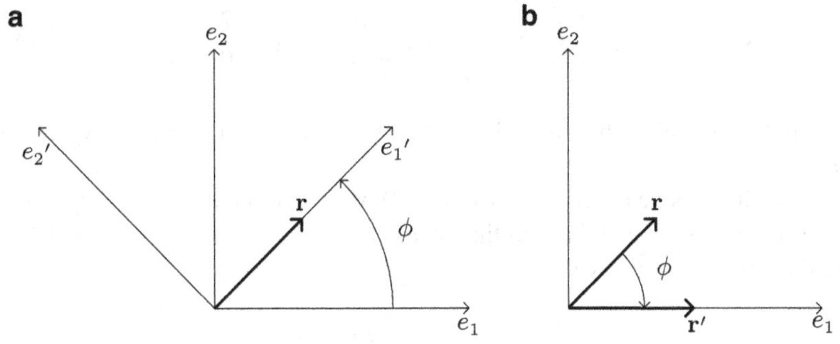

Fig. 3.3 Illustration of the passive and active interpretations of $\mathbf{r}' = A\mathbf{r}$ dimensions. In (**a**) we have a passive transformation, in which the same vector $\mathbf{r}$ is referred to two different bases. The coordinate representation of $\mathbf{r}$ transforms as $\mathbf{r}' = A\mathbf{r}$, though the vector itself does not change. In (**b**) we have an active transformation, where there is only one basis and the vector $\mathbf{r}$ is itself transformed by $\mathbf{r}' = A\mathbf{r}$. In the active case the transformation is opposite that of the passive case

Let $\mathcal{B} = \{e_1, e_2\}$ be the standard basis for $\mathbb{R}^2$, and consider a new basis $\mathcal{B}'$ given by

$$
\begin{aligned}
e_{1'} &\equiv \frac{1}{\sqrt{2}}e_1 + \frac{1}{\sqrt{2}}e_2 \\
e_{2'} &\equiv -\frac{1}{\sqrt{2}}e_1 + \frac{1}{\sqrt{2}}e_2.
\end{aligned}
$$

You can show (as in Example 3.6) that this leads to an orthogonal change of basis matrix given by

$$
A = \begin{pmatrix} \frac{1}{\sqrt{2}} & \frac{1}{\sqrt{2}} \\ -\frac{1}{\sqrt{2}} & \frac{1}{\sqrt{2}} \end{pmatrix} \tag{3.33}
$$

which corresponds to rotating our basis counterclockwise by $\phi = 45°$, see Fig. 3.3a.

Now consider the vector $\mathbf{r} = \frac{1}{2\sqrt{2}}e_1 + \frac{1}{2\sqrt{2}}e_2$, also depicted in the figure. In the standard basis we have

$$
[\mathbf{r}]_{\mathcal{B}} = \begin{pmatrix} \frac{1}{2\sqrt{2}} \\ \frac{1}{2\sqrt{2}} \end{pmatrix}.
$$

What does $\mathbf{r}$ look like in our new basis? From Fig. 3.3a we see that $\mathbf{r}$ is proportional to e_1', and that is indeed what we find; using (3.24a), we have

$$[\mathbf{r}]_{\mathcal{B}'} = A[\mathbf{r}]_{\mathcal{B}} = \begin{pmatrix} \frac{1}{\sqrt{2}} & \frac{1}{\sqrt{2}} \\ -\frac{1}{\sqrt{2}} & \frac{1}{\sqrt{2}} \end{pmatrix} \begin{pmatrix} \frac{1}{2\sqrt{2}} \\ \frac{1}{2\sqrt{2}} \end{pmatrix} = \begin{pmatrix} 1/2 \\ 0 \end{pmatrix} \tag{3.34}$$

as expected. Remember that the column vector at the end of (3.34) is expressed in the *primed* basis.

This was the passive interpretation of (3.32); what about the active interpretation? Taking our matrix A and interpreting it as a linear operator represented in the standard basis we again have

$$[\mathbf{r}']_{\mathcal{B}} = A[\mathbf{r}]_{\mathcal{B}} = \begin{pmatrix} 1/2 \\ 0 \end{pmatrix}$$

except that now the vector $(1/2, 0)$ represents the *new* vector $\mathbf{r}'$ in the *same* basis $\mathcal{B}$. This is illustrated in Fig. 3.3b. As mentioned above, when A is interpreted actively, it corresponds to a *clockwise* rotation, opposite to its interpretation as a passive transformation. □

Exercise 3.12. Verify (3.33) and (3.34).

Example 3.11. *Active transformations and rigid body motion*

Passive transformations are probably the ones encountered most often in classical physics, since a change of cartesian coordinates induces a passive transformation. Active transformations do crop up, though, especially in the case of rigid body motion. In this scenario, one specifies the orientation of a rigid body by the time-dependent orthogonal basis transformation $A(t)$ which relates the space frame K' to the body frame $K(t)$ (we use here the notation of Example 2.12). As we saw above, there corresponds to the time-dependent matrix $A(t)$ a time-dependent linear operator $U(t)$ which satisfies $U(t)(e_{i'}) = e_i(t)$. If K and K' were coincident at $t = 0$ and $\mathbf{r}_0$ is the position vector of a point p of the rigid body at that time (see Fig. 3.4a), then the position of p at a later time is just $\mathbf{r}(t) = U(t)\mathbf{r}_0$ (see Fig. 3.4b), which as a matrix equation in K' would read

$$[\mathbf{r}(t)]_{K'} = A(t)[\mathbf{r}_0]_{K'}. \tag{3.35}$$

In more common and less precise notation this would be written

$$\mathbf{r}(t) = A(t)\mathbf{r}_0.$$

In other words, the position of a specific point on the rigid body at an arbitrary time t is given by the active transformation corresponding to the matrix $A(t)$. □

Example 3.12. *Active and passive transformations and the Schrödinger and Heisenberg pictures*

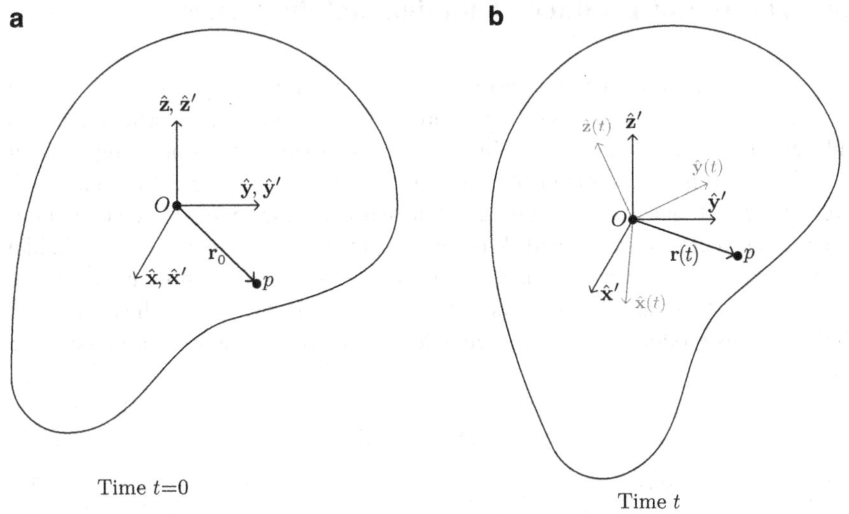

Fig. 3.4 In (**a**) we have the coincident body and space frames at $t = 0$, along with the point p of the rigid body. In (**b**) we have the rotated rigid body, and the vector $\mathbf{r}(t)$ pointing to point p now has different components in the space frame, given by (3.35)

The duality between passive and active transformations is also present in quantum mechanics. In the Schrödinger picture, one considers observables like the momentum or position operator as acting on the state ket while the basis kets remain fixed. This is the active viewpoint. In the Heisenberg picture, however, one considers the state ket to be fixed and considers the *observables* to be time-dependent (recall that (2.18) is the equation of motion for these operators). Since the operators are time-dependent, their eigenvectors (which form a basis[10]) are time-dependent as well, so this picture is the passive one in which the vectors don't change but the basis does. Just as an equation like (3.32) can be interpreted in both the active and passive sense, a quantum-mechanical equation like

$$< \hat{x}(t) > = \langle\psi|\,(U^{\dagger}\hat{x}\,U\,)\,|\psi\rangle \tag{3.36a}$$

$$= (\langle\psi|U^{\dagger})\;\hat{x}\;(U|\psi\rangle), \tag{3.36b}$$

where U is the time-evolution operator for time t, can also be interpreted in two ways: in the active sense of (3.36b), in which the U's act on the vectors and change them into new vectors, and in the passive sense of (3.36a), where the U's act on the operator $\hat{x}$ by a similarity transformation to turn it into a new operator, $\hat{x}(t)$.

[10]For details on why the eigenvectors of Hermitian operators form a basis, at least in the finite-dimensional case, see Hoffman and Kunze [13].

3.4 The Tensor Product: Definition and Properties

Now that we are familiar with tensors and their transformation laws, it is time to introduce the tensor product. The tensor product is one of the most basic operations with tensors and is commonplace in physics, but is often unacknowledged or, at best, dealt with in an ad-hoc fashion. Before we give the precise definition, which takes a little getting used to, we give a rough, heuristic description. Given two finite-dimensional vector spaces V and W (over the same set of scalars C), we would like to construct a product vector space, which we denote $V \otimes W$, whose elements are in some sense "products" of vectors $v \in V$ and $w \in W$. We denote these products by $v \otimes w$. This product, like any respectable product, should be bilinear in the sense that

$$(v_1 + v_2) \otimes w = v_1 \otimes w + v_2 \otimes w \tag{3.37a}$$

$$v \otimes (w_1 + w_2) = v \otimes w_1 + v \otimes w_2 \tag{3.37b}$$

$$c(v \otimes w) = (cv) \otimes w = v \otimes (cw), \quad c \in C. \tag{3.37c}$$

Given these properties, the product of any two arbitrary vectors v and w can then be expanded in terms of bases $\{e_i\}_{i=1\ldots n}$ and $\{f_j\}_{j=1\ldots m}$ for V and W as

$$\begin{aligned} v \otimes w &= (v^i e_i) \otimes (w^j f_j) \\ &= v^i w^j e_i \otimes f_j \end{aligned}$$

so $\{e_i \otimes f_j\}, i = 1 \ldots n,\ j = 1 \ldots m$ should be a basis for $V \otimes W$, which would then have dimension nm. Thus the basis for the product space would be just the product of the basis vectors, and the dimension of the product space would be just the product of the dimensions.

Now let's make this precise. Given two finite-dimensional vector spaces V and W, we define their ***tensor product*** $V \otimes W$ to be the set of all C-valued bilinear functions on $V^* \times W^*$. Such functions do form a vector space, as you can easily check. This definition may seem unexpected or counterintuitive at first, but you will soon see that this definition does yield the vector space described above. Also, given two *vectors* $v \in V$, $w \in W$, we define their ***tensor product*** $v \otimes w$ to be the element of $V \otimes W$ defined as follows:

$$\boxed{(v \otimes w)(h, g) \equiv h(v)w(g) \quad \forall\, h \in V^*,\ g \in W^*.} \tag{3.38}$$

(Remember that an element of $V \otimes W$ is a bilinear function on $V^* \times W^*$, and so is defined by its action on a pair $(h, g) \in V^* \times W^*$). The bilinearity of the tensor product is immediate and you can probably verify it without writing anything down: just check that both sides of Eq. (3.37) are equal when evaluated on any pair of dual vectors. To prove that $\{e_i \otimes f_j\}, i = 1 \ldots n,\ j = 1 \ldots m$ is a basis for $V \otimes W$, let $\{e^i\}_{i=1\ldots n}, \{f^j\}_{i=1\ldots m}$ be the corresponding dual bases and consider an arbitrary $T \in V \otimes W$. Using bilinearity,

$$T(h,g) = h_i g_j T(e^i, f^j) = h_i g_j T^{ij} \tag{3.39}$$

where $T^{ij} \equiv T(e^i, f^j)$. If we consider the expression $T^{ij} e_i \otimes f_j$, then

$$\begin{aligned}(T^{ij} e_i \otimes f_j)(e^k, f^l) &= T^{ij} e_i(e^k) f_j(f^l)\\ &= T^{ij} \delta_i{}^k \delta_j{}^l\\ &= T^{kl}\end{aligned}$$

so $T^{ij} e_i \otimes f_j$ agrees with T on basis vectors, hence on all vectors by bilinearity, so $T = T^{ij} e_i \otimes f_j$. Since T was an arbitrary element of $V \otimes W$, $V \otimes W = \mathrm{Span}\{e_i \otimes f_j\}$. Furthermore, the $e_i \otimes f_j$ are linearly independent as you should check, so $\{e_i \otimes f_j\}$ is actually a basis for $V \otimes W$ and $V \otimes W$ thus has dimension mn.

The tensor product has a couple of important properties besides bilinearity. First, it *commutes with taking duals*, that is

$$(V \otimes W)^* = V^* \otimes W^*.$$

Secondly, and more importantly, the tensor product it is *associative*, i.e. for vector spaces V_i, $i = 1, 2, 3$,

$$(V_1 \otimes V_2) \otimes V_3 = V_1 \otimes (V_2 \otimes V_3).$$

This property allows us to drop the parentheses and write expressions like $V_1 \otimes \cdots \otimes V_n$ without ambiguity. One can think of $V_1 \otimes \cdots \otimes V_n$ as the set of C-valued multilinear functions on $V_1^* \times \cdots \times V_n^*$.

These two properties are both plausible, particularly when thought of in terms of basis vectors, but verifying them rigorously turns out to be slightly tedious. See Warner [21] for proofs and further details.

Exercise 3.13. If $\{e_i\}, \{f_j\}$, and $\{g_k\}$ are bases for V_1, V_2, and V_3 respectively, convince yourself that $\{e_i \otimes f_j \otimes g_k\}$ is a basis for $V_1 \otimes V_2 \otimes V_3$, and hence that $\dim V_1 \otimes V_2 \otimes V_3 = n_1 n_2 n_3$ where $\dim V_i = n_i$. Extend the above to n-fold tensor products.

3.5 Tensor Products of V and V^*

In the previous section we defined the tensor product for two arbitrary vector spaces V and W. Often, though, we'll be interested in just the iterated tensor product of a vector space and its dual, i.e. in tensor products of the form

$$\underbrace{V^* \otimes \cdots \otimes V^*}_{r \text{ times}} \otimes \underbrace{V \otimes \cdots \otimes V}_{s \text{ times}}. \tag{3.40}$$

This space is of particular interest because it is actually identical to $\mathcal{T}_s^r$! From the previous section we know that the vector space in (3.40) can be interpreted as the set of multilinear functions on[11]

$$\underbrace{V \times \cdots \times V}_{r \text{ times}} \times \underbrace{V^* \times \cdots \times V^*}_{s \text{ times}}, \tag{3.41}$$

but these functions are exactly $\mathcal{T}_s^r$! Since the space in (3.40) has basis $\mathcal{B}_s^r = \{e^{i_1} \otimes \cdots \otimes e^{i_r} \otimes e_{j_1} \otimes \cdots \otimes e_{j_s}\}$, we can conclude that $\mathcal{B}_s^r$ is a basis for $\mathcal{T}_s^r$. In fact, we claim that if $T \in \mathcal{T}_s^r$ has components $T_{i_1 \dots i_r}{}^{j_1 \dots j_s}$, then

$$\boxed{T = T_{i_1 \dots i_r}{}^{j_1 \dots j_s} e^{i_1} \otimes \cdots \otimes e^{i_r} \otimes e_{j_1} \otimes \cdots \otimes e_{j_s}} \tag{3.42}$$

is the expansion of T in the basis $\mathcal{B}_s^r$. To prove this, we just need to check that both sides agree when evaluated on an arbitrary set of basis vectors; on the left-hand side we get $T(e_{i_1}, \dots, e_{i_r}, e^{j_1}, \dots, e^{j_s}) = T_{i_1, \dots, i_r}{}^{j_1 \dots j_s}$ by definition, and on the right-hand side we have

$$\begin{aligned}
&(T_{k_1 \dots k_r}{}^{l_1 \dots l_s} e^{k_1} \otimes \cdots \otimes e^{k_r} \otimes e_{l_1} \otimes \cdots \otimes e_{l_s})(e_{i_1}, \dots, e_{i_r}, e^{j_1}, \dots, e^{j_s}) \\
&= T_{k_1 \dots k_r}{}^{l_1 \dots l_s} e^{k_1}(e_{i_1}) \dots e^{k_r}(e_{i_r}) e_{l_1}(e^{j_1}) \dots e_{l_s}(e^{j_s}) \\
&= T_{k_1 \dots k_r}{}^{l_1 \dots l_s} \delta^{k_1}_{i_1} \dots \delta^{k_r}_{i_r} \delta^{j_1}_{l_1} \dots \delta^{j_s}_{l_s} \\
&= T_{i_1, \dots, i_r}{}^{j_1 \dots j_s}
\end{aligned} \tag{3.43}$$

so our claim is true. Thus, for instance, a $(2, 0)$ tensor like the Minkowski metric can be written as $\eta = \eta_{\mu\nu} e^\mu \otimes e^\nu$. Conversely, a tensor product like $f \otimes g = f_i g_j e^i \otimes e^j \in \mathcal{T}_0^2$ thus has components $(f \otimes g)_{ij} = f_i g_j$. **Notice that we now have two ways of thinking about components**:

1. As the values of a tensor on sets of basis vectors, as in (3.5)
2. As the expansion coefficients in a given basis, as in (3.42)

This duplicity of perspective was pointed out in the case of vectors just above Exercise 2.12, and it's essential that you be comfortable thinking about components in either way.

Exercise 3.14. Compute the dimension of $\mathcal{T}_s^r$.

[11] Actually, to interpret (3.40) as the space of multilinear functions on (3.41) also requires the fact that $(V^*)^* \simeq V$. See Problem 2-5 for a proof of this.

Exercise 3.15. Let T_1 and T_2 be tensors of type (r_1, s_1) and (r_2, s_2) respectively on a vector space V. Show that $T_1 \otimes T_2$ can be viewed as an $(r_1 + r_2, s_1 + s_2)$ tensor, so that the tensor product of two tensors is again a tensor, justifying the nomenclature.

One important operation on tensors which we can now discuss is that of ***contraction***, which is the generalization of the trace functional to tensors of arbitrary rank: Given $T \in \mathcal{T}_s^r(V)$ with expansion

$$T = T_{i_1 \dots i_r}{}^{j_1 \dots j_s} e^{i_1} \otimes \cdots \otimes e^{i_r} \otimes e_{j_1} \otimes \cdots \otimes e_{j_s} \tag{3.44}$$

we can define a contraction of T to be any $(r-1, s-1)$ tensor resulting from feeding e^i into one of the arguments, e_i into another, and then summing over i as implied by the summation convention. For instance, if we feed e_i into the rth slot and e^i into the $(r + s)$th slot and sum, we get the $(r - 1, s - 1)$ tensor $\tilde{T}$ defined as

$$\tilde{T}(v_1, \dots, v_{r-1}, f_1, \dots, f_{s-1}) \equiv \sum_i T(v_1, \dots, v_{r-1}, e_i, f_1, \dots, f_{s-1}, e^i).$$

You may be suspicious that $\tilde{T}$ depends on our choice of basis, but Exercise 3.16 shows that contraction is in fact well defined. Notice that the components of $\tilde{T}$ are

$$\tilde{T}_{i_1 \dots i_{r-1}}{}^{j_1 \dots j_{s-1}} = \sum_l T_{i_1 \dots i_{r-1} l}{}^{j_1 \dots j_{s-1} l}.$$

(The summation in the previous two equations is not necessary, since we are assuming the Einstein summation convention, but is written explicitly for emphasis.) Similar contractions can be performed on any two arguments of T provided one argument eats vectors and the other dual vectors. In terms of components, a contraction can be taken with respect to any pair of indices provided that one is covariant and the other contravariant. If we are working on a vector space equipped with a metric g, then we can use the metric to raise and lower indices and so can contract on *any* pair of indices, even if they're both covariant or contravariant. For instance, we can contract a $(2, 0)$ tensor T with components T_{ij} as $\tilde{T} = T_i{}^i = g^{ij} T_{ij}$, which one can interpret as just the trace of the associated linear operator (or (1,1) tensor). For a linear operator or any other rank 2 tensor, this is the only option for contraction. If we have two linear operators A and B, then their tensor product $A \otimes B \in \mathcal{T}_2^2$ has components

$$(A \otimes B)_{ik}{}^{jl} = A_i{}^j B_k{}^l,$$

and contracting on the first and last index gives a $(1, 1)$ tensor AB whose components are

$$(AB)_k{}^j = A_l{}^j B_k{}^l.$$

You should check that this tensor is just the composition of A and B, as our notation suggests. What linear operator do we get if we consider the other contraction $A_i^{\ j} B_j^{\ l}$?

Exercise 3.16. Show that if $\{e_i\}_{i=1\ldots n}$ and $\{e_{i'}\}_{i=1\ldots n}$ are two arbitrary bases that

$$T(v_1,\ldots,v_{r-1},e_i,f_1,\ldots,f_{s-1},e^i) = T(v_1,\ldots,v_{r-1},e_{i'},f_1,\ldots,f_{s-1},e^{i'})$$

so that contraction is well defined.

Example 3.13. $V^* \otimes V$

One of the most important examples of tensor products of the form (3.40) is $V^* \otimes V$, which as we mentioned is the same as $\mathcal{T}_1^1$, the space of linear operators. How does this identification work, explicitly? Well, given $f \otimes v \in V^* \otimes V$, we can define a linear operator by $(f \otimes v)(w) \equiv f(w)v$. More generally, given

$$T_i^{\ j} e^i \otimes e_j \in V^* \otimes V, \tag{3.45}$$

we can define a linear operator T by

$$T(v) = T_i^{\ j} e^i(v) e_j = v^i T_i^{\ j} e_j$$

which is identical to (2.20). This identification of $V^* \otimes V$ and linear operators is actually implicit in many quantum-mechanical expressions. Let $\mathcal{H}$ be a quantum-mechanical Hilbert space and let $\psi, \phi \in \mathcal{H}$ so that $L(\phi) \in \mathcal{H}^*$. The tensor product of $L(\phi)$ and ψ, which we would write as $L(\phi) \otimes \psi$, is written in Dirac notation as $|\psi\rangle\langle\phi|$ (note the transposition of the factors relative to our convention). If we're given an orthonormal basis $\mathcal{B} = \{|i\rangle\}$, the expansion (3.45) of an arbitrary operator H can be written in Dirac notation as

$$H = \sum_{i,j} H_{ij}\ |i\rangle\langle j|,$$

an expression which may be familiar from advanced quantum mechanics texts.[12] In particular, the identity operator can be written as

$$I = \sum_i |i\rangle\langle i|,$$

which is referred to as the ***resolution of the identity*** with respect to the basis $\{|i\rangle\}$.

A word about nomenclature: In quantum mechanics and other contexts the tensor product is often referred to as the ***direct*** or ***outer*** product. This last term is meant

[12]We don't bother here with index positions since most quantum mechanics texts don't employ Einstein summation convention, preferring instead to explicitly indicate summation.

to distinguish it from the inner product, since both the outer and inner products eat a dual vector and a vector (strictly speaking the inner product eats 2 vectors, but remember that with an inner product we may identify vectors and dual vectors) but the outer product yields a linear operator whereas the inner product yields a scalar.

Exercise 3.17. Interpret $e^i \otimes e_j$ as a linear operator, and convince yourself that its matrix representation is

$$[e^i \otimes e_j] = E_{ji}.$$

Recall that E_{ji} is one of the elementary basis matrices introduced way back in Example 2.9, and has a 1 in the jth row and ith column and zeros everywhere else.

3.6 Applications of the Tensor Product in Classical Physics

Example 3.14. *Moment of inertia tensor revisited*

We took an abstract look at the moment of inertia tensor in Example 3.3; now, armed with the tensor product, we can examine the moment of inertia tensor more concretely. Consider a rigid body with a fixed point O, so that it has only rotational degrees of freedom (O need not necessarily be the center of mass). Let O be the origin, pick time-dependent body-fixed axes $K = \{\hat{\mathbf{x}}(t), \hat{\mathbf{y}}(t), \hat{\mathbf{z}}(t)\}$, and let g denote the Euclidean metric. Recall that g allows us to define a map L from vectors to dual vectors. Also, let the ith particle in the rigid body have mass m_i and position vector $\mathbf{r}_i$ with $[\mathbf{r}_i]_K = (x_i, y_i, z_i)$ relative to O, and let $r_i^2 \equiv g(\mathbf{r}_i, \mathbf{r}_i)$. This is illustrated in Fig. 3.5. The $(2, 0)$ moment of inertia tensor is then given by

$$\mathcal{I}_{(2,0)} = \sum_i m_i (r_i^2 g - L(\mathbf{r}_i) \otimes L(\mathbf{r}_i)) \tag{3.46}$$

while the $(1, 1)$ tensor reads

$$\mathcal{I}_{(1,1)} = \sum_i m_i (r_i^2 I - L(\mathbf{r}_i) \otimes \mathbf{r}_i). \tag{3.47}$$

You should check that in components (3.46) reads

$$\mathcal{I}_{jk} = \sum_i m_i (r_i^2 \delta_{jk} - (\mathbf{r}_i)_j (\mathbf{r}_i)_k).$$

Writing a couple of components explicitly yields

$$\mathcal{I}_{xx} = \sum_i m_i (y_i^2 + z_i^2)$$

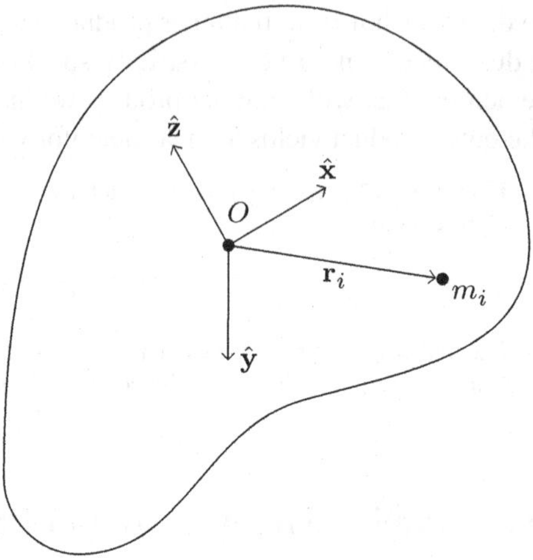

Fig. 3.5 The rigid body with fixed point O and body-fixed axes $K = \{\hat{\mathbf{x}}, \hat{\mathbf{y}}, \hat{\mathbf{z}}\}$, along with ith particle at position $\mathbf{r}_i$ with mass m_i

$$\mathcal{I}_{xy} = -\sum_i m_i x_i y_i, \tag{3.48}$$

expressions which should be familiar from classical mechanics. So long as the basis is orthonormal, the components $\mathcal{I}_j{}^k$ of the $(1, 1)$ tensor in (3.47) will be the same as for the $(2, 0)$ tensor, as remarked earlier. Note that if we had not used body-fixed axes, the components of $\mathbf{r}_i$ [and hence the components of $\mathcal{I}$, by (3.48)] would in general be time-dependent; this is the main reason for using the body-fixed axes in computation.

Example 3.15. *Maxwell Stress Tensor*

In considering the conservation of total momentum (mechanical plus electromagnetic) in classical electrodynamics one encounters the symmetric rank 2 ***Maxwell Stress Tensor***, defined in $(2, 0)$ form as[13]

$$T_{(2,0)} = \mathbf{E} \otimes \mathbf{E} + \mathbf{B} \otimes \mathbf{B} - \frac{1}{2}(\mathbf{E} \cdot \mathbf{E} + \mathbf{B} \cdot \mathbf{B})g,$$

where $\mathbf{E}$ and $\mathbf{B}$ are the dual vector versions of the electric and magnetic field vectors. T can be interpreted in the following way: $T(v, w)$ gives the rate at which momentum in the v-direction flows in the w-direction. In components we have

$$T_{ij} = E_i E_j + B_i B_j - \frac{1}{2}(\mathbf{E} \cdot \mathbf{E} + \mathbf{B} \cdot \mathbf{B})\delta_{ij},$$

which is the expression found in most classical electrodynamics textbooks.

[13]Recall that we've set all physical constants such as c and ϵ_0 equal to 1.

Example 3.16. *The Electromagnetic Field tensor*

As you have probably seen in discussions of relativistic electrodynamics, the electric and magnetic field vectors are properly viewed as components of a rank 2 antisymmetric tensor F, the electromagnetic *field tensor*.[14] To write F in component-free notation requires machinery outside the scope of this text,[15] so we settle for its expression as a matrix in an orthonormal basis, which in $(2,0)$ form is

$$[F_{(2,0)}] = \begin{pmatrix} 0 & -B_z & B_y & -E_x \\ B_z & 0 & -B_x & -E_y \\ -B_y & B_x & 0 & -E_z \\ E_x & E_y & E_z & 0 \end{pmatrix}. \tag{3.49}$$

The Lorentz force law

$$\frac{dp^\mu}{d\tau} = qF^\mu{}_\nu\, v^\nu,$$

where $p = mv$ is the four-momentum of a particle, v is its four-velocity τ its proper time, and q its charge, can be rewritten without components as

$$\frac{dp}{d\tau} = qF_{(1,1)}(v) \tag{3.50}$$

which just says that the Minkowski force $\frac{dp}{d\tau}$ on a particle is given by the action of the field tensor on the particle's 4-velocity!

3.7 Applications of the Tensor Product in Quantum Physics

In this section we'll discuss further applications of the tensor product in quantum mechanics, in particular the oft-unwritten rule that to add degrees of freedom one should take the tensor product of the corresponding Hilbert spaces. Before we get to this, however, we must set up a little more machinery and address an issue that we've so far swept under the rug. The issue is that when dealing with spatial degrees of freedom, as opposed to "internal" degrees of freedom like spin, we often encounter Hilbert spaces like $L^2([-a,a])$ and $L^2(\mathbb{R})$ which are most conveniently described by "basis" vectors which are eigenvectors of either the position operator $\hat{x}$ or the

[14] In this example and the one above we are actually not dealing with tensors but with *tensor fields*, i.e. tensor-valued functions on space and spacetime. For the discussion here, however, we will ignore the spatial dependence, focusing instead on the tensorial properties.

[15] One needs the ***exterior derivative***, a generalization of the curl, divergence and gradient operators from vector calculus. See Schutz [18] for a very readable account.

momentum operator $\hat{p}$. The trouble with these bases is that they are often non-denumerably infinite (i.e., can't be indexed by the integers, unlike all the bases we've worked with so far) and, what's worse, the "basis vectors" don't even belong to the Hilbert space! Consider, for example, $L^2(\mathbb{R})$. The position operator $\hat{x}$ acts on functions $\psi(x) \in L^2(\mathbb{R})$ by

$$\hat{x}\,\psi(x) = x\,\psi(x). \tag{3.51}$$

If we follow the practice of most quantum mechanics texts and treat the Dirac delta functional δ as $L(\delta(x))$ where $\delta(x)$, the "Dirac delta function," is infinite at 0 and 0 elsewhere, you can check (see Exercise 3.18) that

$$\hat{x}\,\delta(x - x_0) = x_0\,\delta(x - x_0)$$

so that $\delta(x - x_0)$ is an "eigenfunction" of $\hat{x}$ with eigenvalue x_0 (in Dirac notation we write the corresponding ket as $|x_0\rangle$). The trouble is that, as we saw in Example 2.25, there is no such $\delta(x) \in L^2(\mathbb{R})$! Furthermore, since the basis $\{\delta(x - x_0)\}_{x_0 \in \mathbb{R}}$ is indexed by $\mathbb{R}$ and not some subset of $\mathbb{Z}$, we must expand $\psi \in L^2(\mathbb{R})$ by integrating instead of summing. Integration, however, is a limiting procedure and one should really worry about what it means for an integral to converge. Rectifying all this in a rigorous manner is possible,[16] but outside the scope of this text, unfortunately. We do wish to work with these objects, however, so we will content ourselves with the traditional approach: ignore the fact that the delta functions are not elements of $L^2(\mathbb{R})$, work without discomfort with the basis $\{\delta(x - x_0)\}_{x_0 \in \mathbb{R}}$,[17] and fearlessly expand arbitrary functions ψ in the basis $\{\delta(x - x_0)\}_{x_0 \in \mathbb{R}}$ as

$$\psi(x) = \int_{-\infty}^{\infty} dx'\,\psi(x')\delta(x - x'), \tag{3.52}$$

where the above equation can be interpreted both as the expansion of ψ and just the definition of the delta function. In Dirac notation (3.52) reads

$$|\psi\rangle = \int_{-\infty}^{\infty} dx'\,\psi(x')|x'\rangle. \tag{3.53}$$

Note that we can think of the numbers $\psi(x)$ as the components of $|\psi\rangle$ with respect to the basis $\{|x\rangle\}_{x \in \mathbb{R}}$. Alternatively, if we define the inner product of our basis vectors to be

$$\langle x|x'\rangle \equiv \delta(x - x')$$

[16] This requires the so-called rigged Hilbert space; see Ballentine [2].

[17] Working with the momentum eigenfunctions e^{ipx} instead doesn't help; though these are legitimate functions, they still are not square-integrable since $\int_{-\infty}^{\infty} |e^{ipx}|^2\,dx = \infty$!

as is usually done, then using (3.53) we have

$$\psi(x) = \langle x|\psi\rangle \tag{3.54}$$

which gives another interpretation of $\psi(x)$. These two interpretations of $\psi(x)$ are just the ones mentioned below (3.43); that is, the components of a vector can be interpreted either as expansion coefficients, as in (3.53), or as the value of a given dual vector on the vector, as in (3.54).

Exercise 3.18. Check/review the following properties of the delta function:

(a) By considering the integral

$$\int_{-\infty}^{\infty} (\hat{x}\,\delta(x-x_0))\, f(x)\, dx$$

(where f is an arbitrary square-integrable function), show formally that

$$\hat{x}\,\delta(x-x_0) = x_0\,\delta(x-x_0).$$

(b) Check that $\{\delta(x-x_0)\}_{x_0\in\mathbb{R}}$ satisfies (2.36).
(c) Verify (3.54).

While we're at it, let's pose the following question: we mentioned in a footnote above that one could use momentum eigenfunctions instead of position eigenfunctions as a basis for $L^2(\mathbb{R})$; what does the corresponding change of basis look like?

Example 3.17. *The Momentum Representation*

As is well known from quantum mechanics, the eigenfunctions of the momentum operator $\hat{p} = -i\frac{d}{dx}$ are the wavefunctions $\{e^{ipx}\}_{p\in\mathbb{R}}$, and these wavefunctions form a basis for $L^2(\mathbb{R})$. In fact, the expansion of an arbitrary function $\psi \in L^2(\mathbb{R})$ in this basis is just the Fourier expansion of ψ, written

$$\psi(x) = \frac{1}{2\pi}\int_{-\infty}^{\infty} dp\,\phi(p)\,e^{ipx}, \tag{3.55}$$

where the component function $\phi(p)$ is known as the ***Fourier transform***[18] of ψ. One could in fact work exclusively with $\phi(p)$ instead of $\psi(x)$, and recast the operators $\hat{x}$ and $\hat{p}$ in terms of their action on $\phi(p)$ (see Exercise 3.19 below); such an approach

[18]The Fourier transform of a function $\psi(x)$ is often alternatively defined as

$$\phi(p) \equiv \int_{-\infty}^{\infty} dx\,\psi(x)e^{-ipx},$$

which in Dirac notation would be written $\langle p|\psi\rangle$. These two equivalent definitions of $\phi(p)$ are totally analogous to the two expressions (3.52) and (3.54) for $\psi(x)$, and are again just the two interpretations of components discussed below (3.43).

is known as the ***momentum representation***. Now, what does it look like when we switch from the position representation to the momentum representation, i.e. when we change bases from $\{\delta(x - x_0)\}_{x_0 \in \mathbb{R}}$ to $\{e^{ipx}\}_{p \in \mathbb{R}}$? Since the basis vectors are indexed by real numbers p and x_0 as opposed to integers i and j, our change of basis will not be given by a matrix with components $A^{i'}_{\ j}$ but rather a *function* $A(x_0, p)$. Using the fact that both bases are orthonormal and assuming that (3.13) extends to infinite dimensions, $A(x_0, p)$ is given by the inner product of $\delta(x - x_0)$ and e^{ipx}. In Dirac notation this would be written as $\langle x_0 | p \rangle$, and we have

$$\langle x_0 | p \rangle = \int_{-\infty}^{\infty} dx\, \delta(x - x_0) e^{ipx} = e^{ipx_0}.$$

This may be a familiar equation, but can now be interpreted as a formula for the change of basis function $A(x_0, p)$. □

Exercise 3.19. Use (3.55) to show that in the momentum representation, $\hat{p}\,\phi(p) = p\,\phi(p)$ and $\hat{x}\,\phi(p) = i\frac{d\phi}{dp}$.

The next issue to address is that of linear operators: having constructed a new Hilbert space[19] $\mathcal{H}_1 \otimes \mathcal{H}_2$ out of two Hilbert spaces $\mathcal{H}_1$ and $\mathcal{H}_2$, can we construct linear operators on $\mathcal{H}_1 \otimes \mathcal{H}_2$ out of the linear operators on $\mathcal{H}_1$ and $\mathcal{H}_2$? Well, given linear operators A_i on $\mathcal{H}_i$, $i = 1, 2$, we can define a linear operator $A_1 \otimes A_2$ on $\mathcal{H}_1 \otimes \mathcal{H}_2$ by

$$(A_1 \otimes A_2)(v \otimes w) \equiv (A_1 v) \otimes (A_2 w). \tag{3.56}$$

You can check that with this definition, $(A \otimes B)(C \otimes D) = AC \otimes BD$. In most quantum-mechanical applications either A_1 or A_2 is the identity, i.e. one considers operators of the form $A_1 \otimes I$ or $I \otimes A_2$. These are often abbreviated as A_1 and A_2 even though they're acting on $\mathcal{H}_1 \otimes \mathcal{H}_2$. We should also mention here that the inner product $(\cdot\,|\,\cdot)_\otimes$ on $\mathcal{H}_1 \otimes \mathcal{H}_2$ is just the product of the inner products on $(\cdot\,|\,\cdot)_i$ on the $\mathcal{H}_i$, that is

$$(v_1 \otimes v_2 | w_1 \otimes w_2)_\otimes \equiv (v_1 | w_1)_1 \cdot (v_2 | w_2)_2.$$

The last subject we should touch upon is that of ***vector operators***, which are defined to be sets of operators that transform as three-dimensional vectors under the adjoint action of the total angular momentum operators J_i. That is, a vector operator is a set of operators $\{B_i\}_{i=1-3}$ (often written collectively as $\mathbf{B}$) that satisfies

[19] You may have noticed that we defined tensor products only for finite-dimensional spaces. The definition can be extended to cover infinite-dimensional Hilbert spaces, but the extra technicalities needed do not add any insight to what we're trying to do here, so we omit them. The theory of infinite-dimensional Hilbert spaces falls under the rubric of *functional analysis*, and details can be found, for example, in Reed and Simon [15].

$$\mathrm{ad}_{J_i}(B_j) = [J_i, B_j] = i\sum_{k=1}^{3}\epsilon_{ijk}B_k, \tag{3.57}$$

where ϵ_{ijk} is the familiar Levi–Civita symbol. The three-dimensional position operator $\hat{\mathbf{r}} = \{\hat{x}, \hat{y}, \hat{z}\}$, momentum operator $\hat{\mathbf{p}} = \{\hat{p}_x, \hat{p}_y, \hat{p}_z\}$, and orbital angular momentum operator $\mathbf{L} = \{L_x, L_y, L_z\}$ are all vector operators, as you can check.

Exercise 3.20. For spinless particles, $\mathbf{J} = \mathbf{L} = \hat{\mathbf{r}}\times\hat{\mathbf{p}}$. Expressions for the components may be obtained by expanding the cross product or referencing Example 2.16 and Exercise 2.11. Use these expressions and the canonical commutation relations $[x_i, p_j] = i\delta_{ij}$ to show that $\hat{\mathbf{r}}$, $\hat{\mathbf{p}}$, and $\mathbf{L}$ are all vector operators.

Now we are finally ready to consider some examples, in which we'll take as an axiom that **adding degrees of freedom is implemented by taking tensor products of the corresponding Hilbert spaces**. You will see that this process reproduces familiar results.

Example 3.18. *Addition of translational degrees of freedom*

Consider a spinless particle constrained to move in one-dimension; the quantum-mechanical Hilbert space for this system is $L^2(\mathbb{R})$ with basis $\{|x\rangle\}_{x\in\mathbb{R}}$. If we consider a second dimension, call it the y dimension, then this degree of freedom has its own Hilbert space $L^2(\mathbb{R})$ with basis $\{|y\rangle\}_{y\in\mathbb{R}}$. If we allow the particle both degrees of freedom, then the Hilbert space for the system is $L^2(\mathbb{R}) \otimes L^2(\mathbb{R})$, with basis $\{|x\rangle \otimes |y\rangle\}_{x,y\in\mathbb{R}}$. An arbitrary ket $|\psi\rangle \in L^2(\mathbb{R}) \otimes L^2(\mathbb{R})$ has expansion

$$|\psi\rangle = \int_{-\infty}^{\infty}\int_{-\infty}^{\infty} dx\, dy\, \psi(x, y)\, |x\rangle \otimes |y\rangle$$

with expansion coefficients $\psi(x, y)$. If we iterate this logic, we get in three-dimensions

$$|\psi\rangle = \int_{-\infty}^{\infty}\int_{-\infty}^{\infty}\int_{-\infty}^{\infty} dx\, dy\, dz\, \psi(x, y, z)\, |x\rangle \otimes |y\rangle \otimes |z\rangle.$$

If we rewrite $\psi(x, y, z)$ as $\psi(\mathbf{r})$ and $|x\rangle \otimes |y\rangle \otimes |z\rangle$ as $|\mathbf{r}\rangle$ where $\mathbf{r} = (x, y, z)$, then we have

$$|\psi\rangle = \int d^3r\, \psi(\mathbf{r})|\mathbf{r}\rangle$$

which is the familiar expansion of a ket in terms of three-dimensional position eigenkets. Such a ket is an element of $L^2(\mathbb{R}) \otimes L^2(\mathbb{R}) \otimes L^2(\mathbb{R})$, which is also denoted as $L^2(\mathbb{R}^3)$.[20]

[20] $L^2(\mathbb{R}^3)$ is actually defined to be the set of all square-integrable functions on $\mathbb{R}^3$, i.e. functions f satisfying

Example 3.19. *Two-particle systems*

Now consider two spinless particles in three-dimensional space, possibly interacting through some sort of potential. The two-body problem with a $1/r$ potential is a classic example of this. The Hilbert space for such a system is then $L^2(\mathbb{R}^3) \otimes L^2(\mathbb{R}^3)$, with basis $\{|\mathbf{r}_1\rangle \otimes |\mathbf{r}_2\rangle\}_{\mathbf{r}_i \in \mathbb{R}^3}$. In many textbooks the tensor product symbol is omitted and such basis vectors are written as $|\mathbf{r}_1\rangle\,|\mathbf{r}_2\rangle$, or even $|\mathbf{r}_1, \mathbf{r}_2\rangle$. A ket $|\psi\rangle$ in this Hilbert space then has expansion

$$|\psi\rangle = \int d^3\mathbf{r}_1 \int d^3\mathbf{r}_2\, \psi(\mathbf{r}_1, \mathbf{r}_2)\, |\mathbf{r}_1, \mathbf{r}_2\rangle$$

which is the familiar expansion of a ket in a two-particle Hilbert space. One can interpret $\psi(\mathbf{r}_1, \mathbf{r}_2)$ as the probability amplitude of finding particle 1 in position $\mathbf{r}_1$ and particle 2 in position $\mathbf{r}_2$ simultaneously.

Example 3.20. *Addition of orbital and spin angular momentum*

Now consider a spin s particle in three-dimensions. As remarked in Example 2.2, the ket space corresponding to the spin degree of freedom is $\mathbb{C}^{2s+1}$, and one usually takes a basis $\{|m\rangle\}_{-s\leq m\leq s}$ of S_z eigenvectors with eigenvalue m. The total Hilbert space for this system is $L^2(\mathbb{R}^3) \otimes \mathbb{C}^{2s+1}$, and we can take as a basis $\{|\mathbf{r}\rangle \otimes |m\rangle\}$ where $\mathbf{r} \in \mathbb{R}^3$ and $-s \leq m \leq s$. Again, the basis vectors are often written as $|\mathbf{r}\rangle|m\rangle$ or even $|\mathbf{r}, m\rangle$. An arbitrary ket $|\psi\rangle$ then has expansion

$$|\psi\rangle = \sum_{m=-s}^{s} \int d^3\mathbf{r}\, \psi_m(\mathbf{r}) |\mathbf{r}, m\rangle,$$

where $\psi_m(\mathbf{r})$ is the probability of finding the particle at position $\mathbf{r}$ and with m units of spin angular momentum in the z-direction. These wavefunctions are sometimes written in column vector form

$$\begin{pmatrix} \psi_s \\ \psi_{s-1} \\ \vdots \\ \psi_{-s+1} \\ \psi_{-s} \end{pmatrix}.$$

The total angular momentum operator $\mathbf{J}$ is given by $\mathbf{L} \otimes I + I \otimes \mathbf{S}$, where $\mathbf{L}$ is the orbital angular momentum operator. One might wonder why $\mathbf{J}$ isn't given by $\mathbf{L} \otimes \mathbf{S}$;

$$\int_{-\infty}^{\infty}\int_{-\infty}^{\infty}\int_{-\infty}^{\infty} dx\, dy\, dz\, |f|^2 < \infty.$$

Not too surprisingly, this space turns out to be identical to $L^2(\mathbb{R}) \otimes L^2(\mathbb{R}) \otimes L^2(\mathbb{R})$.

there is a good answer to this question, but it requires delving into the (fascinating) subject of Lie groups and Lie algebras, which we postpone until Part II. In the meantime, you can get a partial answer by checking (Exercise 3.21 below) that the operators $L_i \otimes S_i$ don't satisfy the angular momentum commutation relations whereas the $L_i \otimes I + I \otimes S_i$ do.

Exercise 3.21. Check that

$$[L_i \otimes I + I \otimes S_i, L_j \otimes I + I \otimes S_j] = \sum_{k=1}^{3} \epsilon_{ijk}(L_k \otimes I + I \otimes S_k).$$

Also show that

$$[L_i \otimes S_i, L_j \otimes S_j] \neq \sum_{k=1}^{3} \epsilon_{ijk} L_k \otimes S_k.$$

Be sure to use the bilinearity of the tensor product carefully.

Example 3.21. *Addition of spin angular momentum*

Next consider two particles of spin s_1 and s_2 respectively, fixed in space so that they have no translational degrees of freedom. The Hilbert space for this system is $\mathbb{C}^{2s_1+1} \otimes \mathbb{C}^{2s_2+1}$, with basis $\{|m_1\rangle \otimes |m_2\rangle\}$ where $-s_i \leq m_i \leq s_i$, $i = 1, 2$. Again, such tensor product kets are usually abbreviated as $|m_1\rangle|m_2\rangle$ or $|m_1, m_2\rangle$. There are several important linear operators on $\mathbb{C}^{2s_1+1} \otimes \mathbb{C}^{2s_2+1}$:

$\mathbf{S}_1 \otimes I$	Vector spin operator on first particle
$I \otimes \mathbf{S}_2$	Vector spin operator on second particle
$\mathbf{S} \equiv \mathbf{S}_1 \otimes I + I \otimes \mathbf{S}_2$	Total vector spin operator
$\mathbf{S}^2 \equiv \sum_i S_i S_i$	Total spin squared operator

(Why aren't $\mathbf{S}_1^2$ and $\mathbf{S}_2^2$ in our list above?) The vectors $|m_1, m_2\rangle$ are clearly eigenvectors of S_{1z} and S_{2z} and hence S_z [we abuse notation as mentioned below (3.57)] but, as you will show in Exercise 3.22, they are *not* necessarily eigenvectors of $\mathbf{S}^2$. However, since the S_i obey the angular momentum commutation relations (as you can check), the general theory of angular momentum tells us that we *can* find a basis for $\mathbb{C}^{2s_1+1} \otimes \mathbb{C}^{2s_2+1}$ consisting of eigenvectors of S_z and $\mathbf{S}^2$. Furthermore, it can be shown that the $\mathbf{S}^2$ eigenvalues that occur are $s(s+1)$ where

$$s = |s_1 - s_2|, \; |s_1 - s_2| + 1, \; \ldots \; , \; s_1 + s_2 \tag{3.58}$$

and for a given s the possible S_z eigenvalues are m where $-s \leq m \leq s$ as usual (see Example 5.23 for further discussion of this). We will write these basis kets as $\{|s, m\rangle\}$ where the above restrictions on s and m are understood, and where we physically interpret $\{|s, m\rangle\}$ as a state with total angular momentum equal to $\sqrt{s(s+1)}$ and with m units of angular momentum pointing along the z-axis. We then have two natural and useful bases for $\mathbb{C}^{2s_1+1} \otimes \mathbb{C}^{2s_2+1}$:

$$\mathcal{B} = \{|m_1, m_2\rangle\} \quad -s_1 \leq m_1 \leq s_1, \quad -s_2 \leq m_2 \leq s_2$$

$$\mathcal{B}' = \{|s, m\rangle\} \quad |s_1 - s_2| \leq s \leq s_1 + s_2, \quad -s \leq m \leq s.$$

What does the transformation between these two bases look like? Well, by their definition, the $A^j_{i'}$ relating the two bases are given by $e^j(e_{i'})$; using s, m collectively in lieu of the primed index and m_1, m_2 collectively in lieu of the unprimed index, we have, in Dirac notation,

$$A^{m_1,m_2}_{s,m} = \langle m_1, m_2 | s, m\rangle. \tag{3.59}$$

These numbers, the notation for which varies widely throughout the literature, are known as ***Clebsch–Gordan Coefficients***, and methods for computing them can be found in any standard quantum mechanics textbook (e.g., Sakurai [17]).

Let us illustrate the foregoing with an example. Take two spin 1 particles, so that $s_1 = s_2 = 1$. The Hilbert space for the first particle is $\mathbb{C}^3$, with S_{1z} eigenvector basis $\{|-1\rangle, |0\rangle, |1\rangle\}$, and so the two-particle system has nine-dimensional Hilbert space $\mathbb{C}^3 \otimes \mathbb{C}^3$ with corresponding basis

$$\begin{aligned}\mathcal{B} &= \{|i\rangle\,|j\rangle \mid i, j = -1, 0, 1\}\\ &= \{|1\rangle\,|1\rangle,\ |1\rangle\,|0\rangle,\ |1\rangle\,|-1\rangle,\ |0\rangle\,|1\rangle, \text{etc.}\}.\end{aligned}$$

There should also be another basis consisting of S_z and $\mathbf{S}^2$ eigenvectors, however. From (3.58) we know that the possible s values are $s = 0, 1, 2$, and it is a standard exercise in angular momentum theory to show that the nine (normalized) S_z and $\mathbf{S}^2$ eigenvectors are

$$\begin{aligned}|s = 2,\ s_z = 2\rangle &= |1\rangle\,|1\rangle\\
|2, 1\rangle &= \frac{1}{\sqrt{2}}(\,|1\rangle\,|0\rangle + |0\rangle\,|1\rangle\,)\\
|2, 0\rangle &= \frac{1}{\sqrt{6}}(\,|1\rangle\,|-1\rangle + 2\,|0\rangle\,|0\rangle + |-1\rangle\,|1\rangle\,)\\
|2, 1\rangle &= \frac{1}{\sqrt{2}}(\,|-1\rangle\,|0\rangle + |0\rangle\,|-1\rangle\,)\\
|2, -2\rangle &= |-1\rangle\,|-1\rangle\\
|1, 1\rangle &= \frac{1}{\sqrt{2}}(\,|1\rangle\,|0\rangle - |0\rangle\,|1\rangle\,)\\
|1, 0\rangle &= \frac{1}{\sqrt{2}}(\,|1\rangle\,|-1\rangle - |-1\rangle\,|1\rangle\,)\\
|1, 1\rangle &= \frac{1}{\sqrt{2}}(\,|0\rangle\,|-1\rangle - |-1\rangle\,|0\rangle\,)\\
|0, 0\rangle &= \frac{1}{\sqrt{3}}(\,|0\rangle\,|0\rangle - |1\rangle\,|-1\rangle - |-1\rangle\,|1\rangle\,).\end{aligned} \tag{3.60}$$

These vectors can be found using standard techniques from the theory of "addition of angular momentum"; for details, see Gasiorowicz [7] or Sakurai [17]. The coefficients appearing on the right-hand side of the above equations are precisely the Clebsch–Gordan coefficients (3.59), as a moment's thought should show.

Exercise 3.22. Show that

$$\mathbf{S}^2 = \mathbf{S}_1^2 \otimes I + I \otimes \mathbf{S}_2^2 + 2\sum_i S_{1i} \otimes S_{2i}.$$

The right-hand side of the above equation is usually abbreviated as $\mathbf{S}_1^2 + \mathbf{S}_2^2 + 2\mathbf{S}_1 \cdot \mathbf{S}_2$. Use this to show that $|m_1, m_2\rangle$ is *not* generally an eigenvector of $\mathbf{S}^2$.

Example 3.22. *Entanglement*

Consider two Hilbert spaces $\mathcal{H}_1$ and $\mathcal{H}_2$ and their tensor product $\mathcal{H}_1 \otimes \mathcal{H}_2$. Only some of the vectors in $\mathcal{H}_1 \otimes \mathcal{H}_2$ can be written as $\psi \otimes \phi$; such vectors are referred to as ***separable states*** or ***product states***. All other vectors must be written as linear combinations of the form $\sum_i \psi_i \otimes \phi_i$, and these vectors are said to be ***entangled***, since in this case the measurement of the degrees of freedom represented by $\mathcal{H}_1$ will influence the measurement of the degrees of freedom represented by $\mathcal{H}_2$. The classic example of an entangled state comes from the previous example of two fixed particles with spin; taking $s_1 = s_2 = 1/2$ and writing the standard basis for $\mathbb{C}^2$ as $\{|+\rangle, |-\rangle\}$, we consider the particular state

$$|+\rangle\,|-\rangle - |-\rangle\,|+\rangle\,. \tag{3.61}$$

If an observer measures the first particle to be spin up, then a measurement of the second particle's spin is guaranteed to be spin-down, and vice-versa, so measuring one part of the system affects what one will measure for the other part. This is the sense in which the system is entangled. For a product state $\psi \otimes \phi$, there is no such entanglement: a particular measurement of the first particle cannot affect what one measures for the second, since the second particle's state will be ϕ no matter what. You will check below that (3.61) is not a product state.

Exercise 3.23. Prove that (3.61) cannot be written as $\psi \otimes \phi$ for any $\psi, \phi \in \mathbb{C}^2$. Do this by expanding ψ and ϕ in the given basis and showing that no choice of expansion coefficients for ψ and ϕ will yield (3.61).

3.8 Symmetric Tensors

Given a vector space V there are certain subspaces of $\mathcal{T}_0^r(V)$ and $\mathcal{T}_r^0(V)$ which are of particular interest: the symmetric and antisymmetric tensors. We'll discuss symmetric tensors in this section and antisymmetric tensors in the next. A ***symmetric*** $(r, 0)$ tensor is an $(r, 0)$ tensor whose value is unaffected by the interchange (or ***transposition***) of any two of its arguments, that is

$$T(v_1, \ldots, v_i, \ldots, v_j, \ldots, v_r) = T(v_1, \ldots, v_j, \ldots, v_i, \ldots, v_r)$$

for any i and j. Symmetric $(0, r)$ tensors are defined similarly. You can easily check that the symmetric $(r, 0)$ and $(0, r)$ tensors each form vector spaces, denoted $S^r(V^*)$ and $S^r(V)$ respectively. For $T \in S^r(V^*)$, the symmetry condition implies that the components $T_{i_1 \ldots i_r}$ are invariant under the transposition of any two indices, hence invariant under *any* rearrangement of the indices (since any rearrangement can be obtained via successive transpositions). Similar remarks apply, of course, to $S^r(V)$. Notice that for rank 2 tensors, the symmetry condition implies $T_{ij} = T_{ji}$ so that $[T]_{\mathcal{B}}$ for any $\mathcal{B}$ is a symmetric matrix. Also note that it doesn't mean anything to say that a linear operator is symmetric, since a linear operator is a $(1, 1)$ tensor and there is no way of transposing the arguments. One might find that the matrix of a linear operator is symmetric in a certain basis, but this won't necessarily be true in other bases. If we have a metric to raise and lower indices, then we can, of course, speak of symmetry by turning our linear operator into a $(2, 0)$ or $(0, 2)$ tensor.

Example 3.23. $S^2(\mathbb{R}^{2*})$

Consider the set $\{e^1 \otimes e^1,\ e^2 \otimes e^2,\ e^1 \otimes e^2 + e^2 \otimes e^1\} \subset S^2(\mathbb{R}^{2*})$ where $\{e^i\}_{i=1,2}$ is the standard dual basis. You can check that this set is linearly independent, and that any symmetric tensor can be written as

$$T = T_{11} e^1 \otimes e^1 + T_{22} e^2 \otimes e^2 + T_{12}(e^1 \otimes e^2 + e^2 \otimes e^1) \tag{3.62}$$

so this set is a basis for $S^2(\mathbb{R}^{2*})$, which is thus three-dimensional. In particular, the Euclidean metric g on $\mathbb{R}^2$ can be written as

$$g = e^1 \otimes e^1 + e^2 \otimes e^2$$

since $g_{11} = g_{22} = 1$ and $g_{12} = g_{21} = 0$. Note that g would *not* take this simple form in a non-orthonormal basis. □

Exercise 3.24. Let $V = \mathbb{R}^n$ with the standard basis $\mathcal{B}$. Convince yourself that

$$[e_i \otimes e_j + e_j \otimes e_i]_{\mathcal{B}} = S_{ij},$$

where S_{ij} is the symmetric matrix defined in Example 2.9.

There are many symmetric tensors in physics, almost all of them of rank 2. Many of them we've met already: the Euclidean metric on $\mathbb{R}^3$, the Minkowski metric on $\mathbb{R}^4$, the moment of inertia tensor, and the Maxwell stress tensor. You should refer to the examples and check that these are all symmetric tensors. We have also met one class of higher rank symmetric tensors: the multipole moments.

Example 3.24. *Multipole moments and harmonic polynomials*

Recall from Example 3.4 that the scalar potential $\Phi(\mathbf{r})$ of a charge distribution $\rho(\mathbf{r}')$ localized around the origin in $\mathbb{R}^3$ can be expanded in a Taylor series in $1/r$ as

$$\Phi(\mathbf{r}) = \frac{1}{4\pi}\left[\frac{Q_0}{r} + \frac{Q_1(\mathbf{r})}{r^3} + \frac{1}{2!}\frac{Q_2(\mathbf{r},\mathbf{r})}{r^5} + \frac{1}{3!}\frac{Q_3(\mathbf{r},\mathbf{r},\mathbf{r})}{r^7}\cdots\right], \tag{3.63}$$

where the Q_l are the symmetric rank l multipole moment tensors. Each symmetric tensor Q_l can be interpreted as a degree l polynomial f_l, just by evaluating on l copies of $\mathbf{r} = (x^1, x^2, x^3)$ as indicated:

$$f_l(\mathbf{r}) \equiv Q_l(\mathbf{r},\ldots,\mathbf{r}) = Q_{i_1\cdots i_l}x^{i_1}\cdots x^{i_l}, \tag{3.64}$$

where the indices i_j are the usual component indices, *not* exponents. Note that the expression $x^{i_1}\cdots x^{i_l}$ in the right-hand side is invariant under any rearrangement of the indices i_j. This is because we fed in l copies of the *same* vector $\mathbf{r}$ into Q_l. This fits in nicely with the symmetry of $Q_{i_1\cdots i_l}$. In fact, the above equation gives a one-to-one correspondence between lth rank symmetric tensors and degree l polynomials; we won't prove this correspondence here, but it shouldn't be too hard to see that (3.64) turns any symmetric tensor into a polynomial, and that, conversely, any fixed degree polynomial can be written in the form of the right-hand side of (3.64) with $Q_{i_1\cdots i_l}$ symmetric. This (roughly) explains why the multipole moments are symmetric tensors: the multipole moments are really just fixed degree polynomials, which in turn correspond to symmetric tensors.

What about the tracelessness of the Q_l, i.e. the fact that $\sum_k Q_{i_1\cdots k\cdots k\cdots i_n} = 0$? Well, $\Phi(\mathbf{r})$ obeys the Laplace equation $\Delta\Phi(\mathbf{r}) = 0$, which means that every term in the expansion (3.63) term is of the form

$$\frac{f_l(\mathbf{r})}{r^{2l+1}}.$$

If we write the polynomial $f_l(\mathbf{r})$ as $r^l Y(\theta,\phi)$ then a quick computation shows that $Y(\theta,\phi)$ must be a spherical harmonic of degree l, and hence f_l must be a harmonic polynomial! Expanding $f_l(\mathbf{r})$ in the form (3.64) and applying the Laplacian then shows that if f_l is harmonic, then Q_l must be traceless. □

Exercise 3.25. What is the polynomial associated with the Euclidean metric tensor $g = \sum_{i=1}^{3} e^i \otimes e^i$? What is the symmetric tensor in $S^3(\mathbb{R}^3)$ associated with the polynomial x^2y?

Exercise 3.26. Substitute $f_l(\mathbf{r}) = r^l Y(\theta,\phi)$ into the equation

$$\Delta\left(\frac{f_l(\mathbf{r})}{r^{2l+1}}\right) = 0$$

and show that $Y(\theta,\phi)$ must be a spherical harmonic of degree l, and thus that f_l is harmonic. Then use (3.64) to show that if f_l is a harmonic polynomial, then the associated symmetric tensor Q_l must be traceless. If you have trouble showing that Q_l is traceless for arbitrary l, try starting with the $l = 2$ (dipole) and $l = 3$ (octopole) cases.

3.9 Antisymmetric Tensors

Now we turn to antisymmetric tensors. An ***antisymmetric*** (or ***alternating***) $(r, 0)$ tensor is one whose value *changes sign* under transposition of any two of its arguments, i.e.

$$T(v_1, \ldots, v_i, \ldots, v_j, \ldots, v_r) = -T(v_1, \ldots, v_j, \ldots, v_i, \ldots, v_r). \tag{3.65}$$

Again, antisymmetric $(0, r)$ tensors are defined similarly and both sets form vector spaces, denoted $\Lambda^r V^*$ and $\Lambda^r V$ (for $r = 1$ we define $\Lambda^1 V^* = V^*$ and $\Lambda^1 V = V$). The following properties of antisymmetric tensors follow directly from (3.65); the first one is immediate, and the second two you will prove in Exercise 3.27 below:

1. $T(v_1, \ldots, v_r) = 0$ if $v_i = v_j$ for any $i \neq j$

$\Longrightarrow$ 2. $T(v_1, \ldots, v_r) = 0$ if $\{v_1, \ldots, v_r\}$ is linearly dependent

$\Longrightarrow$ 3. If $\dim V = n$, then the only tensor in $\Lambda^r V^*$ and $\Lambda^r V$ for $r > n$ is the 0 tensor.

An important operation on antisymmetric tensors is the ***wedge product***: Given $f, g \in V^*$ we define the wedge product of f and g, denoted $f \wedge g$, to be the antisymmetric $(2, 0)$ tensor defined by

$$\boxed{f \wedge g \equiv f \otimes g - g \otimes f.} \tag{3.66}$$

Note that $f \wedge g = -g \wedge f$, and that $f \wedge f = 0$. Expanding (3.66) in terms of the e^i gives

$$f \wedge g = f_i g_j (e^i \otimes e^j - e^j \otimes e^i) = f_i g_j \, e^i \wedge e^j \tag{3.67}$$

so that $\{e^i \wedge e^j\}_{i<j}$ spans all wedge products of dual vectors (note the "$i < j$" stipulation, since $e^i \wedge e^j$ and $e^j \wedge e^i$ are not linearly independent). In fact, you will check in Exercise 3.28 that $\{e^i \wedge e^j\}_{i<j}$ is linearly independent and spans $\Lambda^2 V^*$, hence is a basis for $\Lambda^2 V^*$. The wedge product can be extended to r-fold products of dual vectors as follows: given r dual vectors $f_1, \ldots, f_r$, we define their wedge product $f_1 \wedge \cdots \wedge f_r$ to be the sum of all tensor products of the form $f_{i_1} \otimes \cdots \otimes f_{i_r}$ where each term gets a $+$ or a $-$ sign depending on whether an odd or an even number[21] of transpositions of the factors are necessary to obtain it from $f_1 \otimes \cdots \otimes f_r$; if the number is odd the term is assigned -1, if even a $+1$. Thus,

[21]The number of transpositions required to get a given rearrangement is not unique, of course, but hopefully you can convince yourself that it's always odd or always even. A rearrangement which always decomposes into an odd number of transpositions is an ***odd rearrangement***, and

$$f_1 \wedge f_2 = f_1 \otimes f_2 - f_2 \otimes f_1 \tag{3.68a}$$

$$\begin{aligned} f_1 \wedge f_2 \wedge f_3 = {} & f_1 \otimes f_2 \otimes f_3 + f_2 \otimes f_3 \otimes f_1 + f_3 \otimes f_1 \otimes f_2 \\ & - f_3 \otimes f_2 \otimes f_1 - f_2 \otimes f_1 \otimes f_3 - f_1 \otimes f_3 \otimes f_2 \end{aligned} \tag{3.68b}$$

and so on. You should convince yourself that $\{e^{i_1} \wedge \cdots \wedge e^{i_r}\}_{i_1 < \cdots < i_r}$ is a basis for $\Lambda^r V^*$ (see Example 3.25 below for examples of this). Note that this entire construction can be carried out for vectors as well as dual vectors. Also note that all the comments about symmetry above Example 3.23 apply here as well.

Exercise 3.27. Let $T \in \Lambda^r V^*$. Show that if $\{v_1, \ldots, v_r\}$ is a linearly dependent set then $T(v_1, \ldots, v_r) = 0$. Use the same logic to show that if $\{f_1, \ldots, f_r\} \subset V^*$ is linearly dependent, then $f_1 \wedge \cdots \wedge f_r = 0$. If $\dim V = n$, show that any set of more than n vectors must be linearly dependent, so that $\Lambda^r V = \Lambda^r V^* = 0$ for $r > n$.

Exercise 3.28. Prove that $\{e^i \wedge e^j\}_{i<j}$ is linearly independent by evaluating an arbitrary linear combination on an arbitrary tensor product $e_k \otimes e_l$. Also prove that $\{e^i \wedge e^j\}_{i<j}$ spans $\Lambda^2 V^*$, and hence is a basis for it. You can do this with an argument analogous to that used in and below (3.39).

Example 3.25. *Antisymmetric tensors on* $\mathbb{R}^3$

The algebra of antisymmetric tensors can be a bit intimidating at first, so it may help to warm up with a basic example. By referring to the discussion above, you should convince yourself that the rank r antisymmetric tensors on $\mathbb{R}^3$ have bases as follows:

$$\begin{aligned} \Lambda^1\mathbb{R}^{3*} = \mathbb{R}^{3*} & \text{ has basis } \left\{e^1,\, e^2,\, e^3\right\} \\ \Lambda^2\mathbb{R}^{3*} & \text{ has basis } \left\{e^1 \wedge e^2,\, e^2 \wedge e^3,\, e^1 \wedge e^3\right\}. \\ \Lambda^3\mathbb{R}^{3*} & \text{ has basis } \left\{e^1 \wedge e^2 \wedge e^3\right\}. \end{aligned}$$

See (3.68) for the expansions of the above wedge products in terms of tensor products.

Exercise 3.29. Expand the (2,0) electromagnetic field tensor of (3.49) in the basis $\{e^i \wedge e^j\}$ where $i < j$ and $i, j = 1, 2, 3, 4$.

Exercise 3.30. Let $\dim V = n$. Show that the dimension of $\Lambda^r V^*$ and $\Lambda^r V$ is $\binom{n}{r} = \frac{n!}{(n-r)!\,r!}$.

Example 3.26. *Identical particles*

In quantum mechanics we often consider systems which contain *identical particles*, i.e. particles of the same mass, charge, and spin. For instance, we might consider n non-interacting hydrogen atoms moving in a potential well, or two electrons of an helium atom orbiting around their nucleus. In such cases we would assume that the

---

even rearrangements are defined similarly. We'll discuss this further in Chap. 4, specifically in Example 4.25.

total Hilbert space $\mathcal{H}_{\text{tot}}$ would be just the n-fold tensor product of the single particle Hilbert space $\mathcal{H}$. This is problematic, however. Consider two identical particles in a box, one with state $|\psi\rangle$ and another with state $|\phi\rangle$. Then the above logic implies that the composite state of the two particles is given by $|\psi\rangle \otimes |\phi\rangle \in \mathcal{H}_{\text{tot}} = \mathcal{H} \otimes \mathcal{H}$. The trouble here is that this state is physically indistinguishable from $|\phi\rangle \otimes |\psi\rangle$, since we have no way of labeling the particles so that we know which one is the "first" and which the "second." This makes $\mathcal{H} \otimes \mathcal{H}$ mathematically redundant as a description of the composite system.

To formalize the issue we can introduce the linear "permutation operator" P on $\mathcal{H}_{\text{tot}}$, which just switches the factors in the tensor product, i.e.

$$P(|\psi\rangle \otimes |\phi\rangle) \equiv |\phi\rangle \otimes |\psi\rangle \,. \tag{3.69}$$

Since such a permutation is neither physically executable nor observable, it seems reasonable to require that any vector $v \in \mathcal{H}_{\text{tot}}$ representing a physical state should be invariant under P, at least up to a phase, so that Pv and v represent the same quantum state. In other words, a physical state v should be an eigenvector of P! But by (3.69) it's clear that $P^2 = I$, which when applied to the eigenvector v tells you that $Pv = \pm v$. This means that for our two particles in a box, our options are

$$\tfrac{1}{2}(|\psi\rangle \otimes |\phi\rangle + |\phi\rangle \otimes |\psi\rangle) \in S^2(\mathcal{H})$$
$$\tfrac{1}{2}(|\psi\rangle \otimes |\phi\rangle - |\phi\rangle \otimes |\psi\rangle) \in \Lambda^2(\mathcal{H}),$$

where the P eigenvalues are $+1$ and -1, respectively. But, which of these states should we take? The amazing empirical fact is that nature takes advantage of both states, for different particles. For certain particles (known as ***bosons***) only the state in $S^2(\mathcal{H})$ would be observed, while for other particles (known as ***fermions***) only the state in $\Lambda^2\mathcal{H}$ would be observed. For n-particle systems, bosons would similarly only occupy states in $S^n(\mathcal{H})$ while fermions would only live in $\Lambda^n\mathcal{H}$. This restriction of the total Hilbert space to either $S^n(\mathcal{H})$ or $\Lambda^n\mathcal{H}$ is known as the ***symmetrization postulate***.[22] It is an empirical fact that all known particles are fermions or bosons, in accordance with the symmetrization postulate.

All this has far-reaching consequences. For instance, if we have two fermions, we cannot measure the same values for a complete set of quantum numbers for both particles, since then the state would have to include a term of the form $|\psi\rangle\,|\psi\rangle$ and thus couldn't belong to $\Lambda^2\mathcal{H}$. This fact that two fermions can't be in the same state is

[22] In relativistic quantum field theory, as opposed to non-relativistic quantum mechanics, this fact is no longer an additional postulate but rather an internally deducible fact, known as the ***spin-statistics theorem***. The spin-statistics theorem furthermore states that bosons have integer spin and fermions have half-integral spin. See Zee [24] for a discussion and further references. It is also possible to deduce the symmetrization postulate from our assumption that n-particle states are invariant under permutation operators such as P from (3.69); proving this requires group theory, however, and so is postponed to Sect. 5.3.

known as the ***Pauli Exclusion Principle***. As another example, consider two identical spin 1/2 fermions fixed in space, so that $\mathcal{H}_{\text{tot}} = \Lambda^2\mathbb{C}^2$. $\Lambda^2\mathbb{C}^2$ is one-dimensional with basis vector

$$|0,0\rangle = \left|\frac{1}{2}\right\rangle\left|-\frac{1}{2}\right\rangle - \left|-\frac{1}{2}\right\rangle\left|\frac{1}{2}\right\rangle,$$

where we have used the notation of Example 3.21. If we measure $\mathbf{S}^2$ or S_z for this system we will get 0. This is in marked contrast to the case of two *distinguishable* spin 1/2 fermions; in this case, the Hilbert space is $\mathbb{C}^2 \otimes \mathbb{C}^2$ and we have additional possible state kets

$$\begin{aligned}
|1,1\rangle &= \left|\frac{1}{2}\right\rangle\left|\frac{1}{2}\right\rangle \\
|1,0\rangle &= \left|\frac{1}{2}\right\rangle\left|-\frac{1}{2}\right\rangle + \left|-\frac{1}{2}\right\rangle\left|\frac{1}{2}\right\rangle \\
|1,-1\rangle &= \left|-\frac{1}{2}\right\rangle\left|-\frac{1}{2}\right\rangle
\end{aligned}$$

which yield nonzero values for $\mathbf{S}^2$ and S_z. □

The next three examples are a little more mathematical than physical but they are necessary for the discussion of pseudovectors in the next section. Hopefully you'll also find them of interest in their own right.

Example 3.27. *The Levi–Civita tensor*

Consider $\mathbb{R}^n$ with the standard inner product. Let $\{e_i\}_{i=1\ldots n}$ be an orthonormal basis for $\mathbb{R}^n$ and consider the tensor

$$\boxed{\epsilon \equiv e^1 \wedge \cdots \wedge e^n} \in \Lambda^n \mathbb{R}^{n*}.$$

You can easily check that

$$\epsilon_{i_1\ldots i_n} = \begin{cases} 0 & \text{if } \{i_1,\ldots,i_n\} \text{ contains a repeated index} \\ -1 & \text{if } \{i_1,\ldots,i_n\} \text{ is an odd rearrangement of } \{1,\ldots,n\} \\ +1 & \text{if } \{i_1,\ldots,i_n\} \text{ is an even rearrangement of } \{1,\ldots,n\}. \end{cases}$$

For $n = 3$, we saw in Example 1.1 (and you can also check in Exercise 3.31) that ϵ_{ijk} has the same values as the Levi–Civita symbol, and so ϵ here is an n-dimensional generalization of the three-dimensional Levi–Civita tensor we introduced in Examples 1.1 and 3.2. As in those examples, ϵ should be thought of as eating n vectors and spitting out the n-dimensional volume spanned by those vectors. This can be seen explicitly for $n = 2$ also. Considering two vectors u and v in the $x-y$ plane, we have

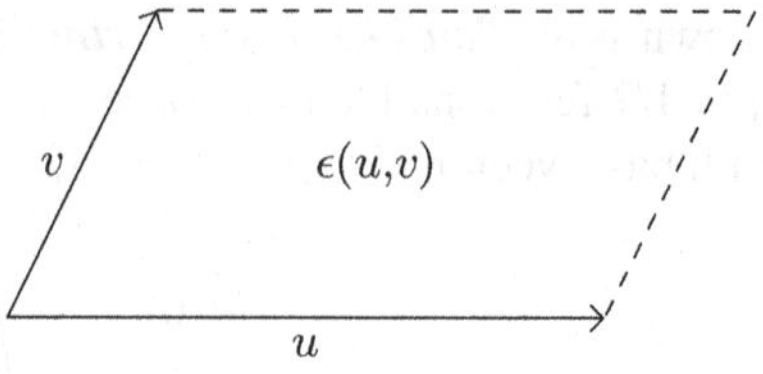

Fig. 3.6 $\epsilon(u, v)$ is just the (signed) area of the parallelogram formed by u and v

$$
\begin{aligned}
\epsilon(u, v) &= \epsilon_{ij} u^i v^j \\
&= u^x v^y - u^y v^x \\
&= (u \times v)^z
\end{aligned}
$$

and we know that this last expression can be interpreted as the (signed) area of the parallelogram spanned by u and v; See Fig. 3.6.

Finally, note that $\Lambda^n \mathbb{R}^{n*}$ is one-dimensional, and that ϵ is the basis for it described under (3.68b).

You may object that our construction of ϵ seems to depend on a choice of metric and orthonormal basis. The former is true: ϵ *does* depend on the metric, and we make no apologies for that. As to whether it depends on a particular choice of orthonormal basis, we must do a little bit of investigating; this will require a brief detour into the subject of determinants.

Exercise 3.31. You may have seen the values of ϵ_{ijk} defined in terms of cyclic and anti-cyclic permutations.[23] The point of this exercise is to make the connection between that definition and ours, and to see to what extent that definition extends to higher dimensions.

(a) Check that the ϵ tensor on $\mathbb{R}^3$ satisfies

$$
\epsilon_{ijk} = \begin{cases} +1 & \text{if } \{i, j, k\} \text{ is a cyclic permutation of } \{1, 2, 3\} \\ -1 & \text{if } \{i, j, k\} \text{ is an anticyclic permutation of } \{1, 2, 3\} \\ 0 & \text{otherwise.} \end{cases}
$$

Thus for three indices the cyclic permutations are the even rearrangements and the anti-cyclic permutations are the odd ones.

(b) Consider ϵ on $\mathbb{R}^4$ with components ϵ_{ijkl}. Are all rearrangements of $\{1, 2, 3, 4\}$ necessarily cyclic or anti-cyclic?

(c) Is it true that $\epsilon_{ijkl} = 1$ if $\{i, j, k, l\}$ is a cyclic permutation of $\{1, 2, 3, 4\}$?

Example 3.28. *The determinant*

You have doubtless encountered determinants before, and have probably seen them defined iteratively; that is, the determinant of a 2×2 square matrix A, denoted $|A|$ (or $\det A$), is defined to be

---

[23]A ***cyclic permutation*** of $\{1, \ldots, n\}$ is any rearrangement of $\{1, \ldots, n\}$ obtained by successively moving numbers from the beginning of the sequence to the end. That is, $\{2, \ldots, n, 1\}, \{3, \ldots, n, 1, 2\}$, and so on are the cyclic permutations of $\{1, \ldots, n\}$. ***Anti-cyclic*** permutations are cyclic permutations of $\{n, n-1, \ldots, 1\}$.

$$|A| \equiv A_{11}A_{22} - A_{12}A_{21} \tag{3.70}$$

and then the determinant of a 3×3 matrix B is defined in terms of this, i.e.

$$|B| \equiv B_{11}\begin{vmatrix} B_{22} & B_{23} \\ B_{32} & B_{33} \end{vmatrix} - B_{12}\begin{vmatrix} B_{21} & B_{23} \\ B_{31} & B_{33} \end{vmatrix} + B_{13}\begin{vmatrix} B_{21} & B_{22} \\ B_{31} & B_{32} \end{vmatrix}. \tag{3.71}$$

This expression is known as the ***cofactor expansion*** of the determinant, and is *not* unique; one can expand about any row (or column), not necessarily (B_{11}, B_{12}, B_{13}).

In our treatment of the determinant we will take a somewhat more sophisticated approach.[24] Take an $n \times n$ matrix A and consider its n columns as n column vectors in $\mathbb{R}^n$, i.e.

$$A = \begin{pmatrix} A_1 & A_2 & \cdots & A_n \\ \vdots & \vdots & \vdots & \vdots \end{pmatrix}.$$

Thus, the first column vector A_1 has ith component A_{i1}, and so on. Then, constructing the ϵ tensor using the standard basis and inner product on $\mathbb{R}^n$, we define the ***determinant of*** A, denoted $|A|$ or $\det A$, to be

$$\boxed{|A| \equiv \epsilon(A_1, \ldots, A_n)} \tag{3.72}$$

or in components

$$\boxed{|A| = \sum_{i_1,\ldots,i_n} \epsilon_{i_1\ldots i_n} A_{i_1 1} \ldots A_{i_n n}.} \tag{3.73}$$

You should check explicitly that this definition reproduces (3.70) and (3.71) for $n = 2, 3$. You can also check in the Problems that many of the familiar properties of determinants (sign change under interchange of columns, invariance under addition of rows, factoring of scalars) follow quite naturally from the definition and the multilinearity and antisymmetry of ϵ.

Since the determinant is defined in terms of the epsilon tensor, which has an interpretation in terms of volume, then perhaps the determinant also has an interpretation in terms of volume. Consider our matrix A as a linear operator on $\mathbb{R}^n$; then, A sends the standard orthonormal basis $\{e_1, \ldots, e_n\}$ to a new, potentially non-orthonormal basis $\{Ae_1, \ldots, Ae_n\}$. If $\{e_1, \ldots, e_n\}$ spans a regular n-cube whose volume is $\epsilon(e_1, \ldots, e_n) = 1$, then the vectors $\{Ae_1, \ldots, Ae_n\}$ span a skewed n-cube with volume given by $\epsilon(Ae_1, \ldots, Ae_n)$. To evaluate this volume, recall from Box 2.4

[24] For a complete treatment, however, you should consult Hoffman and Kunze [13], Chap. 5.

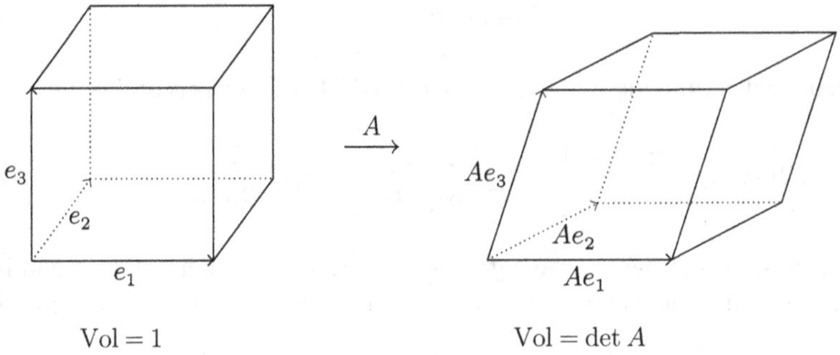

Fig. 3.7 The action of A on the standard cube in $\mathbb{R}^3$. The determinant of A is just the volume of the skew cube spanned by $\{Ae_1, Ae_2, Ae_3\}$

that Ae_1 is nothing but A_1, the first column of A; thus, by the definition (3.72), the volume is just det A! We thus conclude that:

The determinant of a matrix A is the (oriented) volume of the skew n-cube obtained by applying A to the standard n-cube.

This is illustrated for $n = 3$ in Fig. 3.7.

You may have noticed that this volume can be negative, which is why we called the determinant an *oriented* (or *signed*) volume; the interpretation of this is given in the next example.

Example 3.29. *Orientations and the ϵ tensor*

Note: This material is a bit abstract and may be skipped on a first reading.

With the determinant in hand we may now explore to what extent the definition of ϵ depends on our choice of orthonormal basis. Consider another orthonormal basis $\{e_{i'} = A^j_{i'} e_j\}$. If we define an ϵ' in terms of this basis, we find

$$\begin{aligned}
\epsilon' &= e^{1'} \wedge \cdots \wedge e^{n'} \\
&= A^{1'}_{i_1} \ldots A^{n'}_{i_n} e^{i_1} \wedge \cdots \wedge e^{i_n} \\
&= A^{1'}_{i_1} \ldots A^{n'}_{i_n} \epsilon^{i_1 \ldots i_n} e^1 \wedge \cdots \wedge e^n \\
&= |A| \epsilon,
\end{aligned} \tag{3.74}$$

where in the third equality we used the fact that if $e^{i_1} \wedge \cdots \wedge e^{i_n}$ doesn't vanish it can always be rearranged to give $e^1 \wedge \cdots \wedge e^n$, and any resulting sign change is accounted for by the Levi–Civita symbol. Now since both $\{e_i\}$ and $\{e_{i'}\}$ are orthonormal bases, A must be an orthogonal matrix. We can then use the product rule for determinants

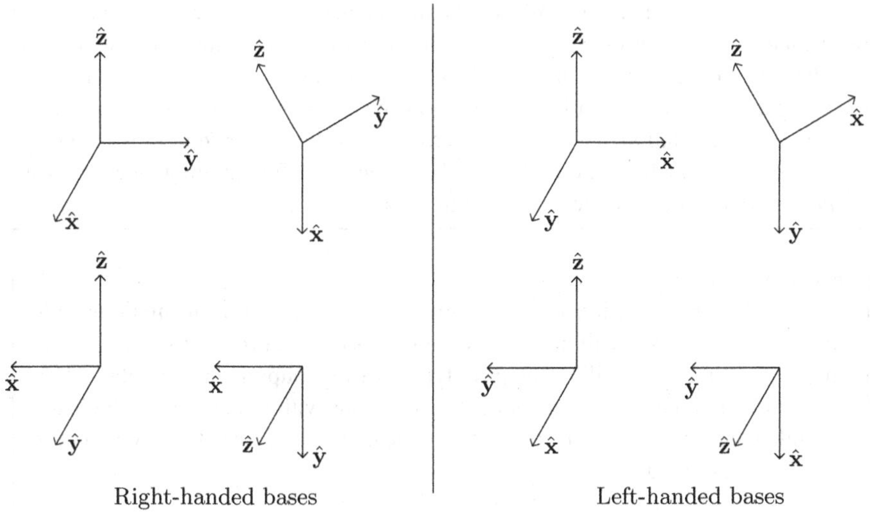

Fig. 3.8 The two orientations of $\mathbb{R}^3$. The *upper-left* most basis is usually considered the standard basis

$|AB| = |A||B|$ (see Problem 3-4 for a simple proof) and the fact that $|A^T| = |A|$ to get

$$1 = |I| = |AA^T| = |A||A^T| = |A|^2$$

which implies $|A| = \pm 1$. Thus by (3.74) $\epsilon' = \epsilon$ if the two orthonormal bases used in their construction are related by an orthogonal transformation A with $|A| = 1$; such a transformation is called a ***rotation***,[25] and two bases related by a rotation, or by any transformation with $|A| > 0$, are said to have the ***same orientation***. If two bases are related by a basis transformation with $|A| < 0$, then the two bases are said to have the ***opposite orientation***. We can then define an ***orientation*** as a maximal[26] set of bases all having the same orientation, and you can show (see Problem 3-6) that $\mathbb{R}^n$ has exactly two orientations. In $\mathbb{R}^3$ these two orientations are the right-handed bases and the left-handed bases, and are depicted schematically in Fig. 3.8. Thus we can say that ϵ doesn't depend on a particular choice of orthonormal basis, but it does depend on a metric and a choice of orientation, where the orientation chosen is the one determined by the standard basis. □

[25]No doubt you are used to thinking about a rotation as a transformation that preserves distances and fixes a line in space (the ***axis of rotation***). This definition of a rotation is particular to $\mathbb{R}^3$, since even in $\mathbb{R}^2$ a rotation can't be considered to be "about an axis" since $\hat{\mathbf{z}} \notin \mathbb{R}^2$. For the equivalence of our general definition and the more intuitive definition in $\mathbb{R}^3$, see Goldstein [8].

[26]i.e., could not be made bigger.

The notion of orientation allows us to understand the interpretation of the determinant as an "oriented" volume: the sign of the determinant just tells us whether or not the orientation of $\{Ae_i\}$ is the same as $\{e_i\}$. Also, for orientation-changing transformations on $\mathbb{R}^3$ one can show that A can be written as $A = A_0(-I)$, where A_0 is a rotation and $-I$ is referred to as the ***inversion*** transformation. The inversion transformation plays a key role in differentiating vectors from *pseudovectors*, which are the subject of the next section.

Box 3.4 *Rotations vs. Translation*

In the last example we introduced rotations as orthogonal transformations with unit determinant. We will have much more to say about rotations in Part II of this book, where they will be the prototype for Lie groups and Lie algebras. For now, though, it may be worth contrasting rotations with *translations*, which are the other familiar symmetry of 3-D Euclidean space. Translation by a vector $w \in \mathbb{R}^3$ is just a map

$$T_w : \mathbb{R}^3 \to \mathbb{R}^3$$
$$v \mapsto v + w.$$

The main contrast with rotations is that whereas rotations can be associated with linear operators (as per our discussion in Sect. 3.3), translations are *non-linear* maps; that is, they do not satisfy the linearity condition (2.16), as you can check. This means their action on $\mathbb{R}^3$ cannot be expressed in terms of matrix multiplication.

We will not have much more to say about translations, though we will discuss the relationship between translations in position and momentum space in Example 4.38.

3.10 Pseudovectors

Note: All indices in this section refer to **orthonormal** bases. The calculations and results below do not apply to non-orthonormal bases, though they can be generalized to such.

A ***pseudovector*** (or ***axial vector***) is a tensor on $\mathbb{R}^3$ whose components transform like vectors under rotations but don't change sign under inversion. Common examples of pseudovectors are the angular velocity vector $\boldsymbol{\omega}$, the magnetic field vector $\mathbf{B}$, as well as all cross products, such as the angular momentum vector $\mathbf{L} = \mathbf{r} \times \mathbf{p}$. It turns out that pseudovectors like these are actually elements of $\Lambda^2\mathbb{R}^3$, which are known as ***bivectors***.

To see the connection, consider the wedge product of two vectors $r, p \in \mathbb{R}^3$. This looks like

$$\begin{aligned} r \wedge p &= (r^1 e_1 + r^2 e_2 + r^3 e_3) \wedge (p^1 e_1 + p^2 e_2 + p^3 e_3) \\ &= (r^1 p^2 - r^2 p^1)\, e_1 \wedge e_2 + (r^3 p^1 - r^1 p^3)\, e_3 \wedge e_1 + (r^2 p^3 - r^3 p^2)\, e_2 \wedge e_3. \end{aligned} \tag{3.75}$$

This looks just like $r \times p$ if we make the identifications

$$\begin{aligned} e_1 \wedge e_2 &\longrightarrow e_3 \\ e_3 \wedge e_1 &\longrightarrow e_2 \\ e_2 \wedge e_3 &\longrightarrow e_1. \end{aligned} \tag{3.76}$$

In terms of matrices, this corresponds to the identification[27]

$$\boxed{\begin{pmatrix} 0 & -z & y \\ z & 0 & -x \\ -y & x & 0 \end{pmatrix} \longrightarrow \begin{pmatrix} x \\ y \\ z \end{pmatrix}.} \tag{3.77}$$

This identification can be embodied in a one-to-one and onto map

$$J : \Lambda^2 \mathbb{R}^3 \to \mathbb{R}^3$$

defined as follows. If $\alpha \in \Lambda^2 \mathbb{R}^3$, then we can expand it as

$$\alpha = \alpha^{23} e_2 \wedge e_3 + \alpha^{31} e_3 \wedge e_1 + \alpha^{12} e_1 \wedge e_2.$$

We then define $J(\alpha)$ via its components $(J(\alpha))^i$ as

$$\boxed{(J(\alpha))^i \equiv \frac{1}{2} \epsilon^i{}_{jk} \alpha^{jk}.} \tag{3.78}$$

You will check below that this definition really does give the identifications written above. Note that J is essentially just a contraction with the epsilon tensor. With this, we see that $r \times p$ is really just $J(r \wedge p)$! Thus:

Cross products are essentially just bivectors.

Exercise 3.32. Check that J, as defined by (3.78), acts on basis vectors as in (3.76). Also check that when written in terms of matrices, J produces the map (3.77).

Exercise 3.33. We can use now give a simple derivation of the BAC-CAB rule of vector algebra. For $A, B, C \in \mathbb{R}^3$, note that $(B \wedge C)(L(A), \cdot)$ is also a vector in $\mathbb{R}^3$ (here $L(A)$ is the metric dual of A; cf Sect. 2.7). Evaluate this vector in two ways: using the definition

[27] To map the components of $e_i \wedge e_j$ to a matrix you'll need the convention discussed in Box 2.4.

(3.66) of the wedge product, and using (3.75). This should yield the BAC-CAB rule, as well as (hopefully) some intuition for it.

Now that we know how to identify bivectors and regular vectors, we must examine what it means for bivectors to transform "like" vectors under rotations, but without a sign change under inversion. On the face of things, it seems like bivectors should transform very differently from vectors; after all, a bivector is a $(0, 2)$ tensor, and you can show that it has matrix transformation law

$$[\alpha]_{\mathcal{B}'} = A[\alpha]_{\mathcal{B}}A^T. \tag{3.79}$$

This looks very different from the transformation law for the associated vector $J(\alpha)$, which is just [cf. (3.24a)]

$$[J(\alpha)]_{\mathcal{B}'} = A[J(\alpha)]_{\mathcal{B}}. \tag{3.80}$$

In particular, the bivector α transforms with two copies of A, and the vector $J(\alpha)$ with just one. How could these transformation laws be "the same"? Well, remember that α isn't any old $(0, 2)$ tensor, but an *antisymmetric* one, and for these a small miracle happens. This is best appreciated by considering an example. Let A be a rotation about the z axis by an angle θ, so that

$$A = \begin{pmatrix} \cos\theta & -\sin\theta & 0 \\ \sin\theta & \cos\theta & 0 \\ 0 & 0 & 1 \end{pmatrix}.$$

Then you can check that, with $\alpha^i \equiv (J(\alpha))^i$,

$$\begin{aligned}
[\alpha]_{\mathcal{B}'} &= A[\alpha]_{\mathcal{B}}A^T \\
&= \begin{pmatrix} \cos\theta & -\sin\theta & 0 \\ \sin\theta & \cos\theta & 0 \\ 0 & 0 & 1 \end{pmatrix} \begin{pmatrix} 0 & -\alpha^3 & \alpha^2 \\ \alpha^3 & 0 & -\alpha^1 \\ -\alpha^2 & \alpha^1 & 0 \end{pmatrix} \begin{pmatrix} \cos\theta & \sin\theta & 0 \\ -\sin\theta & \cos\theta & 0 \\ 0 & 0 & 1 \end{pmatrix} \\
&= \begin{pmatrix} 0 & -\alpha^3 & \alpha^2\cos\theta + \alpha^1\sin\theta \\ \alpha^3 & 0 & \alpha^2\sin\theta - \alpha^1\cos\theta \\ -\alpha^2\cos\theta - \alpha^1\sin\theta & -\alpha^2\sin\theta + \alpha^1\cos\theta & 0 \end{pmatrix} \qquad (3.81a) \\
&\xrightarrow{J} \begin{pmatrix} -\alpha^2\sin\theta + \alpha^1\cos\theta \\ \alpha^2\cos\theta + \alpha^1\sin\theta \\ \alpha^3 \end{pmatrix} \qquad (3.81b)
\end{aligned}$$

which is exactly a rotation of the components α^i of $J(\alpha)$! This seems to suggest that if we transform the components of $\alpha \in \Lambda^2\mathbb{R}^3$ by a rotation *first* and then apply J, or apply J and *then* rotate the components, we get the same thing. In other words, **the**

map J commutes with rotations, and that is what it means for both bivectors and vectors to behave "the same" under rotations.

Exercise 3.34. Derive (3.79). You may need to consult Sect. 3.2. Also, Verify (3.81a) by performing the necessary matrix multiplication.

To *prove* that J commutes with *arbitrary* rotations (the example above just proved it for rotations about the z-axis), we need to show that

$$A^{i'}_{j}\alpha^{j} = \frac{1}{2}\epsilon^{i'}{}_{k'l'}A^{k'}_{m}A^{l'}_{n}\alpha^{mn}, \tag{3.82}$$

where A is a rotation. On the left-hand side J is applied first followed by a rotation, and on the right-hand side the rotation is done first, followed by J.

We now compute:

$$\begin{aligned}\frac{1}{2}\epsilon^{i'}{}_{k'l'}A^{k'}_{m}A^{l'}_{n}\alpha^{mn} &= \frac{1}{2}\epsilon_{p'k'l'}\delta^{i'p'}A^{k'}_{m}A^{l'}_{n}\alpha^{mn}\\ &= \frac{1}{2}\sum_{q}\epsilon_{p'k'l'}A^{i'}_{q}A^{p'}_{q}A^{k'}_{m}A^{l'}_{n}\alpha^{mn}\\ &= \frac{1}{2}\sum_{q}\epsilon_{qmn}|A|A^{i'}_{q}\alpha^{mn}\\ &= \frac{1}{2}|A|\epsilon^{q}{}_{mn}A^{i'}_{q}\alpha^{mn}\\ &= |A|A^{i'}_{q}\alpha^{q},\end{aligned} \tag{3.83}$$

where in the second equality we used a variant of (3.26) which comes from writing out $AA^T = I$ in components, in the third equality we used the easily verified fact that $\epsilon_{p'k'l'}A^{p'}_{q}A^{k'}_{m}A^{l'}_{n} = |A|\epsilon_{qmn}$, and in the fourth equality we raised an index to resume the use of Einstein summation convention and were able to do so because covariant and contravariant components are equal in orthonormal bases. Now, for *rotations* $|A| = 1$ so in this case (3.83) and (3.82) are identical and $J(\alpha)$ *does* transform like a vector. For *inversion*, however, $|A| = |-I| = -1$ so (3.83) tells us that the components of $J(\alpha)$ do *not* change sign under inversion, as those of an ordinary vector would. Another way to see this is to set $A = -I$ in (3.79) (check!).

We have thus shown that **pseudovectors are bivectors**, since bivectors transform like vectors under rotation but don't change sign under inversion. We have also seen that cross products are very naturally interpreted as bivectors. There are other pseudovectors lying around, though, that don't naturally arise as cross products. For instance, what about the angular velocity vector $\boldsymbol{\omega}$?

Example 3.30. *The Angular Velocity Vector*

The angular velocity vector $\boldsymbol{\omega}$ is usually introduced in the context of rigid body rotations. One usually fixes the center of mass of the body, and then the velocity $\mathbf{v}$

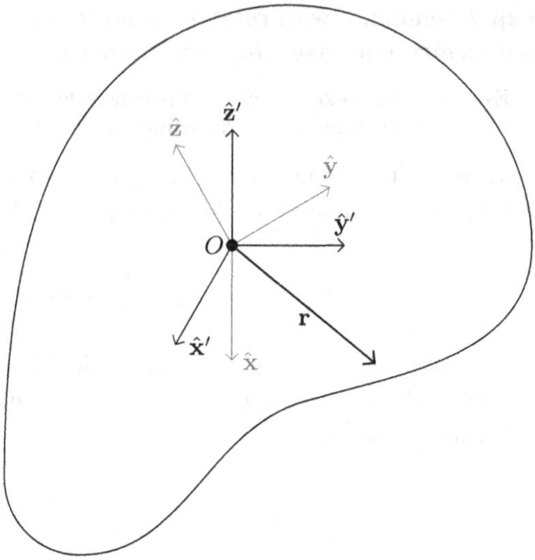

Fig. 3.9 Our rigid body with fixed space frame K' and body frame gray$K$ in gray

of a point of the rigid body is given by

$$\mathbf{v} = \boldsymbol{\omega} \times \mathbf{r}, \tag{3.84}$$

where $\mathbf{r}$ is the position vector of the point as measured from the center of mass. The derivation of this equation usually involves consideration of an angle and axis of rotation, and from these considerations one can argue that $\boldsymbol{\omega}$ is a pseudovector. Here we will take a different approach in which $\boldsymbol{\omega}$ will appear first as an antisymmetric matrix, making the bivector nature of $\boldsymbol{\omega}$ manifest.

Let K and K' be two orthonormal bases for $\mathbb{R}^3$ as in Example 2.12, with K time-dependent. One should think of K as being attached to the rotating rigid body, whereas K' is fixed. We'll refer to K as the ***body*** frame and K' as the ***space*** frame. Both frames have their origin at the center of mass of the rigid body. This is depicted in Fig. 3.9.

Now let $\mathbf{r}$ represent a point of the rigid body; then, $[\mathbf{r}]_K$ will be its coordinates in the body frame, and $[\mathbf{r}]_{K'}$ its coordinates in the space frame. Let A be the (time-dependent) orthogonal matrix of the basis transformation taking K' to K, so that

$$[\mathbf{r}]_{K'} = A\,[\mathbf{r}]_K. \tag{3.85}$$

We'd now like to calculate the velocity $[\mathbf{v}]_{K'}$ of our point in the rigid body relative to the space frame, and compare it to (3.84). This is given by just differentiating $[\mathbf{r}]_{K'}$:

$$[\mathbf{v}]_{K'} = \frac{d}{dt}[\mathbf{r}]_{K'}$$

$$
\begin{aligned}
&= \frac{d}{dt} A\,[\mathbf{r}]_K \quad \text{by (3.85)} \\
&= \frac{dA}{dt}\,[\mathbf{r}]_K \qquad \text{since } [\mathbf{r}]_K \text{ is constant} \\
&= \frac{dA}{dt} A^{-1}\,[\mathbf{r}]_{K'}.
\end{aligned}
\tag{3.86}
$$

So far this doesn't really look like (3.84). What to do? First, observe that $\frac{dA}{dt}A^{-1} = \frac{dA}{dt}A^T$ is actually an antisymmetric matrix:

$$
\begin{aligned}
0 &= \frac{d}{dt}(I) \\
&= \frac{d}{dt}(AA^T) \\
&= \frac{dA}{dt}A^T + A\frac{dA^T}{dt} \\
&= \frac{dA}{dt}A^T + \left(\frac{dA}{dt}A^T\right)^T.
\end{aligned}
\tag{3.87}
$$

We can then define an ***angular velocity bivector*** $\tilde{\boldsymbol{\omega}}$ whose components in the space frame are given by

$$
[\tilde{\boldsymbol{\omega}}]_{K'} = \frac{dA}{dt}A^{-1}.
$$

Then we simply define the ***angular velocity vector*** $\boldsymbol{\omega}$ to be

$$
\boldsymbol{\omega} \equiv J(\tilde{\boldsymbol{\omega}}).
$$

Note that $\boldsymbol{\omega}$ is, in general, time-dependent. It follows from this definition that

$$
\frac{dA}{dt}A^{-1} = \begin{pmatrix} 0 & -\omega^{3'} & \omega^{2'} \\ \omega^{3'} & 0 & -\omega^{1'} \\ -\omega^{2'} & \omega^{1'} & 0 \end{pmatrix}
$$

(we use primed indices since we're working with the K' components), so then

$$
\frac{dA}{dt}A^{-1}[\mathbf{r}]_{K'} = \begin{pmatrix} 0 & -\omega^{3'} & \omega^{2'} \\ \omega^{3'} & 0 & -\omega^{1'} \\ -\omega^{2'} & \omega^{1'} & 0 \end{pmatrix} \begin{pmatrix} x^{1'} \\ x^{2'} \\ x^{3'} \end{pmatrix}
$$

$$= \begin{pmatrix} \omega^{2'} x^{3'} - \omega^{3'} x^{2'} \\ \omega^{3'} x^{1'} - \omega^{1'} x^{3'} \\ \omega^{1'} x^{2'} - \omega^{2'} x^{1'} \end{pmatrix}$$

$$= [\boldsymbol{\omega} \times \mathbf{r}]_{K'} .$$

Combining this with (3.86), we then have

$$[\mathbf{v}]_{K'} = [\boldsymbol{\omega} \times \mathbf{r}]_{K'} \tag{3.88}$$

which is just (3.84) written in the space frame. Thus,

The "pseudovector" $\boldsymbol{\omega}$ in the space frame is nothing more than the vector associated with the antisymmetric matrix $\frac{dA}{dt}A^{-1}$!

□

Exercise 3.35. Use (3.79) to show that the bivector $\tilde{\omega}$ in the body frame is

$$[\tilde{\omega}]_K = A^{-1}\frac{dA}{dt}.$$

Combine this with (3.86) to show that (3.84) is true in the body frame as well.

In the last example we saw that to any time-dependent rotation matrix A we could associate an antisymmetric matrix $\frac{dA}{dt}A^{-1}$, which we can identify with the angular velocity vector which represents "infinitesimal" rotations. This association between finite transformations and their infinitesimal versions, which in the case of rotations takes us from orthogonal matrices to antisymmetric matrices, is precisely the relationship between a *Lie group* and its *Lie algebra*. We turn our attention to these objects in the next part of this book.

Chapter 3 Problems

Note: Problems marked with an "$*$" tend to be longer, and/or more difficult, and/or more geared towards the completion of proofs and the tying up of loose ends. Though these problems are still worthwhile, they can be skipped on a first reading.

3-1. In this problem we explore the properties of $n \times n$ orthogonal matrices. This is the set of real invertible matrices A satisfying $A^T = A^{-1}$, and is denoted $O(n)$.

(a) Is $O(n)$ a vector subspace of $M_n(\mathbb{R})$?
(b) Show that the product of two orthogonal matrices is again orthogonal, that the inverse of an orthogonal matrix is again orthogonal, and that the identity matrix is orthogonal. These properties show that $O(n)$ is

a ***group***, i.e. a set with an associative multiplication operation and identity element such that the set is closed under multiplication and every element has a multiplicative inverse. Groups are the subject of Chap. 4.

(c) Show that the columns of an orthogonal matrix A, viewed as vectors in $\mathbb{R}^n$, are mutually orthogonal under the usual inner product. Show the same for the rows. Show that for an active transformation, i.e.

$$[e_{i'}]_{\mathcal{B}} = A[e_i]_{\mathcal{B}},$$

where $\mathcal{B} = \{e_i\}_{i=1\ldots n}$ so that

$$[e_i]_{\mathcal{B}}^T = (0, \ldots, \underbrace{1}_{i\text{th slot}}, \ldots, 0),$$

the columns of A *are* the $[e_{i'}]_{\mathcal{B}}$. In other words, the components of the *new* basis vectors in the *old* basis are just the columns of A. This also shows that for a passive transformation, where

$$[e_i]_{\mathcal{B}'} = A[e_i]_{\mathcal{B}}$$

the columns of A are the components of the *old* basis vectors in the *new* basis.

(d) Show that the orthogonal matrices A with $|A| = 1$, the rotations, form a subgroup unto themselves, denoted $SO(n)$. Do the matrices with $|A| = -1$ also form a subgroup?

3-2. In this problem we'll compute the dimension of the space of $(0, r)$ symmetric tensors $S^r(V)$. This is slightly more difficult to compute than the dimension of the space of $(0, r)$ antisymmetric tensors $\Lambda^r V$, which was Exercise 3.30.

(a) Let $\dim V = n$ and $\{e_i\}_{i=1\ldots n}$ be a basis for V. Argue that $\dim S^r(V)$ is given by the number of ways you can choose r (possibly repeated) vectors from the basis $\{e_i\}_{i=1\ldots n}$.

(b) We've now reduced the problem to a combinatorics problem: how many ways can you choose r objects from a set of n objects, where any object can be chosen more than once? The answer is

$$\dim S^r(V) = \binom{n+r-1}{n-1} \equiv \frac{(n+r-1)!}{r!(n-1)!}. \tag{3.89}$$

Try to derive this on your own. If you need help, the solution to this is known as the "stars and bars" or "balls and walls" method; you can also refer to Sternberg [19], Chap. 5.

3-3. Prove the following basic properties of the determinant directly from the definition (3.72). We will restrict our discussion to operations with columns, though it can be shown that all the corresponding statements for rows are true as well.

(a) Any matrix with a column of zeros has $|A| = 0$.
(b) Multiplying a column by a scalar c multiplies the whole determinant by c.
(c) The determinant changes sign under interchange of any two columns.
(d) Adding two columns together, i.e. sending $A_i \rightarrow A_i + A_j$ for any i and j, doesn't change the value of the determinant.

3-4. One can extend the definition of determinants from matrices to more general linear operators as follows: We know that a linear operator T on a vector space V (equipped with an inner product and orthonormal basis $\{e_i\}_{i=1\ldots n}$) can be extended to an operator on the p-fold tensor product $\mathcal{T}^0_p(V)$ by

$$T(v_1 \otimes \cdots \otimes v_p) = (Tv_1) \otimes \cdots \otimes (Tv_p)$$

and thus, since $\Lambda^n V \subset \mathcal{T}^0_n(V)$, the action of T extends to $\Lambda^n V$ similarly by

$$T(v_1 \wedge \cdots \wedge v_n) = (Tv_1) \wedge \cdots \wedge (Tv_n).$$

Consider then the action of T on the contravariant version of ϵ, the tensor $\tilde{\epsilon} \equiv e_1 \wedge \cdots \wedge e_n$. We know from Exercise 3.30 that $\Lambda^n V$ is one-dimensional, so that $T(\tilde{\epsilon}) = (Te_1) \wedge \cdots \wedge (Te_n)$ is proportional to $\tilde{\epsilon}$. We then define the ***determinant*** of T to be this proportionality constant, so that

$$(Te_1) \wedge \cdots \wedge (Te_n) \equiv |T| \, e_1 \wedge \cdots \wedge e_n. \tag{3.90}$$

(a) Show by expanding the left-hand side of (3.90) in components that this more general definition reduces to the old one of (3.73) in the case of $V = \mathbb{R}^n$.
(b) Use this definition of the determinant to show that for two linear operators B and C on V,

$$|BC| = |B||C|.$$

In particular, this result holds when B and C are square matrices.
(c) Use (b) to show that the determinant of a matrix is invariant under similarity transformations (see Example 3.8). Conclude that we could have defined the determinant of a linear operator T as the determinant of its matrix in *any* basis.

3-5. Let V be a vector space with an inner product and orthonormal basis $\{e_i\}_{i=1...n}$. Prove that a linear operator T is invertible if and only if $|T| \neq 0$, as follows:

(a) Show that T is invertible if and *only if* $\{T(e_i)\}_{i=1...n}$ is a linearly independent set (see Exercise 2.9 for the "if" part of the statement).
(b) Show that $|T| \neq 0$ if and only if $\{T(e_i)\}_{i=1...n}$ is a linearly independent set.
(c) This is not a problem, just a comment. In Example 3.28 we interpreted the determinant of a matrix A as the oriented volume of the n-cube determined by $\{Ae_i\}$. As you just showed, if A is not invertible then the Ae_i are linearly dependent, hence span a space of dimension less than n and thus yield an n-dimensional volume of 0. Thus, the geometrical picture is consistent with the results you just obtained!

3-6. Let $\mathcal{B}$ be the standard basis for $\mathbb{R}^n$, O the set of all bases related to $\mathcal{B}$ by a basis transformation with $|A| > 0$, and O' the set of all bases related to $\mathcal{B}$ by a transformation with $|A| < 0$.

(a) Using what we've learned in the preceding problems, show that a basis transformation matrix A cannot have $|A| = 0$.
(b) O is by definition an orientation. Show that O' is also an orientation, and conclude that $\mathbb{R}^n$ has exactly two orientations. Note that both O and O' contain orthonormal and non-orthonormal bases.
(c) For what n is $A = -I$ an orientation-changing transformation?

Part II
Group Theory

Chapter 4
Groups, Lie Groups, and Lie Algebras

In physics we are often interested in how a particular object behaves under a particular set of transformations; for instance, one often reads that under rotations a dipole moment transforms like a vector, and a quadropole moment like a (second rank) tensor. Electric and magnetic fields supposedly transform like independent vectors under rotations, but collectively like a second rank antisymmetric tensor under Lorentz transformations. Similarly, in quantum mechanics one is often interested in the "spin" of a ket (which specifies how it transforms under rotations), or its behavior under the time-reversal or space inversion (parity) transformations. This knowledge is particularly useful as it leads to the many famous "selection rules" which greatly simplify evaluation of matrix elements. Transformations are also crucial in quantum mechanics because, as we'll see, all physical observables can be considered as "infinitesimal generators" of particular transformations; for example, the angular momentum operators "generate" rotations (as we discussed briefly in Problem 2-4) and the momentum operator "generates" translations.

Like tensors, this material is usually treated in a somewhat ad-hoc way, which facilitates computation but obscures the underlying mathematical structures. These underlying structures are known to mathematicians as *group theory*, *Lie theory*, and *representation theory*, and are known collectively to physicists as just "group theory". In this second half of the book we'll present the basic facts of this theory, along with many physical applications. As in Part I, the aim is to clarify and unify the diverse phenomena in physics that this mathematics underlies. Furthermore, the mathematics of group theory is in some senses a natural extension and application of what we learned in Part I.

Before we discuss how particular objects transform, however, we must discuss the transformations themselves, in both their "finite" and "infinitesimal" form. That discussion is the subject of the present chapter. As with Part I, we begin with a heuristic introduction which hopefully conveys some of the essential points, as well as motivates the precise discussion that follows.

N. Jeevanjee, *An Introduction to Tensors and Group Theory for Physicists*,
DOI 10.1007/978-3-319-14794-9_4

4.1 Invitation: Lie Groups and Infinitesimal Generators

In studying physics you may have encountered the terms "group" and "Lie group", and if you've studied quantum mechanics you most likely have heard of "infinitesimal generators". These terms can sound quite mysterious, and physics students are often left with the sense that something profound but just out of reach is going on underneath the routine homework problems. Our goal in this section is to demystify these subjects by introducing their basic ideas, along with familiar examples.

We begin with groups. Though group theory is a vast, profound, and often abstract subject, the basic idea of a group is simple: it is just a set of transformations that are *composable* and *invertible*. We'll define this more precisely in the next section, but for now we'll illustrate what that means with an example.

Example 4.1. *2-D Rotations*

Consider a position vector $\mathbf{r} = (x, y)$ in the plane, and rotate it counterclockwise by an angle θ, as illustrated in Fig. 4.1. Then the rotated vector $\mathbf{r}'$ has coordinates

$$\begin{aligned} x' &= x\cos\theta - y\sin\theta \\ y' &= x\sin\theta + y\cos\theta . \end{aligned}$$

This transformation can be re-written in terms of matrix multiplication as

$$\begin{pmatrix} x' \\ y' \end{pmatrix} = \begin{pmatrix} \cos\theta & -\sin\theta \\ \sin\theta & \cos\theta \end{pmatrix} \begin{pmatrix} x \\ y \end{pmatrix}.$$

The 2×2 matrix

$$R(\theta) \equiv \begin{pmatrix} \cos\theta & -\sin\theta \\ \sin\theta & \cos\theta \end{pmatrix} \tag{4.1}$$

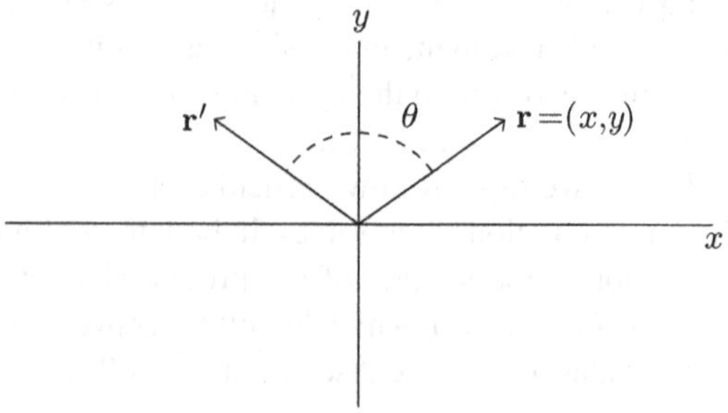

Fig. 4.1 An arbitrary vector $\mathbf{r}$ in the plane, along with its rotated version $\mathbf{r}'$

is what we mean by a "rotation"; it's what a rotation *is*. We can then consider the set of all such 2-D rotations, which we'll denote[1] $SO(2)$:

$$SO(2) \equiv \left\{ \begin{pmatrix} \cos\theta & -\sin\theta \\ \sin\theta & \cos\theta \end{pmatrix} \middle| \; \theta \in [0, 2\pi) \right\}.$$

It should be clear from our geometric picture of rotations in Fig. 4.1 that composing two rotations yields a third rotation; you can also check analytically that

$$R(\theta) \cdot R(\phi) = R(\theta + \phi). \tag{4.2}$$

Thus, rotations are *composable*. Or, put another way, $SO(2)$ is *closed* under matrix multiplication.

It's hopefully also clear from our picture of rotations that

$$R(\theta)^{-1} = R(-\theta), \tag{4.3}$$

and this is also easily checked analytically. Thus, every rotation has an inverse that is also a rotation, so rotations are *invertible*. Thus, rotations are a set of transformations that are both composable and invertible, and that makes $SO(2)$ a *group*. □

Exercise 4.1. Use (4.1) to verify (4.2) and (4.3).

Note that $R(\theta)$ from (4.1) is completely and uniquely determined by the angle θ; we then say that $SO(2)$ is *parameterized* by θ. Any such group, which can be smoothly parameterized by one or more continuous variables, is known as a ***Lie group***. Lie groups stand in contrast to ***discrete groups***, which don't accommodate such a parameterization. (We will meet examples of these in the next section.) Also, note that $SO(2)$ is a group composed of matrices. Thus, $SO(2)$ is a ***matrix Lie group***. For many applications in physics this particular class of groups is the most important, and we will spend much time in Part II studying these groups.

Another mysterious concept that one encounters in physics is that of an "infinitesimal transformation" or "infinitesimal generator". This concept only applies to Lie groups, since a smooth, continuous parameterization is required to make a transformation "infinitesimal". As with groups, it is possible to give infinitesimal generators a one-sentence description: they are just *derivatives of Lie group elements with respect to their parameters.* We again illustrate with some examples.

Example 4.2. *Infinitesimal generators of 2-D rotations*

If an infinitesimal generator is just a derivative, then let's differentiate $R(\theta)$ from (4.1) and evaluate at $\theta = 0$:

[1]This notation may be familiar if you did Problem 3-1 d). If not, it will be explained in the next section.

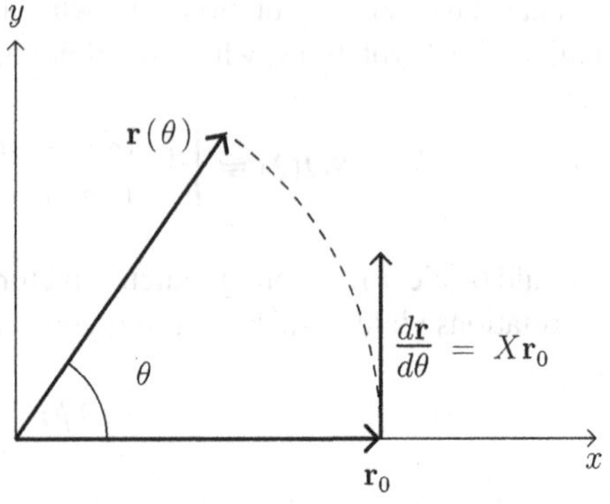

Fig. 4.2 The *dashed line* is the parameterized curve $\mathbf{r}(\theta)$ from (4.4). The tangent vector $\frac{d\mathbf{r}}{d\theta}$ at $\theta = 0$ is just given by $X\mathbf{r}_0$

$$\left.\frac{dR}{d\theta}\right|_{\theta=0} = \begin{pmatrix} 0 & -1 \\ 1 & 0 \end{pmatrix} \equiv X.$$

The matrix X is called a "generator" because it generates rotations, in the following sense. Consider the vector $\mathbf{r}_0 = (1, 0)$ in the plane, as well as the curve

$$\mathbf{r}(\theta) \equiv R(\theta) \cdot \mathbf{r}_0 \tag{4.4}$$

traced out by the vector as it rotates; see Fig. 4.2. Then the tangent vector to this curve at $\theta = 0$ is given by

$$\left.\frac{d\mathbf{r}}{d\theta}\right|_{\theta=0} = \left.\frac{dR}{d\theta}\right|_{\theta=0} \cdot \mathbf{r}_0 = X\mathbf{r}_0 = \begin{pmatrix} 0 & -1 \\ 1 & 0 \end{pmatrix}\begin{pmatrix} 1 \\ 0 \end{pmatrix} = \begin{pmatrix} 0 \\ 1 \end{pmatrix}.$$

We can thus interpret X as the matrix which turns $\mathbf{r}_0$ into $\left.\frac{d\mathbf{r}}{d\theta}\right|_{\theta=0}$; it takes in a position vector and produces *the direction in which the position vector will change as a rotation is applied.* This is the sense in which X "generates" rotations, and is also illustrated in Fig. 4.2.

Example 4.3. *Infinitesimal generators of 3-D rotations*

In three-dimensions we now have three different axes about which we can rotate. The matrices corresponding to rotations about the x, y, and z axes are analogous to (4.1) and are given, respectively, by:

$$R_x(\theta) \equiv \begin{pmatrix} 1 & 0 & 0 \\ 0 & \cos\theta & -\sin\theta \\ 0 & \sin\theta & \cos\theta \end{pmatrix}$$

$$R_y(\theta) \equiv \begin{pmatrix} \cos\theta & 0 & \sin\theta \\ 0 & 1 & 0 \\ -\sin\theta & 0 & \cos\theta \end{pmatrix}$$

$$R_z(\theta) \equiv \begin{pmatrix} \cos\theta & -\sin\theta & 0 \\ \sin\theta & \cos\theta & 0 \\ 0 & 0 & 1 \end{pmatrix}.$$

Note that $R_x(\theta)$ leaves the x-axis invariant, as a rotation about the x-axis should. Similarly for $R_y(\theta)$ and $R_z(\theta)$. The corresponding generators are

$$L_x \equiv \left.\frac{dR_x}{d\theta}\right|_{\theta=0} = \begin{pmatrix} 0 & 0 & 0 \\ 0 & 0 & -1 \\ 0 & 1 & 0 \end{pmatrix}$$

$$L_y = \begin{pmatrix} 0 & 0 & 1 \\ 0 & 0 & 0 \\ -1 & 0 & 0 \end{pmatrix}$$

$$L_z = \begin{pmatrix} 0 & -1 & 0 \\ 1 & 0 & 0 \\ 0 & 0 & 0 \end{pmatrix}. \tag{4.5}$$

(You may recognize L_z from (2.23), up to a factor of i. The appearance here of L_z and the discrepancy of a factor of i will be explained in Chap. 5). Note that

$$\mathrm{Span}\{L_x, L_y, L_z\} = \{\text{anti-symmetric } 3\times 3 \text{ matrices}\} \equiv \mathfrak{so}(3)$$

where $\mathfrak{so}(3)$ denotes the *vector space* of 3×3 antisymmetric matrices.[2] Note that we have again uncovered a close connection between rotations and antisymmetric matrices, just as we found at the very end of Part I in Example 3.30.

If you read that example, then much of the following will be familiar. It turns out that *any* element of $\mathfrak{so}(3)$, and not just L_x, L_y, or L_z, can be interpreted as a generator of a rotation. If $\omega \in \mathfrak{so}(3)$, then ω is antisymmetric and can be written as

$$\omega = \begin{pmatrix} 0 & -\omega_z & \omega_y \\ \omega_z & 0 & -\omega_x \\ -\omega_y & \omega_x & 0 \end{pmatrix}.$$

[2]The notation $\mathfrak{so}(3)$ may seem like it has a deeper meaning, and it does, but we can't explain that until Sect. 4.6. Until then, just consider $\mathfrak{so}(3)$ as odd notation for the vector space of 3×3 antisymmetric matrices.

(The reason for the strange labeling of the components of ω will be clear in a moment.) Now, ω generates rotations just as the matrix X from the last example does: if we matrix multiply a position vector $\mathbf{r} = (x, y, z)$ by ω, we get the direction in which $\mathbf{r}$ is changing as the rotation is applied. To figure out what rotation that is, you can check that:

$$\begin{aligned}\omega\,\mathbf{r} &= \begin{pmatrix} 0 & -\omega_z & \omega_y \\ \omega_z & 0 & -\omega_x \\ -\omega_y & \omega_x & 0 \end{pmatrix}\begin{pmatrix} x \\ y \\ z \end{pmatrix} \\ &= \begin{pmatrix} \omega_y z - \omega_z y \\ \omega_z x - \omega_x z \\ \omega_x y - \omega_y x \end{pmatrix} \\ &= \boldsymbol{\omega} \times \mathbf{r} \end{aligned} \tag{4.6}$$

where

$$\boldsymbol{\omega} \equiv (\omega_x, \omega_y, \omega_z)$$

is a "vector" (really a pseudovector; see Sect. 3.10) associated with the antisymmetric matrix ω. Now, carefully distinguishing ω the matrix from $\boldsymbol{\omega}$ the vector, we set $\mathbf{r} = \boldsymbol{\omega}$ in (4.6) to get

$$\omega\,\boldsymbol{\omega} = \boldsymbol{\omega} \times \boldsymbol{\omega} = 0.$$

This means that $\boldsymbol{\omega}$ is unchanged by the rotation, so must lie along the axis of rotation. Furthermore, the expression (4.6) is just the expression from classical mechanics for the velocity of a point $\mathbf{r}$ on a rigid body rotating with angular velocity vector $\boldsymbol{\omega}$. We can thus conclude that

The generator $\omega \in \mathfrak{so}(3)$ generates a rotation along the $\boldsymbol{\omega}$ axis, where the pseudovector $\boldsymbol{\omega}$ is the familiar angular velocity vector.

To conclude this example, we note that the matrices L_x, L_y, and L_z have interesting interrelationships via the commutator. For instance:

$$\begin{aligned}[L_x, L_y] &= L_x L_y - L_y L_x \\ &= \begin{pmatrix} 0 & 0 & 0 \\ 0 & 0 & -1 \\ 0 & 1 & 0 \end{pmatrix}\begin{pmatrix} 0 & 0 & 1 \\ 0 & 0 & 0 \\ -1 & 0 & 0 \end{pmatrix} - \begin{pmatrix} 0 & 0 & 1 \\ 0 & 0 & 0 \\ -1 & 0 & 0 \end{pmatrix}\begin{pmatrix} 0 & 0 & 0 \\ 0 & 0 & -1 \\ 0 & 1 & 0 \end{pmatrix} \\ &= \begin{pmatrix} 0 & 0 & 0 \\ 1 & 0 & 0 \\ 0 & 0 & 0 \end{pmatrix} - \begin{pmatrix} 0 & 1 & 0 \\ 0 & 0 & 0 \\ 0 & 0 & 0 \end{pmatrix}\end{aligned}$$

$$= \begin{pmatrix} 0 & -1 & 0 \\ 1 & 0 & 0 \\ 0 & 0 & 0 \end{pmatrix}$$

$$= L_z! \tag{4.7}$$

Similarly, you can check that

$$\begin{aligned} [L_y, L_z] &= L_x \\ [L_z, L_x] &= L_y. \end{aligned} \tag{4.8}$$

Taken together, (4.7) and (4.8) are, of course, just the familiar angular momentum commutation relations from quantum mechanics. Note that they are derived here not in the abstract setting of Hermitian operators on a Hilbert space, but rather from simple 3×3 matrices that generate rotations! Furthermore, (4.7) and (4.8) tell us that $\mathfrak{so}(3)$ is a vector space which is *closed under commutators*, meaning that a commutator of two $\mathfrak{so}(3)$ elements yields another $\mathfrak{so}(3)$ element. Such a vector space is known as a ***Lie algebra***, and we have thus seen here that:

The vector space of infinitesimal generators of a matrix Lie group forms a Lie algebra.

Matrix Lie groups and their associated Lie algebras will be the central (but not exclusive) focus of Part II of this book.

Exercise 4.2. Verify (4.8).

4.2 Groups: Definition and Examples

Now that we have a little bit of a feel for what groups are, it is time to give their precise definition. As with vector spaces, we will give an abstract, axiomatic definition, where the axioms are just meant to embody the most important properties of sets of transformations, including the composability and invertibility we just discussed. As we proceed, you may find it helpful to carry the examples of 2-D and 3-D rotations from the previous section in the back of your mind. However, as with vector spaces, the utility in taking the abstract, axiomatic approach is that what we learn about abstract groups will then apply to objects that look nothing like rotations in Euclidean space, so long as those objects satisfy the axioms. After laying down the axioms and establishing some basic properties of groups, we'll proceed directly to concrete examples.

That said, a ***group*** is a set G together with a "multiplication" operation, denoted $\cdot$, that satisfies the following axioms:

1. (**Closure**) $g, h \in G$ implies $g \cdot h \in G$.
2. (**Associativity**) For $g, h, k \in G$, $g \cdot (h \cdot k) = (g \cdot h) \cdot k$.

3. (**Existence of the identity**) There exists an element $e \in G$ such that $g \cdot e = e \cdot g = g \;\; \forall g \in G$.
4. (**Existence of inverses**) $\forall \; g \in G$ there exists an element $h \in G$ such that $g \cdot h = h \cdot g = e$.

If we think of a group as a set of transformations, as we usually do in physics, then the multiplication operation is obviously just composition; that is, if R and S are three-dimensional rotations, for instance, then $R \cdot S$ is just S followed by R. Note that we don't necessarily have $R \cdot S = S \cdot R$ for all rotations R and S; in cases such as this, G is said to be ***non-commutative*** (or ***non-abelian***). If we did have $S \cdot R = R \cdot S$ for all $R, S \in G$, then we would say that G is ***commutative*** (or ***abelian***).

There are several important properties of groups that follow almost immediately from the definition. Firstly, the *identity is unique*, for if e and f are both elements satisfying axiom 3 then we have

$$\begin{aligned} e &= e \cdot f \quad \text{since } f \text{ is an identity} \\ &= f \quad \text{since } e \text{ is an identity.} \end{aligned}$$

Secondly, *inverses are unique*: Let $g \in G$ and let h and k both be inverses of g. Then

$$g \cdot h = e$$

so multiplying both sides on the left by k gives

$$\begin{aligned} k \cdot (g \cdot h) &= k, \\ (k \cdot g) \cdot h &= k \quad \text{by associativity,} \\ e \cdot h &= k \quad \text{since } k \text{ is an inverse of } g, \\ h &= k. \end{aligned}$$

We henceforth denote the unique inverse of an element g as g^{-1}.

Thirdly, if $g \in G$ and h is merely a *right* inverse for g, i.e.

$$g \cdot h = e, \tag{4.9}$$

then h is also a *left* inverse for g and is hence the unique inverse g^{-1}. This is seen as follows:

$$\begin{aligned} h \cdot g &= (g^{-1} \cdot g) \cdot (h \cdot g) \\ &= (g^{-1} \cdot (g \cdot h)) \cdot g \quad \text{by associativity} \end{aligned}$$

$$
\begin{aligned}
&= (g^{-1} \cdot e) \cdot g \quad \text{by (4.9)} \\
&= g^{-1} \cdot g \\
&= e,
\end{aligned}
$$

so $h = g^{-1}$.

The last few properties concern inverses and can be verified immediately:

$$
\begin{aligned}
(g^{-1})^{-1} &= g \\
(g \cdot h)^{-1} &= h^{-1} \cdot g^{-1} \\
e^{-1} &= e.
\end{aligned}
$$

Exercise 4.3. Prove the *cancellation laws* for groups, i.e. that

$$
\begin{aligned}
g_1 \cdot h = g_2 \cdot h &\Longrightarrow g_1 = g_2 \\
h \cdot g_1 = h \cdot g_2 &\Longrightarrow g_1 = g_2.
\end{aligned}
$$

Before we get to some examples, we should note that axioms 2 and 3 in the definition above are usually obviously satisfied and one rarely needs to check them explicitly. The important thing in showing that a set is a group is verifying that it is closed under multiplication and contains all its inverses. Also, as a matter of notation, from now on we will usually omit the $\cdot$ when writing a product, and simply write gh for $g \cdot h$.

Example 4.4. $\mathbb{R}$: *The real numbers as an additive group*

Consider the real numbers $\mathbb{R}$ with the group "multiplication" operation given by regular addition, i.e.

$$x \cdot y \equiv x + y \quad x, y \in \mathbb{R}.$$

It may seem counterintuitive to define "multiplication" as addition, but the definition of a group is rather abstract so there is nothing that prevents us from doing this, and this point of view will turn out to be useful. With addition as the product, $\mathbb{R}$ becomes an abelian group: The first axiom to verify is closure, and this is satisfied since the sum of two real numbers is always a real number. The associativity axiom is also satisfied, since it is a fundamental property of real numbers that addition is associative. The third axiom, dictating the existence of the identity, is satisfied since $0 \in \mathbb{R}$ fits the bill. The fourth axiom, which dictates the existence of inverses, is satisfied since for any $x \in \mathbb{R}$, $-x$ is its (additive) inverse. Thus $\mathbb{R}$ is a group under addition, and is in fact an abelian group since $x + y = y + x \;\; \forall\, x, y \in \mathbb{R}$.

Note that $\mathbb{R}$ is *not* a group under regular multiplication, since 0 has no multiplicative inverse. If we remove 0, though, then we do get a group, the multiplicative group of nonzero real numbers, denoted $\mathbb{R}^*$. We leave it to you to verify that $\mathbb{R}^*$ is a group. You should also verify that this entire discussion goes through for $\mathbb{C}$ as well, so that $\mathbb{C}$ under addition and $\mathbb{C}^* \equiv \mathbb{C}\backslash\{0\}$ under multiplication are both abelian groups.

Example 4.5. *Vector spaces as additive groups*

The previous example of $\mathbb{R}$ and $\mathbb{C}$ as additive groups can be generalized to the case of vector spaces, which are abelian groups under vector space addition. The group axioms follow directly from the vector space axioms, as you should check, with 0 as the identity. While viewing vector spaces as additive groups means we ignore the crucial feature of scalar multiplication, we'll see that this perspective will occasionally prove useful.

Example 4.6. *$GL(V)$, $GL(n,\mathbb{R})$, and $GL(n,\mathbb{C})$: The general linear groups*

The ***general linear group*** of a vector space V, denoted $GL(V)$, is defined to be the subset of $\mathcal{L}(V)$ consisting of all *invertible* linear operators on V. We can easily verify that $GL(V)$ is a group: to verify closure, note that for any $T, U \in GL(V)$, TU is linear and $(TU)^{-1} = U^{-1}T^{-1}$, so TU is invertible. To verify associativity, note that for any $T, U, V \in GL(V)$ and $v \in V$, we have

$$(T(UV))(v) = T(U(V(v))) = ((TU)V)(v)$$

(careful unraveling the meaning of the parentheses!) so that $T(UV) = (TU)V$. To verify the existence of the identity, just note that I is invertible and linear, hence in $GL(V)$. To verify the existence of inverses, note that for any $T \in GL(V)$, T^{-1} exists and is invertible and linear, hence is in $GL(V)$ also. Thus $GL(V)$ is a group.

Let V have scalar field C and dimension n. If we pick a basis for V, then for each $T \in GL(V)$ we get an invertible matrix $[T] \in M_n(C)$. Just as all the invertible $T \in \mathcal{L}(V)$ form a group, so do the corresponding invertible matrices in $M_n(C)$; this group is denoted as $GL(n, C)$, and the group axioms can be readily verified for it.[3] When $C = \mathbb{R}$ we get $GL(n,\mathbb{R})$, the ***real general linear group in** n **dimensions***, and when $C = \mathbb{C}$ we get $GL(n,\mathbb{C})$, the ***complex general linear group in** n **dimensions***. □

While neither $GL(V)$, $GL(n,\mathbb{R})$, nor $GL(n,\mathbb{C})$ occur explicitly very often in physics, they have many important ***subgroups***, i.e. subsets which themselves are groups. The most important of these arise when we have a vector space V equipped with a non-degenerate Hermitian form $(\cdot\,|\,\cdot)$. In this case, we can consider the set of ***isometries*** $\mathrm{Isom}(V)$, consisting of those operators T which "preserve" $(\cdot\,|\,\cdot)$ in the sense that

$$\boxed{(Tv|Tw) = (v|w)\ \forall\, v, w, \in V.} \tag{4.10}$$

[3] You may recall having met $GL(n,\mathbb{R})$ at the end of Sect. 2.1. There we asked why it isn't a vector space, and now we know—it's more properly thought of as a group!

If $(\cdot\,|\,\cdot)$ can be interpreted as giving the "length" of vectors, then an isometry T can be thought of as an operator that preserves lengths.[4] Note that any such T is invertible (why?), hence $\mathrm{Isom}(V) \subset GL(V)$. $\mathrm{Isom}(V)$ is in fact a subgroup of $GL(V)$, as we'll now verify. First off, for any $T, U \in \mathrm{Isom}(V)$,

$$\begin{aligned} ((TU)v|(TU)w) &= (T(Uv)|T(Uw)) \\ &= (Uv|Uw) \quad \text{since } T \text{ is an isometry} \\ &= (v|w) \qquad\quad \text{since } U \text{ is an isometry} \end{aligned}$$

so TU is an isometry as well, hence $\mathrm{Isom}(V)$ is closed under multiplication. As for associativity, this axiom is automatically satisfied since $\mathrm{Isom}(V) \subset GL(V)$ and multiplication in $GL(V)$ is associative, as we proved above. As for the existence of the identity, just note that the identity operator I is trivially an isometry. To verify the existence of inverses, note that T^{-1} exists and is an isometry since

$$\begin{aligned} (T^{-1}v|T^{-1}w) &= (TT^{-1}v|TT^{-1}w) \quad \text{since } T \in \mathrm{Isom}(V) \\ &= (v|w). \end{aligned}$$

Thus $\mathrm{Isom}(V)$ is a group. Why is it of interest? Well, as we'll show in the next few examples, the matrix representations of $\mathrm{Isom}(V)$ actually turn out to be the orthogonal matrices, the unitary matrices, and the Lorentz transformations, depending on whether or not V is real or complex and whether or not $(\cdot\,|\,\cdot)$ is positive-definite. Our discussion here shows[5] that all of these sets of matrices are groups, and that *they can all be thought of as representing linear operators which preserve the relevant non-degenerate Hermitian form.*

Example 4.7. *The orthogonal group $O(n)$*

Let V be an n-dimensional real inner product space. The isometries of V can be thought of as operators which preserve lengths *and* angles, since the formula

$$\cos\theta = \frac{(v|w)}{||v||\,||w||}$$

for the angle between v and w is defined purely in terms of the inner product. Now, if T is an isometry and we write out $(v, w) = (Tv|Tw)$ in components referred to an orthonormal basis $\mathcal{B}$, we find that

$$\begin{aligned} [v]^T[w] &= (v|w) \\ &= (Tv|Tw) \end{aligned}$$

[4] Hence the term "iso-metry" = "same length."

[5] We are glossing over some subtleties with this claim. See Example 4.20 for the full story.

$$
\begin{aligned}
&= \delta_{ij}(Tv)^i(Tw)^j \\
&= \delta_{ij}T^i_{\ k}\, v^k T^j_{\ l}\, w^l \\
&= \sum_i v^k T^i_{\ k}\, T^i_{\ l}\, w^l \\
&= [v]^T[T]^T[T][w] \;\; \forall\, v, w, \in V. \qquad (4.11)
\end{aligned}
$$

As you will show in Exercise 4.4, this is true if and only if $[T]^T[T] = I$, or

$$[T]^T = [T]^{-1}. \qquad (4.12)$$

This is the familiar orthogonality condition, and the set of all orthogonal matrices (which, as we know from the above discussion and from Problem 3-1, form a group) is known as the ***orthogonal group*** $O(n)$. We will consider $O(n)$ in detail for $n = 2, 3$ in the next section.

Exercise 4.4. Show that

$$[v]^T[w] = [v]^T[T]^T[T][w] \;\; \forall\, v, w \in V$$

if and only if $[T]^T[T] = I$. One direction is easy; for the other, let $v = e_i,\ w = e_j$ where $\{e_i\}_{i=1\ldots n}$ is orthonormal.

Exercise 4.5. Verify directly that $O(n)$ is a group, by using the defining condition (4.12). This is the same as Problem 3-1 b.

Example 4.8. *The unitary group* $U(n)$

Now let V be a complex inner product space. Recall from Problem 2-7 that we can define the ***adjoint*** $T^\dagger$ of a linear operator T by the equation

$$\boxed{(T^\dagger v|w) \equiv (v|Tw).} \qquad (4.13)$$

If T is an isometry, then we can characterize it in terms of its adjoint, as follows: first, we have

$$
\begin{aligned}
(v|w) &= (Tv|Tw) \\
&= (T^\dagger T v|w) \;\; \forall\, v, w \in V.
\end{aligned}
$$

Calculations identical to those of Exercise 4.4 then show that this can be true if and only if $T^\dagger T = TT^\dagger = I$, which is equivalent to the more familiar condition

$$\boxed{T^\dagger = T^{-1}.} \qquad (4.14)$$

Such an operator is said to be ***unitary***. Thus every isometry of a complex inner product space is unitary, and vice-versa. Now, if V has dimension n and we choose an orthonormal basis for V, then we have $[T^\dagger] = [T]^\dagger$ (cf. Problems 2-7), and so (4.14) implies

$$[T]^\dagger = [T]^{-1}. \tag{4.15}$$

Thus, in an orthonormal basis, a unitary operator is represented by a unitary matrix! (Note that this is NOT necessarily true in a non-orthonormal basis.) By the discussion preceding Example 4.7, the set of all unitary matrices forms a group, denoted $U(n)$. We won't discuss $U(n)$ in depth in this text, but we will discuss one of its cousins, $SU(2)$ (to be defined below), extensively.

Note that there is nothing in the above discussion that *requires* V to be complex, so we can actually use the same definitions (of adjoints and unitarity) to define unitary operators on *any* inner product space, real or complex. Thus, a unitary operator is just an isometry of a real or complex inner product space.[6] In the case of a real vector space, the unitary matrix condition (4.15) reduces to the orthogonality condition (4.12), as you might expect.

Exercise 4.6. Verify directly that $U(n)$ is a group, using the defining condition (4.15).

Example 4.9. *The Lorentz group $O(n-1,1)$*

Now let V be a real vector space with a ***Minkowski metric*** η, which is defined, as in Example 2.20, as a symmetric, non-degenerate (2,0) tensor whose matrix in an orthonormal basis has the form

$$[\eta] = \begin{pmatrix} 1 & & & & \\ & 1 & & & \\ & & \cdots & & \\ & & & 1 & \\ & & & & -1 \end{pmatrix} \tag{4.16}$$

with zeros on all the off-diagonals. This is to be compared with (2.33), which is just (4.16) with $n = 4$. Now, since η is a non-degenerate Hermitian form, we can consider its group of isometries. If $T \in \text{Isom}(V)$, then in analogy to the computation leading to (4.11), we have (in an arbitrary basis $\mathcal{B}$),

$$\begin{aligned} [v]^T[\eta][w] &= \eta(v,w) \\ &= \eta(Tv,Tw) \\ &= [v]^T[T]^T[\eta][T][w] \quad \forall\, v,w \in V. \end{aligned} \tag{4.17}$$

[6]In fact, the only reason for speaking of both "isometries" and "unitary operators" is that unitary operators act solely on inner product spaces, whereas isometries can act on spaces with non-degenerate Hermitian forms that are not necessarily positive-definite, such as $\mathbb{R}^4$ with the Minkoswki metric.

Again, the same argument as you used in Exercise 4.4 shows that the above holds if and only if

$$[T]^T[\eta][T] = [\eta] \tag{4.18}$$

which in components reads

$$T_\mu{}^\rho T_\nu{}^\sigma \eta_{\rho\sigma} = \eta_{\mu\nu}. \tag{4.19}$$

If $\mathcal{B}$ is orthonormal, (4.18) becomes

$$[T]^T \begin{pmatrix} 1 & & & & \\ & 1 & & & \\ & & \cdots & & \\ & & & 1 & \\ & & & & -1 \end{pmatrix} [T] = [\eta]$$

which you will recognize from (3.29) as the definition of a Lorentz transformation, though now we are working in an arbitrary dimension n rather than just dimension four. We thus see that the set of all Lorentz transformations forms a group, known as the ***Lorentz group*** and denoted by $O(n-1,1)$ [the notation just refers to the number of positive and negative 1's present in the matrix form of η given in (4.16)]. The Lorentz transformations lie at the heart of special relativity, and we will take a close look at these matrices for $n = 4$ in the next section. □

Exercise 4.7. Verify directly that $O(n-1,1)$ is group, using the defining condition (4.18).

Box 4.1 *Active and Passive Interpretations of Isometries*
You may recall that we originally defined orthogonal matrices, unitary matrices, and Lorentz transformations as those matrices which implement a basis change from one orthonormal basis to another (on vector spaces with real inner products, Hermitian inner products, and Minkowski metrics, respectively). In the preceding examples, however, we've seen that these matrices can alternatively be defined as those which represent (in an orthonormal basis) operators which preserve a non-degenerate Hermitian form. These two definitions correspond to the active and passive viewpoints of transformations: our first definition of these matrices (as those which implement orthonormal basis changes) gives the passive viewpoint, while the second definition (as those matrices which represent isometries) gives the active viewpoint.

Example 4.10. *The special unitary and orthogonal groups $SU(n)$ and $SO(n)$*

The groups $O(n)$ and $U(n)$ have some very important subgroups, the ***special unitary*** and ***special orthogonal*** groups, denoted $SU(n)$ and $SO(n)$ respectively, which are defined as those matrices in $U(n)$ and $O(n)$ that have determinant equal to 1. You will verify below that these are subgroups of $U(n)$ and $O(n)$. These groups are basic in mathematics and (for certain n) fundamental in physics: as we'll see, $SO(n)$ is the group of rotations in n dimensions, $SU(2)$ is crucial in the theory of angular momentum in quantum mechanics, and (though we won't discuss it here) $SU(3)$ is fundamental in particle physics, especially in the mathematical description of quarks. □

Exercise 4.8. Show that $SO(n)$ and $SU(n)$ are subgroups of $O(n)$ and $U(n)$.

Before moving on to a more detailed look at some specific instances of the groups described above, we switch gears for a moment and consider groups that aren't subsets of $GL(n, C)$. These groups have a very different flavor than the groups we've been considering, but are useful in physics nonetheless. We'll make more precise the sense in which they differ from the previous examples when we get to Sect. 4.5.

Example 4.11. $\mathbb{Z}_2$ *The group with two elements*

Consider the set $\mathbb{Z}_2 \equiv \{+1, -1\} \subset \mathbb{Z}$ with the product being just the usual multiplication of integers. You can easily check that this is a group, in fact an abelian group. Though this group may seem trivial and somewhat abstract, it pops up in a few places in physics, as we'll see in Sect. 4.4.

Example 4.12. S_n *The symmetric group on n letters*

This group does not usually occur explicitly in physics but is intimately tied to permutation symmetry, the physics of identical particles, and much of the mathematics we discussed in Sect. 3.8. The ***symmetric group on*** n ***letters*** (also known as the ***permutation group***), denoted S_n, is defined to be the set of all one-to-one and onto maps of the set $\{1, 2, \ldots, n\}$ to itself, where the product is just the composition of maps. The maps are known as ***permutations***. You should check that any composition of permutations is again a permutation and that permutations are invertible, so that S_n is a group. This verification is simple, and just relies on the fact that permutations are, by definition, one-to-one and onto.

Any permutation σ is specified by the n numbers $\sigma(i)$, $i = 1 \ldots n$, and can conveniently be notated as

$$\begin{pmatrix} 1 & 2 & \cdots & n \\ \sigma(1) & \sigma(2) & \cdots & \sigma(n) \end{pmatrix}.$$

In such a scheme, the identity in S_3 would just look like

$$\begin{pmatrix} 1\,2\,3 \\ 1\,2\,3 \end{pmatrix}$$

while the cyclic permutation σ_1 given by $1 \to 2$, $2 \to 3$, $3 \to 1$ would look like

$$\sigma_1 = \begin{pmatrix} 1\,2\,3 \\ 2\,3\,1 \end{pmatrix}.$$

The transposition σ_2 which switches 1 and 2 and leaves 3 alone would look like

$$\sigma_2 = \begin{pmatrix} 1\,2\,3 \\ 2\,1\,3 \end{pmatrix}.$$

How do we take products of permutations? Well, the product $\sigma_1 \cdot \sigma_2$ would take on the following values:

$$\begin{aligned}(\sigma_1 \cdot \sigma_2)(1) = \sigma_1(\sigma_2(1)) = \sigma_1(2) = 3 \\ (\sigma_1 \cdot \sigma_2)(2) = \sigma_1(1) = 2 \\ (\sigma_1 \cdot \sigma_2)(3) = \sigma_1(3) = 1\end{aligned} \tag{4.20}$$

so we have

$$\sigma_1 \cdot \sigma_2 = \begin{pmatrix} 1\,2\,3 \\ 2\,3\,1 \end{pmatrix} \cdot \begin{pmatrix} 1\,2\,3 \\ 2\,1\,3 \end{pmatrix} = \begin{pmatrix} 1\,2\,3 \\ 3\,2\,1 \end{pmatrix}. \tag{4.21}$$

You should take the time to inspect (4.21) and understand how to take such a product of permutations without having to write out (4.20).

Though a proper discussion of the applications of S_n to physics must wait until Sect. 4.4, we point out here that if we have a vector space V and consider its n-fold tensor product $\mathcal{T}_n^0(V)$, then S_n acts on product states by

$$\sigma(v_1 \otimes v_2 \otimes \ldots \otimes v_n) = v_{\sigma(1)} \otimes v_{\sigma(2)} \otimes \ldots \otimes v_{\sigma(n)}.$$

(Note that this generalizes the permutation operator introduced in Example 3.26.) A generic element of $\mathcal{T}_n^0(V)$ will be a sum of such product states, and the action of $\sigma \in S_n$ on these more general states is determined by imposing the linearity condition. In the case of n identical particles in quantum mechanics, where the total Hilbert space is naively the n-fold tensor product $\mathcal{T}_n^0(\mathcal{H})$ of the single-particle Hilbert space $\mathcal{H}$, this action effectively *interchanges particles*, and we will later restate the symmetrization postulate from Example 3.26 in terms of this action of S_n on $\mathcal{T}_n^0(\mathcal{H})$.

Exercise 4.9. Show that S_n has $n!$ elements.

4.3 The Groups of Classical and Quantum Physics

We are now ready for a detailed look at some of the specific groups which arise in physics.

Example 4.13. $SO(2)$ *Special orthogonal group in two dimensions*

As discussed above, $SO(2)$ is the group of all orthogonal 2×2 matrices with determinant equal to 1. You will check in Exercise 4.10 that $SO(2)$ is abelian and that the general form of an element of $SO(2)$ is

$$\begin{pmatrix} \cos\theta & -\sin\theta \\ \sin\theta & \cos\theta \end{pmatrix}. \tag{4.22}$$

You will recognize that such a matrix represents a counterclockwise rotation of θ radians in the $x - y$ plane, as discussed in Example 4.1. Though we won't discuss $SO(2)$ very much, it serves as a nice warmup for the next example, which is ubiquitous in physics and will be discussed throughout the text.

Exercise 4.10. Consider an arbitrary matrix

$$A = \begin{pmatrix} a & b \\ c & d \end{pmatrix}$$

and impose the orthogonality condition, as well as $|A| = 1$. Show that (4.22) is the most general solution to these constraints. Then, verify explicitly that $SO(2)$ is a group (even though we already know it is by Exercise 4.8) by showing that the product of two matrices of the form (4.22) is again a matrix of the form (4.22). This will also show that $SO(2)$ is abelian.

Example 4.14. $SO(3)$ *Special orthogonal group in three-dimensions*

This group is of great importance in physics, as it is the group of all rotations in three-dimensional space! For that statement to mean anything, however, we must carefully define what a "rotation" is. One commonly used definition is the following:

Definition. A ***rotation*** in n dimensions is any linear operator R which can be *obtained continuously from the identity*[7] and takes orthonormal bases to orthonormal bases. This means that for any orthonormal basis $\{e_i\}_{i=1\dots n}$, $\{Re_i\}_{i=1\dots n}$ must also be an orthonormal basis.

You will show in Problem 4-1 that this definition is equivalent to saying $R \in SO(n)$.

Given that $SO(3)$ really is the group of three-dimensional rotations, then, can we find a general form for an element of $SO(3)$? As you may know from classical mechanics courses, an arbitrary rotation can be described in terms of the ***Euler***

[7]Meaning that there exists a continuous map $\gamma : [0,1] \to GL(n,\mathbb{R})$ such that $\gamma(0) = I$ and $\gamma(1) = R$. In other words, there is a path of *invertible* matrices connecting R to I.

angles, which tell us how to rotate a given orthonormal basis into another of the same orientation (or handedness). In classical mechanics texts,[8] it is shown that this can be achieved by rotating the given axes by an angle ϕ around the original z-axis, then by an angle θ around the *new* x-axis, and finally by an angle ψ around the *new* z-axis. If we take the passive point of view, these three rotations take the form

$$\begin{pmatrix} \cos\psi & \sin\psi & 0 \\ -\sin\psi & \cos\psi & 0 \\ 0 & 0 & 1 \end{pmatrix}, \quad \begin{pmatrix} 1 & 0 & 0 \\ 0 & \cos\theta & \sin\theta \\ 0 & -\sin\theta & \cos\theta \end{pmatrix}, \quad \begin{pmatrix} \cos\phi & \sin\phi & 0 \\ -\sin\phi & \cos\phi & 0 \\ 0 & 0 & 1 \end{pmatrix}$$

so multiplying them together gives a general form for $R \in SO(3)$:

$$\begin{pmatrix} \cos\psi\cos\phi - \cos\theta\sin\phi\sin\psi & \cos\psi\sin\phi + \cos\theta\cos\phi\sin\psi & \sin\psi\sin\theta \\ -\sin\psi\cos\phi - \cos\theta\sin\phi\cos\psi & -\sin\psi\sin\phi + \cos\theta\cos\phi\cos\psi & \cos\psi\sin\theta \\ \sin\theta\sin\phi & -\sin\theta\cos\phi & \cos\theta \end{pmatrix}. \tag{4.23}$$

Another general form for $R \in SO(3)$ is that of a rotation by an arbitrary angle θ about an arbitrary axis $\hat{\mathbf{n}}$; you will see in Sect. 4.7 that this is given by

$$\begin{pmatrix} n_x^2(1-\cos\theta)+\cos\theta & n_xn_y(1-\cos\theta)-n_z\sin\theta & n_xn_z(1-\cos\theta)+n_y\sin\theta \\ n_yn_x(1-\cos\theta)+n_z\sin\theta & n_y^2(1-\cos\theta)+\cos\theta & n_yn_z(1-\cos\theta)-n_x\sin\theta \\ n_zn_x(1-\cos\theta)-n_y\sin\theta & n_zn_y(1-\cos\theta)+n_x\sin\theta & n_z^2(1-\cos\theta)+\cos\theta \end{pmatrix} \tag{4.24}$$

where $\hat{\mathbf{n}} = (n_x, n_y, n_z)$ and the components of $\hat{\mathbf{n}}$ are not all independent since $n_x^2 + n_y^2 + n_z^2 = 1$. This constraint, along with the three components of $\hat{\mathbf{n}}$ and the angle θ, gives us three free parameters with which to describe an arbitrary rotation, just as with the Euler angles. For a nice geometric interpretation of the above matrix, see Problem 4-3.

Example 4.15. $O(3)$ *Orthogonal group in three-dimensions*

If $SO(3)$ is the group of all three-dimensional rotations, then what are we to make of $O(3)$, the group of all orthogonal 3×3 matrices without the restriction on the determinant? Well, as we pointed out in Example 3.29, the orthogonality condition actually implies[9] that $|R| = \pm 1$, so in going from $SO(3)$ to $O(3)$ we are just adding all the orthogonal matrices with $|R| = -1$. These new matrices are sometimes referred to as ***improper rotations***, as opposed to the elements with $|R| = 1$ which

[8] Such as Goldstein [8].

[9] This fact can be understood geometrically: since orthogonal matrices preserve distances and angles, they should preserve volumes as well. As we learned in Example 3.28, the determinant measures how volume changes under the action of a linear operator, so any volume preserving operator should have determinant ± 1. The sign is determined by whether or not the orientation is reversed.

are known as ***proper rotations***. Now, amongst the improper rotations is our old friend the ***inversion*** transformation

$$-I = \begin{pmatrix} -1 & 0 & 0 \\ 0 & -1 & 0 \\ 0 & 0 & -1 \end{pmatrix}.$$

Any improper rotation can be written as the product of a proper rotation and the inversion transformation, as $R = (-I)(-R)$ (note that if R is an improper rotation, then $-R$ is a proper rotation). Thus, an improper rotation can be thought of as a proper rotation followed[10] by the inversion transformation.

One important feature of $O(3)$ is that its two parts, the proper and improper rotations, are ***disconnected***, in the sense that one cannot continuously go from matrices with $|R| = 1$ to matrices with $|R| = -1$. (If one *can* continuously go from one group element to any other, then the group is said to be ***connected***. It is disconnected if it is not connected.) One can, however, multiply by $-I$ to go between the two components. This is represented schematically in Fig. 4.3. Note that the stipulation in our definition that a rotation must be continuously obtainable from the identity excludes all the improper rotations, as it should.

Example 4.16. *$SU(2)$ Special unitary group in two complex dimensions*

As mentioned in Example 4.10, $SU(2)$ is the group of all 2×2 complex matrices A which satisfy $|A| = 1$ and

$$A^\dagger = A^{-1}.$$

You can check (see Exercise 4.11 below) that a generic element of $SU(2)$ looks like

$$\begin{pmatrix} \alpha & \beta \\ -\bar{\beta} & \bar{\alpha} \end{pmatrix} \quad \alpha, \beta \in \mathbb{C}, \quad |\alpha|^2 + |\beta|^2 = 1. \tag{4.25}$$

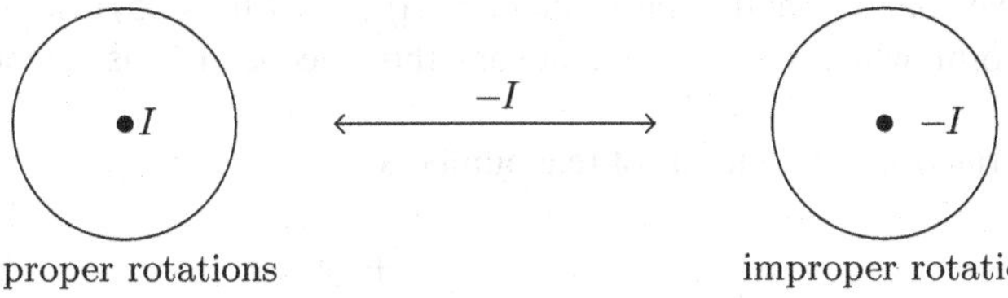

Fig. 4.3 The two components of $O(3)$. The proper rotations are just $SO(3)$. Multiplying by the inversion transformation $-I$ takes one back and forth between the two components

[10] One can actually think of the inversion as following *or* preceding the proper rotation, since $-I$ commutes with all matrices.

We could also use three real parameters with no conditions rather than two complex parameters with a constraint; one such parametrization is

$$\begin{pmatrix} e^{i(\psi+\phi)/2}\cos\frac{\theta}{2} & ie^{i(\psi-\phi)/2}\sin\frac{\theta}{2} \\ ie^{-i(\psi-\phi)/2}\sin\frac{\theta}{2} & e^{-i(\psi+\phi)/2}\cos\frac{\theta}{2} \end{pmatrix} \tag{4.26}$$

where we have used the same symbols for our parameters as we did for the Euler angles. Another real parameterization is

$$\begin{pmatrix} \cos(\theta/2) - in_z\sin(\theta/2) & (-in_x - n_y)\sin(\theta/2) \\ (-in_x + n_y)\sin(\theta/2) & \cos(\theta/2) + in_z\sin(\theta/2) \end{pmatrix}, \quad n_x^2 + n_y^2 + n_z^2 = 1 \tag{4.27}$$

where $\hat{\mathbf{n}} \equiv (n_x, n_y, n_z)$ is a unit vector and θ is an arbitrary angle. The similarities between the $SU(2)$ parameterizations (4.26) and (4.27) and the $SO(3)$ parameterizations (4.23) and (4.24) is no accident, as there is a close relationship between $SU(2)$ and $SO(3)$, which we will discuss in detail in the next section. This relationship underlies the appearance of $SU(2)$ in quantum mechanics, where rotations are implemented on spin $1/2$ particles by elements of $SU(2)$. In fact, we will see that a rotation with Euler angles ϕ, θ, and ψ is implemented by the matrix (4.26), and a rotation of angle θ about the axis $\hat{\mathbf{n}}$ is implemented by the matrix (4.27)!

Exercise 4.11. Consider an arbitrary complex matrix

$$\begin{pmatrix} \alpha & \beta \\ \gamma & \delta \end{pmatrix}$$

and impose the unit determinant and unitary conditions. Show that (4.25) is the most general solution to these constraints. Then show that any such solution can also be written in the form (4.26).

Box 4.2 *SU(2) as the 3-sphere*

You may have noticed that the constraint $|\alpha|^2 + |\beta|^2 = 1$ in (4.25) has a simple, symmetric form, which seems to hint at something deeper. Let us elaborate on this.

If we write α and β in terms of real numbers as

$$\alpha = u + iv, \quad \beta = x + iy$$

then $|\alpha|^2 + |\beta|^2 = 1$ becomes

$$u^2 + v^2 + x^2 + y^2 = 1. \tag{4.28}$$

This equation is clearly analogous to the algebraic equations $x^2 + y^2 = 1$ for the unit circle and $x^2 + y^2 + z^2 = 1$ for the surface of the unit sphere.

Equation (4.28) is, in fact, the equation of the ***3-sphere***, which you can think of as a three-dimensional "spherical" space embedded in four-dimensions, just as the unit circle (or "1-sphere") is a one-dimensional space embedded in 2-D, and the surface of the unit sphere (or "2-sphere") is a two-dimensional space embedded in 3-D. Equations (4.25) and (4.28) tell us that

***SU(2)* is the 3-sphere.**

What's more, you can think of $\pm I \in SU(2)$, coordinatized by $(u, v, x, y) = \pm(1, 0, 0, 0)$, as two "poles" of the 3-sphere, in analogy with the North and South poles $(x, y, z) = \pm(0, 0, 1)$ of the 2-sphere. This is schematically illustrated in Fig. 4.4.

Though one can take this much further and analyze the "spherical" geometry of the 3-sphere and its relation to the $SU(2)$ group structure (see Frankel [6], section 21.4), our point here is that $SU(2)$ can be thought of not just as a parametrized set of matrices *but as a highly symmetric multi-dimensional space in its own right*. This is also true for $SO(2)$, which by (4.22) is clearly just the unit circle. It's also true of other, more complicated groups like $SO(3)$ and $SO(3,1)_o$ (see the next example), but in these cases there is no simple analogy between the structure of these groups and that of more familiar spaces, so they are harder to visualize. However, it's important to remember that such interpretations of these groups do exist, and we will come back to this way of thinking in Sect. 4.5.

Example 4.17. $SO(3,1)_o$ *The restricted Lorentz group*

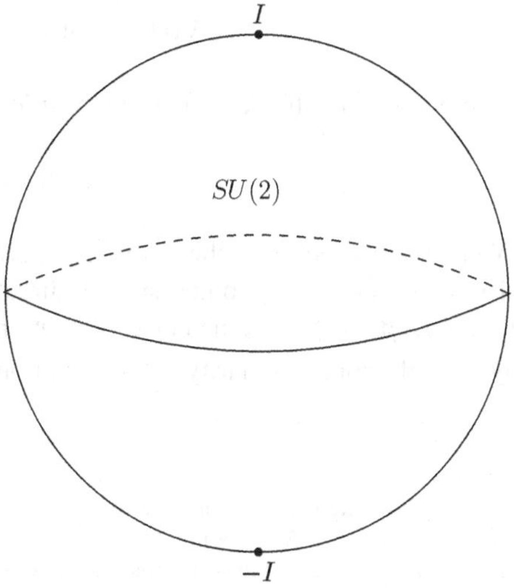

Fig. 4.4 The Lie group $SU(2)$ represented as a multidimensional spherical object. One can identify $SU(2)$ with the 3-sphere, as per (4.25) and (4.28), but this is difficult to visualize (especially on paper), so we schematically represent it here as the 2-sphere. The point is to think of matrix Lie groups not just as parametrized sets of matrices, but as multidimensional spaces in their own right

The ***restricted Lorentz group*** $SO(3,1)_o$ is defined to be the set of all $A \in O(3,1)$ which satisfy $|A| = 1$ as well as $A_{44} > 1$. You will verify in Problem 4-4 that $SO(3,1)_o$ is a subgroup of $O(3,1)$. Where does its definition come from? Well, just as $O(3)$ can be interpreted physically as the set of all orthonormal coordinate transformations, $O(3,1)$ can be interpreted as the set of all transformations between inertial reference frames. However, we often are interested in restricting those transformations to those which preserve the orientation of time and space, which is what the additional conditions $|A| = 1$ and $A_{44} > 1$ do. The condition $A_{44} > 0$ means that A doesn't reverse the direction of time, so that clocks in the new coordinates aren't running backwards. This, together with $|A| = 1$, then implies that A does not reverse the orientation of the space axes. Such transformations are known as ***restricted Lorentz transformations***.

The most familiar such transformation is probably

$$L = \begin{pmatrix} 1 & 0 & 0 & 0 \\ 0 & 1 & 0 & 0 \\ 0 & 0 & \gamma & -\beta\gamma \\ 0 & 0 & -\beta\gamma & \gamma \end{pmatrix} \quad -1 < \beta < 1,\ \gamma \equiv \frac{1}{\sqrt{1-\beta^2}} \tag{4.29}$$

which is interpreted passively[11] as a coordinate transformation to a new reference frame that is unrotated relative to the old frame but is moving uniformly along the z-axis with relative velocity β.[12] Such a transformation is often referred to as a ***boost*** along the z-axis, and is also sometimes written as

$$L = \begin{pmatrix} 1 & 0 & 0 & 0 \\ 0 & 1 & 0 & 0 \\ 0 & 0 & \cosh u & -\sinh u \\ 0 & 0 & -\sinh u & \cosh u \end{pmatrix}, \qquad u \in \mathbb{R} \tag{4.30}$$

where u is a quantity known as the ***rapidity*** and is related to β by

$$\tanh u = \beta.$$

(You should check that the above L matrices really are in $SO(3,1)_o$.) We could also boost along any other spatial direction; if the relative velocity vector is $\boldsymbol{\beta} \equiv (\beta_x, \beta_y, \beta_z)$, then the corresponding matrix should be obtainable from (4.29) by an orthogonal similarity transformation that takes $\hat{\mathbf{z}}$ into $\hat{\boldsymbol{\beta}}$. You will show in

[11] It's worth noting that, in contrast to rotations, Lorentz transformations are pretty much *always* interpreted passively. A vector in $\mathbb{R}^4$ is considered an event, and it doesn't make much sense to start moving that event around in spacetime (the active interpretation), though it does make sense to ask what a different observer's coordinates for that particular event would be (passive interpretation).

[12] Note that β is measured in units of the speed of light, hence the restriction $-1 < \beta < 1$.

Exercise 4.12 below that this yields

$$L = \begin{pmatrix} \frac{\beta_x^2(\gamma-1)}{\beta^2}+1 & \frac{\beta_x\beta_y(\gamma-1)}{\beta^2} & \frac{\beta_x\beta_z(\gamma-1)}{\beta^2} & -\beta_x\gamma \\ \frac{\beta_y\beta_x(\gamma-1)}{\beta^2} & \frac{\beta_y^2(\gamma-1)}{\beta^2}+1 & \frac{\beta_y\beta_z(\gamma-1)}{\beta^2} & -\beta_y\gamma \\ \frac{\beta_z\beta_x(\gamma-1)}{\beta^2} & \frac{\beta_z\beta_y(\gamma-1)}{\beta^2} & \frac{\beta_z^2(\gamma-1)}{\beta^2}+1 & -\beta_z\gamma \\ -\beta_x\gamma & -\beta_y\gamma & -\beta_z\gamma & \gamma \end{pmatrix}. \tag{4.31}$$

If we generalize the relation between u and β to three-dimensions as

$$\boldsymbol{\beta} = \frac{\tanh u}{u}\mathbf{u}, \quad u = |\mathbf{u}| \tag{4.32}$$

then you can check that this arbitrary boost can also be written as

$$L = \begin{pmatrix} \frac{u_x^2(\cosh u-1)}{u^2}+1 & \frac{u_xu_y(\cosh u-1)}{u^2} & \frac{u_xu_z(\cosh u-1)}{u^2} & -\frac{u_x}{u}\sinh u \\ \frac{u_yu_x(\cosh u-1)}{u^2} & \frac{u_y^2(\cosh u-1)}{u^2}+1 & \frac{u_yu_z(\cosh u-1)}{u^2} & -\frac{u_y}{u}\sinh u \\ \frac{u_zu_x(\cosh u-1)}{u^2} & \frac{u_zu_y(\cosh u-1)}{u^2} & \frac{u_z^2(\cosh u-1)}{u^2}+1 & -\frac{u_z}{u}\sinh u \\ -\frac{u_x}{u}\sinh u & -\frac{u_y}{u}\sinh u & -\frac{u_z}{u}\sinh u & \cosh u \end{pmatrix}. \tag{4.33}$$

Note that the set of boosts is *not* closed under matrix multiplication; you could check this directly (and laboriously) using (4.31) or (4.33), but we'll prove it more simply and elegantly in Example 4.33.

Now we know what boosts look like, but how about an arbitrary restricted Lorentz transformation? Well, the nice thing about $SO(3,1)_o$ is that any element A can be decomposed as $A = LR'$, where

$$R' = \begin{pmatrix} R & \\ & 1 \end{pmatrix} \quad R \in SO(3) \tag{4.34}$$

and L is of the form (4.31). This is the usual decomposition of an arbitrary restricted Lorentz transformation into a rotation and a boost, which you will perform in Problem 4-5. Note that L has three arbitrary parameters, so that our arbitrary restricted Lorentz transformation LR' has six parameters total.

Exercise 4.12. Construct an orthogonal matrix A which implements an orthonormal change of basis from the standard basis $\{\hat{\mathbf{x}}, \hat{\mathbf{y}}, \hat{\mathbf{z}}\}$ to one of the form $\{\mathbf{r}_1, \mathbf{r}_2, \hat{\boldsymbol{\beta}}\}$ where the $\mathbf{r}_i$ are any two vectors mutually orthonormal with $\hat{\boldsymbol{\beta}}$ and each other. Embed A in $SO(3,1)_o$ as in (4.34) and use this to obtain (4.31) by performing a similarity transformation on (4.29). Parts of Problem 3-1 may be useful here.

Exercise 4.13. (Properties of Boosts)

(a) Check that L in (4.31) really does represent a boost of velocity $\boldsymbol{\beta}$ as follows: Use L as a passive transformation to obtain new coordinates (x', y', z', t') from the old ones by

$$\begin{pmatrix} x' \\ y' \\ z' \\ t' \end{pmatrix} = L \begin{pmatrix} x \\ y \\ z \\ t \end{pmatrix}.$$

Show that the spatial origin of the unprimed frame, defined by $x = y = z = 0$, moves with velocity $-\boldsymbol{\beta}$ in the primed coordinate system, which tells us that the primed coordinate system moves with velocity $+\boldsymbol{\beta}$ with respect to the unprimed system.

(b) A photon traveling in the $+z$ direction has energy-momentum 4-vector[13] given by $(E/c, 0, 0, E)$, where E is the photon energy and we have restored the speed of light c. By applying a boost in the z direction [as given by (4.29)] and using the quantum-mechanical relation $E = h\nu$ between a photon's energy and its frequency ν, derive the ***relativistic doppler shift***

$$\nu' = \sqrt{\frac{1-\beta}{1+\beta}}\,\nu.$$

Example 4.18. $O(3,1)$ *The extended Lorentz group*

In the previous example we restricted our changes of inertial reference frame to those which preserved the orientation of space and time. This is sufficient in classical mechanics, but in quantum mechanics we are often interested in the effects of space and time inversion on the various Hilbert spaces we're working with. If we add spatial inversion, also called ***parity*** and represented by the matrix

$$P = \begin{pmatrix} -1 & 0 & 0 & 0 \\ 0 & -1 & 0 & 0 \\ 0 & 0 & -1 & 0 \\ 0 & 0 & 0 & 1 \end{pmatrix},$$

as well as time-reversal, represented by

$$T = \begin{pmatrix} 1 & 0 & 0 & 0 \\ 0 & 1 & 0 & 0 \\ 0 & 0 & 1 & 0 \\ 0 & 0 & 0 & -1 \end{pmatrix}, \tag{4.35}$$

to the restricted Lorentz group, we actually recover $O(3,1)$, which is thus known as the ***improper*** or ***extended*** Lorentz group. You should verify that $P, T \in O(3,1)$, but $P, T \notin SO(3,1)_o$. In fact, $|P| = |T| = -1$, which is no accident; as in the case of the orthogonal group, the defining Eq. (4.18) restricts the determinant, and in fact implies that $|A| = \pm 1$. In this case, however, the group has four

[13] See Griffiths [10], Ch. 12.2.

components instead of two! Obviously those matrices with $|A| = 1$ must be disconnected from those with $|A| = -1$, but those which reverse the orientation of the space axes must also be disconnected from those which do not, and those which reverse the orientation of time must be disconnected from those which do not. This is represented schematically in Fig. 4.5. Note that, as in the case of $O(3)$, multiplication by the transformations P and T takes us to and from the various different components.

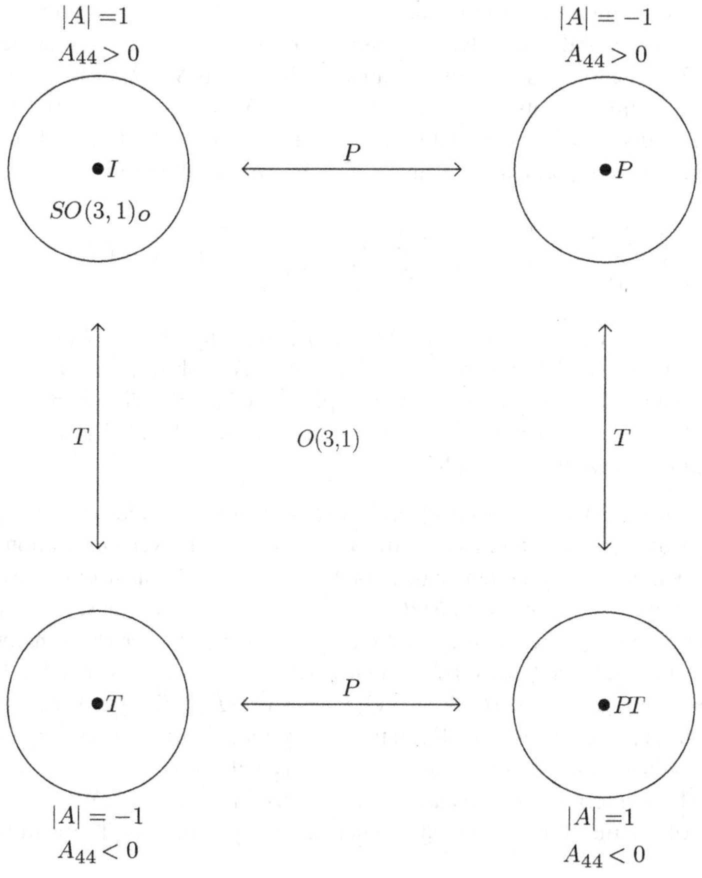

Fig. 4.5 The four components of $O(3, 1)$. The proper Lorentz transformations in the *upper-left corner* are just $SO(3, 1)_o$. Note that the transformations in the *lower-right hand corner* change both the orientation of the space axes and time, and so must be disconnected from $SO(3, 1)_o$ even though they have $|A| = 1$. Also note that multiplying by the parity and time-reversal operators P and T take one back and forth between the various components

Example 4.19. $SL(2,\mathbb{C})$ *Special linear group in two complex dimensions*

This group cannot be viewed as a group of isometries, but it is important in physics nonetheless. $SL(2,\mathbb{C})$ is defined to be the set of all 2×2 complex matrices A with $|A| = 1$. By now it should be apparent that this set is a group. The general form of $A \in SL(2,\mathbb{C})$ is

$$A = \begin{pmatrix} a & b \\ c & d \end{pmatrix} \qquad a, b, c, d \in \mathbb{C},\ ad - bc = 1.$$

The unit determinant constraint means that A is determined by three complex parameters or six real parameters, just as for $SO(3,1)_o$. This is no coincidence; in fact, $SL(2,\mathbb{C})$ bears the same relationship to $SO(3,1)_o$ as $SU(2)$ bears to $SO(3)$, in that $SL(2,\mathbb{C})$ implements restricted Lorentz transformations on spin 1/2 particles! This will be discussed in the next section. You will also show later[14] that a boost with rapidity $\mathbf{u}$ is implemented by an $SL(2,\mathbb{C})$ matrix of the form

$$\tilde{L} = \begin{pmatrix} \cosh\frac{u}{2} + \frac{u_z}{u}\sinh\frac{u}{2} & -\frac{1}{u}(u_x - i u_y)\sinh\frac{u}{2} \\ -\frac{1}{u}(u_x + i u_y)\sinh\frac{u}{2} & \cosh\frac{u}{2} - \frac{u_z}{u}\sinh\frac{u}{2} \end{pmatrix}, \quad \mathbf{u} \in \mathbb{R}^3, \tag{4.36}$$

and it can be shown,[15] just as for $SO(3,1)_o$, that any $A \in SL(2,\mathbb{C})$ can be decomposed as $A = \tilde{L}\tilde{R}$, where $\tilde{R} \in SU(2)$ and $\tilde{L}$ is as above. This, together with the facts that an arbitrary rotation can be implemented by $\tilde{R} \in SU(2)$ parametrized as in (4.26), yields the general form $\tilde{L}\tilde{R}$ for an element of $SL(2,\mathbb{C})$ in terms of the same parameters we used for $SO(3,1)_o$. □

Now that we have discussed several groups of physical relevance, it's a good time to try and organize them somehow. If you look back over this section, you'll see that each group implements either rotations or Lorentz transformations, but in different "flavors". For instance, $SO(3)$ are just the proper rotations, whereas $O(3)$ includes the improper rotations, and $SU(2)$ implements proper rotations but only for quantum-mechanical spin 1/2 particles. Analogous remarks, but for Lorentz transformations, apply to $SO(3,1)_o$, $O(3,1)$, and $SL(2,\mathbb{C})$. Furthermore, since proper rotations are a subset of all rotations, and since rotations are just a restricted class of Lorentz transformations, there are various inclusion relations amongst these groups. These interrelationships are all summarized in Table 4.1. We will add one more set of relationships to this table when we study homomorphisms in the next section.

[14] See Problem 4-8.

[15] See Problem 4-6.

Table 4.1 The interrelationships between the groups discussed in this section

Rotations		Lorentz transformations	Description
$SU(2)$	$\subset$	$SL(2,\mathbb{C})$	Quantum-mechanical
$SO(3)$	$\subset$	$SO(3,1)_o$	Proper
$\cap$		$\cap$	–
$O(3)$	$\subset$	$O(3,1)$	Improper

4.4 Homomorphism and Isomorphism

In the last section we claimed that there is a close relationship between $SU(2)$ and $SO(3)$, as well as between $SL(2,\mathbb{C})$ and $SO(3,1)_o$. We now make this relationship precise, and show that a similar relationship exists between S_n and $\mathbb{Z}_2$. We will also define what it means for two groups to be "the same", which will then tie into our somewhat abstract discussion of $\mathbb{Z}_2$ in the last section.

Given two groups G and H, a ***homomorphism*** from G to H is a map $\Phi : G \to H$ such that

$$\boxed{\Phi(g_1 g_2) = \Phi(g_1)\Phi(g_2)\ \ \forall\, g_1, g_2 \in G.} \tag{4.37}$$

Note that the product in the left-hand side of (4.37) takes place in G, whereas the product on the right-hand side takes place in H. A homomorphism should be thought of as a map from one group to another *which preserves the multiplicative structure*. Note that Φ need not be one-to-one or onto; if it is onto, then Φ is said to be a ***homomorphism onto*** H, and if in addition it is one-to-one, then we say Φ is an ***isomorphism***. If Φ is an isomorphism, then it is invertible and thus sets up a one-to-one correspondence which preserves the group structure, so we can then regard G and H as "the same" group, just with different labels for the elements. When two groups G and H are isomorphic we write $G \simeq H$.

Exercise 4.14. Let $\Phi : G \to H$ be a homomorphism, and let e be the identity in G and e' the identity in H. Show that

$$\begin{aligned} \Phi(e) &= e' \\ \Phi(g^{-1}) &= \Phi(g)^{-1}\ \ \forall\, g \in G. \end{aligned}$$

Example 4.20. *Isometries and the orthogonal, unitary, and Lorentz groups*

A nice example of a group isomorphism is when we have an n-dimensional vector space V (over some scalars C) and a basis $\mathcal{B}$, hence a map

$$\begin{aligned} GL(V) &\to GL(n, C) \\ T &\mapsto [T]_{\mathcal{B}}. \end{aligned}$$

It's easily checked that this map is one-to-one and onto. Furthermore, it is a homomorphism since $[TU] = [T][U]$, a fact you proved in Exercise 2.10. Thus this map is an isomorphism and $GL(V) \simeq GL(n, C)$. If V has a non-degenerate Hermitian form $(\cdot\,|\,\cdot)$, then we can restrict this map to $\mathrm{Isom}(V) \subset GL(V)$, which yields

$$\mathrm{Isom}(V) \simeq O(n)$$

when V is real and $(\cdot\,|\,\cdot)$ is positive-definite,

$$\mathrm{Isom}(V) \simeq U(n)$$

when V is complex and $(\cdot\,|\,\cdot)$ is positive-definite, and

$$\mathrm{Isom}(V) \simeq O(n-1, 1)$$

when V is real and $(\cdot\,|\,\cdot)$ is a Minkoswki metric. These isomorphisms were implicit in the discussion of Examples 4.6–4.9, where we identified the operators in $\mathrm{Isom}(V)$ with their matrix representations in the corresponding matrix group. □

Example 4.21. *Linear maps as homomorphisms*

A ***linear map*** from a vector space V to a vector space W is a map $\Phi : V \to W$ that satisfies the usual linearity condition

$$\Phi(cv_1 + v_2) = c\Phi(v_1) + \Phi(v_2). \tag{4.38}$$

(A linear *operator* is then just the special case in which $V = W$.) In particular we have $\Phi(v_1 + v_2) = \Phi(v_1) + \Phi(v_2)$, which just says that Φ is a homomorphism between the additive groups V and W! (cf. Example 4.5). If Φ is one-to-one and onto, then it is an isomorphism, and in particular we refer to it as a ***vector space isomorphism***.

The notions of linear map and vector space isomorphism are basic ones, and could have been introduced much earlier (as they are in standard linear algebra texts), but because of our specific goals in this book we haven't needed them yet. These objects will start to play a role soon, though, and will recur throughout the rest of the book.

Exercise 4.15. Use an argument similar to that of Exercise 2.8 to prove that a linear map $\phi : V \to W$ is an isomorphism if and only if $\dim V = \dim W$ *and* ϕ satisfies

$$\phi(v) = 0 \implies v = 0.$$

Exercise 4.16. Before moving on to more complicated examples, let's get some practice by acquainting ourselves with a few more basic homomorphisms.

(a) First, show that the map

$$\begin{aligned} \exp : \mathbb{R} &\to \mathbb{R}^* \\ x &\mapsto e^x, \end{aligned}$$

from the additive group of real numbers to the multiplicative group of nonzero real numbers, is a homomorphism. Is it an isomorphism? Why or why not?

(b) Repeat the analysis from (a) for $\exp : \mathbb{C} \to \mathbb{C}^*$.

(c) Show that the map

$$\begin{aligned} \det : GL(n, C) &\to C^* \\ A &\mapsto \det A \end{aligned}$$

is a homomorphism for both $C = \mathbb{R}$ and $C = \mathbb{C}$. Is it an isomorphism in either case? Would you expect it to be?

Exercise 4.17. Recall that $U(1)$ is the group of 1×1 unitary matrices. Show that this is just the set of complex numbers z with $|z| = 1$, and that $U(1)$ is isomorphic to $SO(2)$.

Suppose Φ is a homomorphism but not one-to-one. Is there a way to quantify how far it is from being one-to-one? Define the ***kernel*** of Φ to be the set

$$K \equiv \{g \in G \,|\, \Phi(g) = e'\},$$

where e' is the identity in H. In words, K is the set of all elements of G that get sent to e' under Φ; see Fig. 4.6. Note that $e \in K$ by Exercise 4.14. Furthermore, if Φ is one-to-one, then $K = \{e\}$, since there can be only one element in G that maps to e'. If Φ is not one-to-one, then the size of K tells us how far it is from being so. Also, if we have $\Phi(g_1) = \Phi(g_2) = h \in H$, then

$$\Phi(g_1 {g_2}^{-1}) = \Phi(g_1)\Phi(g_2)^{-1} = hh^{-1} = e$$

so $g_1 {g_2}^{-1}$ is in the kernel of Φ, i.e. $g_1 {g_2}^{-1} = k \in K$. Multiplying this on the right by g_2 then gives $g_1 = kg_2$, so we see that any two elements of G that give the same element of H under Φ are related by left multiplication by an element of K. Conversely, if we are given $g \in G$ and $\Phi(g) = h$, then for all $k \in K$,

$$\Phi(kg) = \Phi(k)\Phi(g) = e'\Phi(g) = \Phi(g) = h.$$

Thus if we define[16] $Kg \equiv \{kg \,|\, k \in K\}$, then Kg are *precisely* those elements (no more, and no less) of G which get sent to h. This is also depicted in Fig. 4.6. Thus,

[16]We could also proceed by defining $gK \equiv \{gk \,|\, k \in K\}$, but this turns out to be the same as Kg. A subgroup K with this property, that $Kg = gK \; \forall g \in G$, is called a ***normal*** subgroup. For more on this, see Herstein [12].

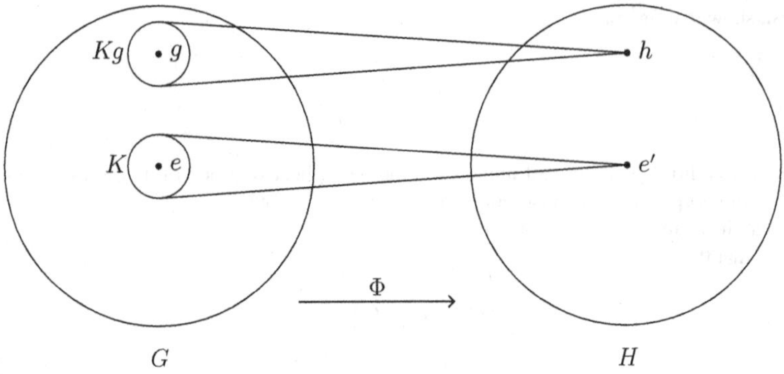

Fig. 4.6 Schematic depiction of the subsets K and Kg of G. They are mapped respectively to e' and h in H by the homomorphism Φ

the size of K tells us how far Φ is from being one-to-one, and the elements of K tell us exactly which elements of G will map to a specific element $h \in H$.

Homomorphisms and isomorphisms are ubiquitous in mathematics and occur frequently in physics, as we'll see in the examples below.

Exercise 4.18. Show that the kernel K of any homomorphism $\Phi : G \to H$ is a sub*group* of G. Then determine the kernels of the maps exp and det of Exercise 4.16.

Exercise 4.19. Suppose $\Phi : V \to W$ is a linear map between vector spaces, hence a homomorphism between abelian groups. Conclude from the previous exercise that the kernel K of Φ is a sub*space* of V, also known as the ***null space*** of Φ. The dimension of K is known as the ***nullity*** of K. Also show that the range of Φ is a subspace of W, whose dimension is known as the ***rank*** of Φ. Finally, prove the ***rank-nullity*** theorem of linear algebra, which states that

$$\text{rank}(\Phi) + \text{nullity}(\Phi) = \dim V. \tag{4.39}$$

(Hint: Take a basis $\{e_1, \cdots, e_k\}$ for K and complete it to a basis $\{e_1, \cdots, e_n\}$ for V, where $n = \dim V$. Then show that $\{\Phi(e_{k+1}), \cdots, \Phi(e_n)\}$ is a basis for the range of Φ.)

Example 4.22. *$SU(2)$ and $SO(3)$*

In most physics textbooks the relationship between $SO(3)$ and $SU(2)$ is described in terms of the "infinitesimal generators" of these groups. We will discuss infinitesimal transformations in the next section and make contact with the standard physics presentation then; here, we present the relationship in terms of a group homomorphism $\rho : SU(2) \to SO(3)$, defined as follows: consider the vector space (check!) of all 2×2 traceless anti-Hermitian matrices, denoted as $\mathfrak{su}(2)$ (for reasons we will explain later). You can check that an arbitrary element $X \in \mathfrak{su}(2)$ can be written as

$$X = \frac{1}{2}\begin{pmatrix} -iz & -y - ix \\ y - ix & iz \end{pmatrix} \quad x, y, z \in \mathbb{R}. \tag{4.40}$$

If we take as basis vectors

$$
\begin{aligned}
S_x &\equiv -\frac{i}{2}\sigma_x = \frac{1}{2}\begin{pmatrix} 0 & -i \\ -i & 0 \end{pmatrix} \\
S_y &\equiv -\frac{i}{2}\sigma_y = \frac{1}{2}\begin{pmatrix} 0 & -1 \\ 1 & 0 \end{pmatrix} \\
S_z &\equiv -\frac{i}{2}\sigma_z = \frac{1}{2}\begin{pmatrix} -i & 0 \\ 0 & i \end{pmatrix}
\end{aligned}
\tag{4.41}
$$

then we have

$$X = xS_x + yS_y + zS_z$$

so the column vector corresponding to X in the basis $\mathcal{B} = \{S_x, S_y, S_z\}$ is

$$[X] = \begin{pmatrix} x \\ y \\ z \end{pmatrix}.$$

Note that

$$\det X = \frac{1}{4}(x^2 + y^2 + z^2) = \frac{1}{4}||[X]||^2$$

so the determinant of $X \in \mathfrak{su}(2)$ is proportional to the norm squared of $[X] \in \mathbb{R}^3$ with the usual Euclidean metric. Now, you will check below that $A \in SU(2)$ acts on $X \in \mathfrak{su}(2)$ by the map $X \mapsto AXA^\dagger$, and that this map is linear. Thus, this map is a linear operator on $\mathfrak{su}(2)$, and can be represented in the basis $\mathcal{B}$ by a 3×3 matrix which we'll call $\rho(A)$, so that

$$[AXA^\dagger] = \rho(A)[X]$$

where $\rho(A)$ acts on $[X]$ by the usual matrix multiplication. Furthermore,

$$||\rho(A)[X]||^2 = ||[AXA^\dagger]||^2 = 4\det(AXA^\dagger) = 4\det X = ||[X]||^2 \tag{4.42}$$

so $\rho(A)$ preserves the norm of X. This implies (see Exercise 4.21 below) that $\rho(A) \in O(3)$, and one can in fact show[17] that $\det \rho(A) = 1$, so that $\rho(A) \in SO(3)$. In fact,

[17]See Problem 4-7 of this chapter, or consider the following rough (but correct) argument: ρ : $SU(2) \to O(3)$ as defined above is a continuous map, and so the composition

$$\det \circ \rho : SU(2) \to \mathbb{R}$$

if A is as in (4.26), then $\rho(A)$ is just (4.23) (Problem 4-7). Thus we may construct a map

$$\rho : SU(2) \to SO(3)$$
$$A \mapsto \rho(A).$$

Furthermore, ρ is a homomorphism, since

$$\rho(AB)[X] = [(AB)X(AB)^\dagger] = [ABXB^\dagger A^\dagger] = \rho(A)[BXB^\dagger] = \rho(A)\rho(B)[X] \tag{4.43}$$

and hence $\rho(AB) = \rho(A)\rho(B)$. Is ρ an isomorphism? One can show[18] that ρ is onto but not one-to-one, and in fact has kernel $K = \{I, -I\}$. From the discussion preceding this example, we then know that $\rho(A) = \rho(-A)\ \forall A \in SU(2)$ (this fact is also clear from the definition of ρ), so for every rotation $R \in SO(3)$ there correspond exactly two matrices in $SU(2)$ which map to R under ρ. Thus, when trying to implement a rotation R on a spin 1/2 particle we have two choices for the $SU(2)$ matrix we use, and it is sometimes said that the map ρ^{-1} from $SO(3)$ to $SU(2)$ is *double-valued*. In mathematical terms one doesn't usually speak of functions with multiple-values, though, so instead we say that $SU(2)$ is the ***double-cover*** of $SO(3)$, since the map ρ is onto ("cover") and two-to-one ("double").

Box 4.3 *Interpreting the Map ρ*
Note that the S_i of Eq. (4.41) are, up to a factor of i, just the spin 1/2 angular momentum matrices. Then, noting that $A^\dagger = A^{-1}$, the map $X \mapsto AXA^\dagger = AXA^{-1}$ may remind you of a rotation of spin operators, which is just what it is. This connection will be explained when we discuss the Ad homomorphism in Example 4.43.

Exercise 4.20. Let $A \in SU(2)$, $X \in \mathfrak{su}(2)$. Show that $AXA^\dagger \in \mathfrak{su}(2)$ and note that $A(X+Y)A^\dagger = AXA^\dagger + AYA^\dagger$, so that the map $X \to AXA^\dagger$ really is a linear operator on $\mathfrak{su}(2)$.

Exercise 4.21. Let V be a real vector space with a metric g and let $R \in \mathcal{L}(V)$ preserve norms on V, i.e. $g(Rv, Rv) = g(v, v)\ \forall v \in V$. Show that this implies that

$$g(Rv, Rw) = g(v, w)\ \forall v, w \in V,$$

i.e. that R is an isometry. Hint: consider $g(v+w, v+w)$ and use the bilinearity of g.

$$A \mapsto \det(\rho(A))$$

is also continuous. Since $SU(2)$ is connected any continuous function must itself vary continuously, so $\det \circ \rho$ can't jump between 1 and -1, which are its only possible values. Since $\det(\rho(I)) = 1$, we can then conclude that $\det(\rho(A)) = 1\ \forall A \in SU(2)$.

[18] See Problem 4-7 again.

Example 4.23. $SL(2,\mathbb{C})$ *and* $SO(3,1)_o$

Just as there is a two-to-one homomorphism from $SU(2)$ to $SO(3)$, there is a two-to-one homomorphism from $SL(2,\mathbb{C})$ to $SO(3,1)_o$ which is defined similarly. Consider the vector space $H_2(\mathbb{C})$ of 2×2 Hermitian matrices. As we saw in Example 2.10, this is a four-dimensional vector space with basis $\mathcal{B} = \{\sigma_x, \sigma_y, \sigma_z, I\}$, and an arbitrary element $X \in H_2(\mathbb{C})$ can be written as

$$X = \begin{pmatrix} t+z & x-iy \\ x+iy & t-z \end{pmatrix} = x\sigma_x + y\sigma_y + z\sigma_z + tI \tag{4.44}$$

so that

$$[X] = \begin{pmatrix} x \\ y \\ z \\ t \end{pmatrix}.$$

Now, $SL(2,\mathbb{C})$ acts on $H_2(\mathbb{C})$ in the same way that $SU(2)$ acts on $\mathfrak{su}(2)$: by sending $X \to AXA^\dagger$ where $A \in SL(2,\mathbb{C})$. You can again check that this is actually a linear map from $H_2(\mathbb{C})$ to itself, and hence can be represented in components by a matrix which we'll again call $\rho(A)$. You can also check that

$$\det X = t^2 - x^2 - y^2 - z^2 = -\eta([X],[X])$$

so the determinant of $X \in H_2(\mathbb{C})$ gives minus the norm squared of $[X]$ in the *Minkowski* metric on $\mathbb{R}^4$. As before, the action of $\rho(A)$ on $[X]$ preserves this norm [by a calculation identical to (4.42)], and you will show in Problem 4-8 that $\det \rho(A) = 1$ and $\rho(A)_{44} > 1$, so $\rho(A) \in SO(3,1)_o$. Thus we can again construct a map

$$\begin{aligned} \rho : SL(2,\mathbb{C}) &\to SO(3,1)_o \\ A &\mapsto \rho(A) \end{aligned}$$

and it can be shown[19] that ρ is onto. Furthermore, ρ is a homomorphism, by a calculation identical to (4.43). The kernel of ρ is again $K = \{I, -I\}$ so $SL(2,\mathbb{C})$ is a double-cover of $SO(3,1)_o$.

Exercise 4.22. The elements of $SL(2,\mathbb{C})$ that correspond to rotations should fix timelike vectors, i.e. leave them unchanged. Identify the elements of $H_2(\mathbb{C})$ that correspond to timelike vectors, and show that it is precisely the subgroup $SU(2) \subset SL(2,\mathbb{C})$ which leaves them unchanged, as expected.

[19] See Problem 4-8 again.

Now that we have met the homomorphisms $\rho : SU(2) \to SO(3)$ and $\rho : SL(2,\mathbb{C}) \to SO(3,1)_o$, we can complete Table 4.1 by including those maps. This yields Table 4.2 below.

Table 4.2 As in Table 4.1, but with the homomorphisms $\rho : SU(2) \to SO(3)$ and $\rho : SL(2,\mathbb{C}) \to SO(3,1)_o$

Rotations		Lorentz transformations	Description
$SU(2)$	$\subset$	$SL(2,\mathbb{C})$	Quantum-mechanical
$\rho\downarrow$		$\rho\downarrow$	
$SO(3)$	$\subset$	$SO(3,1)_o$	Proper
$\cap$		$\cap$	–
$O(3)$	$\subset$	$O(3,1)$	Improper

This table summarizes the interrelationships between the matrix Lie groups introduced so far

Example 4.24. $\mathbb{Z}_2$, *parity, and time-reversal*

Consider the set $\{I, P\} \subset O(3,1)$. This is an abelian group of two elements with P^2=I, and so looks just like $\mathbb{Z}_2$. In fact, if we define a map $\Phi : \{I, P\} \to \mathbb{Z}_2$ by

$$\Phi(I) = 1$$
$$\Phi(P) = -1$$

then Φ is a homomorphism since

$$\Phi(P \cdot P) = \Phi(I) = 1 = (-1)^2 = \Phi(P)\Phi(P).$$

Φ is also clearly one-to-one and onto, so Φ is in fact an isomorphism! We could also consider the two-element group $\{I, T\} \in O(3,1)$; since $T^2 = I$, we could define a similar isomorphism from $\{I, T\}$ to $\mathbb{Z}_2$. Thus,

$$\mathbb{Z}_2 \simeq \{I, P\} \simeq \{I, T\}.$$

In fact, you will show below that *all* two element groups are isomorphic. That is why we chose to present $\mathbb{Z}_2$ somewhat abstractly; there are many groups in physics that are isomorphic to $\mathbb{Z}_2$, so it makes sense to use an abstract formulation so that we can talk about all of them at the same time without having to think in terms of a particular representation like $\{I, P\}$ or $\{I, T\}$. We will see the advantage of this in the next example.

Exercise 4.23. Show that any group G with only two elements e and g must be isomorphic to $\mathbb{Z}_2$. To do this you must define a map $\Phi : G \to \mathbb{Z}_2$ which is one-to-one, onto, and which satisfies (4.37). Note that S_2, the symmetric group on two letters, has only two elements. What is the element in S_2 that corresponds to $-1 \in \mathbb{Z}_2$?

Example 4.25. *S_n, $\mathbb{Z}_2$, and the sgn homomorphism*

In Sect. 3.8 we discussed rearrangements, transpositions, and the evenness or oddness of a rearrangement in terms of the number of transpositions needed to obtain it. We are now in a position to make this much more precise, which will facilitate neater descriptions of the ϵ tensor, the determinant of a matrix, and the symmetrization postulate.

We formally define a ***transposition*** in S_n to be any permutation τ which switches two numbers i and j and leaves all the others alone. (We'll usually denote transpositions as τ and more general permutations as σ.) You can check that σ_2 and $\sigma_1 \cdot \sigma_2$ from Example 4.12 are both transpositions. It is a fact that *any* permutation can be written (non-uniquely) as a product of transpositions. Though we won't prove this here,[20] the following argument should make this fact plausible: a permutation σ corresponds to a rearrangement $\{\sigma(1), \sigma(2), \dots, \sigma(n)\}$ of the numbers $\{1, 2, \dots, n\}$. A transposition corresponds to switching any two numbers in the ordered list $\{1, 2, \dots, n\}$. Given a particular rearrangement $\{\sigma(1), \sigma(2), \dots, \sigma(n)\}$, we can build it by starting from $\{1, 2, \dots, n\}$, repeatedly switching the "1" with its neighbor to the right until it gets to the right place, then switching the "2" with its neighbor to the right until it gets to the right place, and repeating until we arrive at $\{\sigma(1), \sigma(2), \dots, \sigma(n)\}$. In this way $\{\sigma(1), \sigma(2), \dots, \sigma(n)\}$ can be written as a product of transpositions. As an example, you can check that

$$\sigma_1 = \begin{pmatrix} 1\,2\,3 \\ 2\,3\,1 \end{pmatrix} = \begin{pmatrix} 1\,2\,3 \\ 3\,2\,1 \end{pmatrix} \begin{pmatrix} 1\,2\,3 \\ 2\,1\,3 \end{pmatrix}$$

is a decomposition of σ_1 into transpositions.

Though the decomposition of a given permutation is far from unique (for instance, the identity can be decomposed as a product of any transposition σ and its inverse), the *evenness* or *oddness* of the number of transpositions in a decomposition is invariant. For instance, even though we could write the identity as

$$\begin{aligned} e &= e \\ e &= \tau_1 {\tau_1}^{-1} \\ e &= \tau_1 {\tau_1}^{-1} \tau_2 {\tau_2}^{-1} \end{aligned}$$

and so on for any transpositions $\tau_1, \tau_2, \dots$, every decomposition will consist of an *even* number of transpositions. Likewise, any decomposition of the transposition $\begin{pmatrix} 1\,2\,3 \\ 2\,1\,3 \end{pmatrix}$ will consist of an odd number of transpositions. A general proof of this fact is relegated to Problem 4-9. What this allows us to do, though, is define a homomorphism sgn: $S_n \to \mathbb{Z}_2$ by

[20] See Herstein [12] for a proof and nice discussion.

$$\text{sgn}(\sigma) = \begin{cases} +1 \text{ if } \sigma \text{ consists of an even number of transpositions} \\ -1 \text{ if } \sigma \text{ consists of an odd number of transpositions.} \end{cases}$$

You should be able to verify with a moment's thought that

$$\text{sgn}(\sigma_1\sigma_2) = \text{sgn}(\sigma_1)\text{sgn}(\sigma_2),$$

so that sgn is actually a homomorphism. If $\text{sgn}(\sigma) = +1$ we say σ is ***even***, and if $\text{sgn}(\sigma) = -1$ we say σ is ***odd***.

With the sgn homomorphism in hand we can now tidy up several definitions from Sect. 3.8. First of all, we can now define the wedge product of r dual vectors $f_i,\ i = 1, \ldots, r$ to be

$$f_1 \wedge \ldots \wedge f_r \equiv \sum_{\sigma \in S_r} \text{sgn}(\sigma) f_{\sigma(1)} \otimes f_{\sigma(2)} \otimes \ldots \otimes f_{\sigma(r)}.$$

You should compare this with the earlier definition and convince yourself that the two definitions are equivalent. Also, from our earlier definition of the ϵ tensor it should be clear that the nonzero components of ϵ are

$$\epsilon_{i_1 \ldots i_n} = \text{sgn}(\sigma) \text{ where } \sigma = \begin{pmatrix} 1 & \cdots & n \\ i_1 & \cdots & i_n \end{pmatrix}$$

and so the definition of the determinant, (3.73), becomes

$$|A| = \sum_{\sigma \in S_n} \text{sgn}(\sigma) A_{1\sigma(1)} \ldots A_{n\sigma(n)}.$$

Finally, at the end of Example 4.12 we described how S_n acts on an n-fold tensor product $\mathcal{T}_n^0(V)$ by

$$\sigma(v_1 \otimes v_2 \otimes \ldots \otimes v_n) = v_{\sigma(1)} \otimes v_{\sigma(2)} \otimes \ldots \otimes v_{\sigma(n)}, \tag{4.45}$$

and extending linearly. If we have a totally symmetric tensor $T = T^{i_1 \ldots i_n} e_{i_1} \otimes \ldots \otimes e_{i_n} \in S^n(V)$, we then have

$$\begin{aligned}
\sigma(T) &= T^{i_1 \ldots i_n} e_{\sigma(i_1)} \otimes \ldots \otimes e_{\sigma(i_n)} \\
&= T^{\sigma^{-1}(j_1) \ldots \sigma^{-1}(j_n)} e_{j_1} \otimes \ldots \otimes e_{j_n} && \text{where we relabel indices using } j_k \equiv \sigma(i_k) \\
&= T^{j_1 \ldots j_n} e_{j_1} \otimes \ldots \otimes e_{j_n} && \text{by total symmetry of } T^{j_1 \ldots j_n} \\
&= T
\end{aligned}$$

so all elements of $S^n(V)$ are fixed by the action of S_n. If we now consider a totally antisymmetric tensor $T = T^{i_1 \dots i_n} e_{i_1} \otimes \dots \otimes e_{i_n} \in \Lambda^n(V)$, then the action of $\sigma \in S_n$ on it is given by

$$\begin{aligned}
\sigma(T) &= T^{i_1 \dots i_n} e_{\sigma(i_1)} \otimes \dots \otimes e_{\sigma(i_n)} \\
&= T^{\sigma^{-1}(j_1) \dots \sigma^{-1}(j_n)} e_{j_1} \otimes \dots \otimes e_{j_n} \\
&= \mathrm{sgn}(\sigma)\, T^{j_1 \dots j_n} e_{j_1} \otimes \dots \otimes e_{j_n} \qquad \text{by antisymmetry of } T^{j_1 \dots j_n} \\
&= \mathrm{sgn}(\sigma)\, T
\end{aligned}$$

so if σ is odd then T changes sign under it, and if σ is even then T is invariant. Thus, we can restate the symmetrization postulate as follows:

> **Symmetrization Postulate II**: Any state of an n-particle system is either *invariant* under a permutation of the particles (in which case the particles are known as ***bosons***), or *changes sign* depending on whether the permutation is even or odd (in which case the particles are known as ***fermions***).

Switching now to Dirac notation, if we furthermore want to comply with the symmetrization postulate and construct a totally symmetric/anti-symmetric n-particle state $|\psi\rangle$ out of n states $|\psi_1\rangle, \ |\psi_2\rangle, \ \cdots |\psi_n\rangle$, we can simply write

$$|\psi\rangle = \sum_{\sigma \in S_n} |\psi_{\sigma(1)}\rangle |\psi_{\sigma(2)}\rangle \cdots |\psi_{\sigma(n)}\rangle \in S^n(V)$$

or

$$|\psi\rangle = \sum_{\sigma \in S_n} \mathrm{sgn}(\sigma) |\psi_{\sigma(1)}\rangle |\psi_{\sigma(2)}\rangle \cdots |\psi_{\sigma(n)}\rangle \in \Lambda^n(V).$$

4.5 From Lie Groups to Lie Algebras

You may have noticed that the examples of groups we met in the last two sections had a couple of different flavors: there were the matrix groups like $SU(2)$ and $SO(3)$ which were parametrizable by a certain number of real parameters, and then there were the "discrete" groups like $\mathbb{Z}_2$ and the symmetric groups S_n that had a finite number of elements and were described by discrete labels rather than continuous parameters. The first type of group forms an extremely important subclass of groups known as ***Lie Groups***, named after the Norwegian mathematician Sophus Lie who was among the first to study them systematically in the late 1800s. Besides their ubiquity in math and physics, Lie groups are important because their continuous nature means that we can study group elements that are "infinitely close" to the identity; these are known to physicists as the "infinitesimal transformations" or "generators" of the group, and to mathematicians as the ***Lie algebra*** of the group. As we make this notion precise, we'll see that Lie algebras are vector spaces and as such are sometimes simpler to understand than the "finite" group elements. Also, the "generators" in some Lie algebras are taken in quantum mechanics to represent certain physical observables, and in fact almost all observables can be built out of elements of certain Lie algebras. We will see that many familiar objects and structures in physics can be understood in terms of Lie algebras.

Before we can study Lie algebras, however, we should make precise what we mean by a Lie group. Here we run into a snag, because the proper and most general definition requires machinery well outside the scope of this text.[21] We do wish to be precise, though, so we follow Hall [11] and use a restricted definition which doesn't really capture the essence of what a Lie group is, but which will get the job done and allow us to discuss Lie algebras without having to wave our hands.

That said, we define a ***matrix Lie group*** to be a subgroup $G \subset GL(n, \mathbb{C})$ which is ***closed***, in the following sense: for any sequence of matrices $A_n \in G$ which converges to a limit matrix A, either $A \in G$ or $A \notin GL(n, \mathbb{C})$. All this says is that a limit of matrices in G must either itself be in G, or otherwise be noninvertible. As remarked above, this definition is technical and doesn't provide much insight into what a Lie group really is, but it will provide the necessary hypotheses in proving the essential properties of Lie algebras.

Let's now prove that some of the groups we've encountered above are indeed matrix Lie groups. We'll verify this explicitly for one class of groups, the orthogonal groups, and leave the rest as problems for you. The orthogonal group $O(n)$ is defined by the equation $R^{-1} = R^T$, or $R^T R = I$. Let's consider the function from $GL(n, \mathbb{R})$ to itself defined by $f(A) = A^T A$. Each entry of the matrix $f(A)$ is easily seen to be a continuous function of the entries of A, so f is continuous. Consider now a sequence R_i in $O(n)$ that converges to some limit matrix R. We then have

$$\begin{aligned} f(R) &= f\left(\lim_{i\to\infty} R_i\right) \\ &= \lim_{i\to\infty} f(R_i) \quad \text{since } f \text{ is continuous} \\ &= \lim_{i\to\infty} I \\ &= I \end{aligned}$$

so $R \in O(n)$. Thus $O(n)$ is a matrix Lie group. The unitary and Lorentz groups, as well as their cousins with unit determinant, are similarly defined by continuous functions, and can analogously be shown to be matrix Lie groups. For an example of a subgroup of $GL(n, \mathbb{C})$ which is *not* closed, hence not a matrix Lie group, see Problem 4-10.

[21] The necessary machinery being the theory of differentiable manifolds; in this context, a Lie group is essentially a group that is also a differentiable manifold. See Schutz [18] or Frankel [6] for very readable introductions for physicists, and Warner [21] for a systematic but terse account.

We remarked that the above definition doesn't really capture the essence of what a Lie group is. What is that essence? As mentioned before, one should think of Lie groups as groups which can be *parametrized* in terms of a certain number of real variables. This number is known as the ***dimension*** of the Lie group, and we will see that this number is also the usual (vector space) dimension of its corresponding Lie algebra. This parameterization is totally analogous to the way one parameterizes the surface of a sphere with polar and azimuthal angles in vector calculus in order to compute surface integrals. In fact, we already showed in Box 4.2 that one can (and should) think of a Lie group as a kind of multidimensional space (like the surface of a sphere) that also has a group structure. Unlike the surface of a sphere, though, a Lie group has a distinguished point, the identity e. Furthermore, as we mentioned above, studying transformations "close to" the identity will lead us to Lie algebras.

For the sake of completeness, we should point out here that there are Lie groups out there which are *not* matrix Lie groups, i.e. which cannot be described as a subset of $GL(n,\mathbb{C})$ for some n. Their relevance for basic physics has not been established, however, so we don't consider them here.[22]

Now that we have a better sense of what Lie groups are, we'd like to zoom in to Lie algebras by considering group elements that are "close" to the identity. For concreteness consider the rotation group $SO(3)$. An arbitrary rotation about the z axis looks like

$$R_z(\theta) = \begin{pmatrix} \cos\theta & -\sin\theta & 0 \\ \sin\theta & \cos\theta & 0 \\ 0 & 0 & 1 \end{pmatrix}$$

and if we take our rotation angle to be $\epsilon \ll 1$, we can approximate $R_z(\epsilon)$ by expanding to first order, which yields

$$R_z(\epsilon) \approx R_z(0) + \epsilon \left.\frac{dR_z}{d\theta}\right|_{\theta=0} = I + \epsilon L_z \tag{4.46}$$

where you may recall from Sect. 4.1 that

$$L_z \equiv \left.\frac{dR_z}{d\theta}\right|_{\theta=0} = \begin{pmatrix} 0 & -1 & 0 \\ 1 & 0 & 0 \\ 0 & 0 & 0 \end{pmatrix}.$$

Now, we should be able to describe a finite rotation through an angle θ as an n-fold iteration of smaller rotations through an angle θ/n. As we take n larger, then θ/n becomes smaller and the approximation (4.46) for the smaller rotation through $\epsilon = \theta/n$ becomes better. Thus, we expect

---

[22] See Hall [11] for further information and references.

$$R_z(\theta) = [R_z(\theta/n)]^n \approx \left(I + \frac{\theta L_z}{n}\right)^n$$

to become an equality in the limit $n \to \infty$. However, you should check, if it is not already a familiar fact, that

$$\lim_{n\to\infty}\left(I + \frac{X}{n}\right)^n = \sum_{n=0}^{\infty} \frac{X^n}{n!} = e^X \tag{4.47}$$

for any real number or matrix X. Thus, we can write

$$R_z(\theta) = e^{\theta L_z} \tag{4.48}$$

where from here on out the exponential of a matrix or linear operator is *defined* by the power series in (4.47). Notice that the set $\{R_z(\theta) = e^{\theta L_z} | \theta \in \mathbb{R}\}$ is a subgroup of $SO(3)$; this can be seen by explicitly checking that

$$R_z(\theta_1)R_z(\theta_2) = R_z(\theta_1 + \theta_2), \tag{4.49}$$

or using the property[23] of exponentials that $e^X e^Y = e^{X+Y}$, or recognizing intuitively that any two successive rotations about the z-axis yields another rotation about the z-axis. Notice that (4.49) says that if we consider $R_z(\theta)$ to be the image of θ under a map

$$R_z : \mathbb{R} \to SO(3), \tag{4.50}$$

then R_z is a homomorphism! Any such continuous homomorphism from the additive group $\mathbb{R}$ to a matrix Lie group G is known as a ***one-parameter subgroup***. One-parameter subgroups are actually very familiar to physicists; the set of rotations in $\mathbb{R}^3$ about *any* particular axis (not just the z-axis) is a one-parameter subgroup (where the parameter can be taken to be the rotation angle), as is the set of all boosts in a particular direction (in which case the parameter can be taken to be the absolute value u of the rapidity). We'll see that translations along a particular direction in both momentum and position space are one-parameter subgroups as well.

If we have a matrix X such that $e^{tX} \in G \;\; \forall t \in \mathbb{R}$, then the map

$$\begin{aligned} \exp : \mathbb{R} &\to G \\ t &\mapsto e^{tX} \end{aligned} \tag{4.51}$$

is a one-parameter subgroup, by the abovementioned property of exponentials. Conversely, if we have a one-parameter subgroup $\gamma : \mathbb{R} \to G$, then we know that $\gamma(0) = I$ (since γ is a homomorphism), and, defining

[23]This is actually only true when X and Y commute; more on this later.

$$X \equiv \frac{d\gamma}{dt}(0), \tag{4.52}$$

we have

$$\begin{aligned}\frac{d\gamma}{dt}(t) &= \lim_{h\to 0}\frac{\gamma(h+t)-\gamma(t)}{h}\\ &= \lim_{h\to 0}\frac{\gamma(h)\gamma(t)-\gamma(t)}{h}\\ &= \lim_{h\to 0}\frac{\gamma(h)-I}{h}\gamma(t)\\ &= \lim_{h\to 0}\frac{\gamma(h)-\gamma(0)}{h}\gamma(t)\\ &= X\gamma(t).\end{aligned}$$

You should recall[24] that the first order linear matrix differential equation $\frac{d\gamma}{dt}(t) = X\gamma(t)$ has the unique solution $\gamma(t) = e^{tX}$. Thus, every one-parameter subgroup is of the form (4.51), so we have a one-to-one correspondence between one-parameter subgroups and matrices X such that $e^{tX} \in G\ \forall t \in \mathbb{R}$. The matrix X is sometimes said to "generate" the corresponding one-parameter subgroup[25] [e.g., L_z "generates" rotations about the z-axis, according to (4.48)], and to each X there corresponds an "infinitesimal transformation" $I + \epsilon X$. Really, though, X is best thought of as a derivative, as given by (4.52) and emphasized in Sect. 4.1.

Any X that generates a one-parameter subgroup also carries a geometric interpretation, which we won't make precise[26] but describe here as a useful heuristic. If we think of G as a multidimensional space, and our one-parameter subgroup $\gamma(t) = e^{tX}$ as a parameterized curve in that space, then (4.52) tells us that X is a *tangent vector to $\gamma(t)$ at the identity*. We can then interpret *all* such X this way, and it turns out that these X actually compose the entire tangent space to G at e, pictured in Fig. 4.7. We are thus led to a *vector space* of matrices corresponding to one-parameter subgroups, and this is precisely the ***Lie algebra*** of the matrix Lie group.

---

[24]See any standard text on linear algebra and linear ODEs.

[25]From (4.52) it's hopefully also clear that any such X also "generates" the transformations corresponding to the one-parameter subgroup, in the sense described in Sect. 4.1.

[26]See Frankel [6] for a detailed discussion.

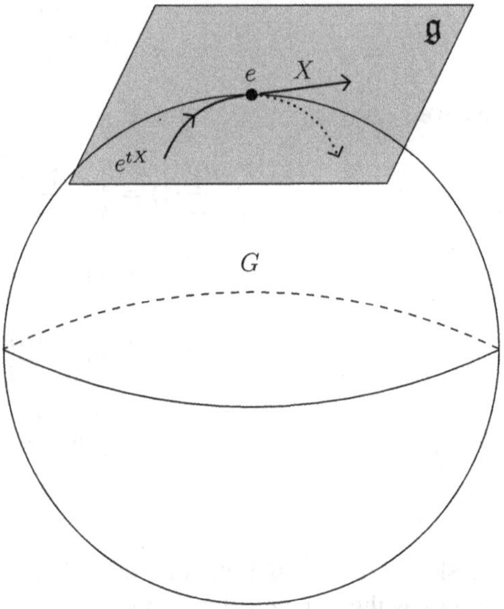

Fig. 4.7 Schematic representation of a Lie group G, depicted here as a sphere, along with its Lie algebra $\mathfrak{g}$, given by the plane tangent to the identity e. A one-parameter subgroup e^{tX} is shown running through e, and its tangent vector at e is X

4.6 Lie Algebras: Definition, Properties, and Examples

In accordance with the previous section, we define the ***Lie algebra*** $\mathfrak{g}$ of a given matrix Lie group $G \subset GL(n, \mathbb{C})$ as :

$$\boxed{\mathfrak{g} \equiv \{X \in M_n(\mathbb{C}) \mid e^{tX} \in G \ \forall t \in \mathbb{R}\}.} \tag{4.53}$$

In this section we'll establish some of the basic properties of the Lie algebras of the various isometry groups we've met—$O(n)$, $U(n)$, $O(n-1, 1)$, $SO(n)$, etc. We'll find that these Lie algebras have certain properties in common, and we'll prove that in fact *all* Lie algebras of matrix Lie groups have these properties. The discussion in this section will also set us up for a detailed look in Sect. 4.7 at the Lie algebras of our specific groups from Sect. 4.3.

Before looking at isometries, let's warm up by considering the more general group of invertible linear operators $GL(n, \mathbb{C})$, and its Lie algebra $\mathfrak{gl}(n, \mathbb{C})$. How can we describe $\mathfrak{gl}(n, \mathbb{C})$? If X is *any* element of $M_n(\mathbb{C})$, then e^{tX} is invertible for all t, as its inverse is simply e^{-tX}. Thus

$$\mathfrak{gl}(n, \mathbb{C}) = M_n(\mathbb{C}). \tag{4.54}$$

Likewise,

$$\mathfrak{gl}(n, \mathbb{R}) = M_n(\mathbb{R}).$$

Thus, the Lie algebra of the group of *invertible* linear operators is simply the space of *all* linear operators!

Now, consider a finite-dimensional vector space V equipped with a non-degenerate Hermitian form $(\cdot\,|\,\cdot)$. How can we describe the Lie algebra associated with the isometry group Isom(V)?[27] If X is in this Lie algebra, then (4.53) and (4.10) tell us that

$$(e^{tX}v|e^{tX}w) = (v|w) \quad \forall\, v, w \in V,\ t \in \mathbb{R}.$$

Differentiating with respect to t yields

$$(Xe^{tX}v|w) + (v|Xe^{tX}w) = 0 \quad \forall\, v, w \in V,\ t \in \mathbb{R}.$$

Evaluating at $t = 0$ and rearranging gives

$$\boxed{(Xv|w) = -(v|Xw) \quad \forall\, v, w \in V.} \tag{4.55}$$

This is the fundamental description of any Lie algebra of isometries, and will take various concrete forms as we proceed through the examples below. The minus sign, in particular, will play a key role.

Example 4.26. $\mathfrak{o}(n)$ *The Lie algebra of* $O(n)$

Let $V = \mathbb{R}^n$ with the Euclidean dot product $(\cdot\,|\,\cdot)$, and denote the Lie algebra of $O(n)$ by $\mathfrak{o}(n)$. If we take an orthonormal basis and use X to denote both the matrix $X \in \mathfrak{o}(n)$ and the corresponding linear operator on $\mathbb{R}^n$, (4.55) becomes

$$\begin{aligned} (X[v])^T[w] &= -[v]^T X[w] \quad \forall\, v, w \in \mathbb{R}^n \\ \Longleftrightarrow \quad [v]^T X^T[w] &= -[v]^T X[w] \quad \forall\, v, w \in \mathbb{R}^n \\ \Longleftrightarrow \quad X^T &= -X \end{aligned} \tag{4.56}$$

where the last line follows from the same logic as in Exercise 4.4. Thus:

The Lie algebra of *O(n)* is the set of *n* x *n* antisymmetric matrices.

You can check that the matrices A_{ij} from Example 2.9 form a basis for $\mathfrak{o}(n)$, so that $\dim \mathfrak{o}(n) = \frac{n(n-1)}{2}$. As we'll see in detail in the $n = 3$ case, the A_{ij} can be interpreted as generating rotations in the $i - j$ plane.

Example 4.27. $\mathfrak{u}(n)$ *The Lie algebra of* $U(n)$

Now let $V = \mathbb{C}^n$ with the Hermitian inner product $(\cdot\,|\,\cdot)$, and denote the Lie algebra of $U(n)$ by $\mathfrak{u}(n)$. For $X \in \mathfrak{u}(n)$ consider its Hermitian adjoint $X^\dagger$, which

[27] We will freely use the identification between finite-dimensional linear operators and matrices here, so that we may think of Isom(V) as a matrix Lie group and hence consider its Lie algebra. The exponential of linear operators is defined by (4.47), just as for matrices.

satisfies (2.49). Again identifying matrices and linear operators via an orthonormal basis (4.55) yields

$$\begin{aligned} (v|X^\dagger w) &= -(v|Xw) \\ \iff \qquad X^\dagger &= -X \end{aligned} \tag{4.57}$$

again by the same logic as in Exercise 4.4. In other words:

The Lie algebra of *U(n)* is the set of $n \times n$ anti-Hermitian matrices.

In Exercise 2.6 you constructed a basis for $H_n(\mathbb{C})$, and as we mentioned in Example 2.4, multiplying a Hermitian matrix by i yields an anti-Hermitian matrix. Thus, if we multiply the basis from Exercise 2.6 by i we get a basis for $\mathfrak{u}(n)$, and it then follows that $\dim \mathfrak{u}(n) = n^2$. Note that a real anti-symmetric matrix can also be considered an anti-Hermitian matrix, so that $\mathfrak{o}(n) \subset \mathfrak{u}(n)$. □

Box 4.4 *The Physicist's Definition of a Lie Algebra*
Before meeting some other Lie algebras, we have a loose end to tie up. We claimed earlier that most physical observables are best thought of as elements of Lie algebras (we'll justify this statement in Sect. 4.8) . However, we know that those observables must be represented by Hermitian operators (so that their eigenvalues are real), and yet we just saw that the elements of $\mathfrak{o}(n)$ and $\mathfrak{u}(n)$ are *anti*-Hermitian! This is where our mysterious factor of i, which we mentioned in the last section, comes into play. Strictly speaking, $\mathfrak{o}(n)$ and $\mathfrak{u}(n)$ are real vector spaces (check!) whose elements are anti-Hermitian matrices. However, if we permit ourselves to ignore the real vector space structure and treat these matrices just as elements of $M_n(\mathbb{C})$, we can multiply them by i to get Hermitian matrices, which we can then take to represent physical observables. In the physics literature, one usually turns this around and *defines* generators of transformations to be these Hermitian observables, and then multiplies by i to get an anti-Hermitian operator, which can then be exponentiated into an isometry. So one could say that the physicist's definition of the Lie algebra of a matrix Lie group G is

$$\mathfrak{g}_{\text{physics}} = \{X \in M_n(\mathbb{C}) |\, e^{itX} \in G \ \forall t \in \mathbb{R}\}.$$

In the rest of this book we will stick with our original definition of the Lie algebra of a matrix Lie group, but you should be aware that the physicist's definition is often implicit in the physics literature.

Example 4.28. $\mathfrak{o}(n-1,1)$ *The Lie algebra of* $O(n-1,1)$

Now let $V = \mathbb{R}^n$ with the Minkowski metric η, and let $\mathfrak{o}(n-1,1)$ denote the Lie algebra of the Lorentz group $O(n-1,1)$. For $X \in \mathfrak{o}(n-1,1)$, and letting $[\eta] = \text{Diag}(1,1\dots,1,-1)$ (see (4.60) below), Eq. (4.55) becomes

$$
\begin{aligned}
& [v]^T X^T[\eta][w] = -[v]^T[\eta]X[w] \quad \forall\, v,w \in \mathbb{R}^n \\
\Longleftrightarrow \quad & X^T[\eta] = -[\eta]X. \qquad (4.58)
\end{aligned}
$$

Writing X out in block form with a $(n-1)\times(n-1)$ matrix X' for the spatial components (i.e., the X_{ij} where $i,j<n$), and vectors $\mathbf{a}$ and $\mathbf{b}$ for the components X_{in} and X_{ni}, $i<n$, this reads

$$
\begin{pmatrix} X'^T & -\mathbf{b} \\ \mathbf{a} & -X_{nn} \end{pmatrix} = -\begin{pmatrix} X' & \mathbf{a} \\ -\mathbf{b} & -X_{nn} \end{pmatrix}.
$$

This implies that X has the form

$$
\boxed{X = \begin{pmatrix} X' & \mathbf{a} \\ \mathbf{a} & 0 \end{pmatrix}, \quad X' \in \mathfrak{o}(n-1), \quad \mathbf{a} \in \mathbb{R}^{n-1}.}
$$

One can think of X' as generating rotations in the $n-1$ spatial dimensions, and $\mathbf{a}$ as generating a boost along the direction it points in $\mathbb{R}^{n-1}$. We will discuss this in detail in the case $n=4$ in the next section. □

We have now described the Lie algebras of the isometry groups $O(n)$, $U(n)$, and $O(n-1,1)$. What about their cousins $SO(n)$ and $SU(n)$? Since these groups are defined by the additional condition that they have unit determinant, examining their Lie algebras will require that we know how to evaluate the determinant of an exponential. This can be accomplished via the following beautiful and useful formula:

Proposition 4.1. *For any finite-dimensional matrix* X,

$$
\det e^X = e^{\text{Tr}\, X}. \qquad (4.59)
$$

A general proof of this is postponed to Example 4.44 and Problem 4-14, but we can gain some insight into the formula by proving it in the case where X is diagonalizable. Recall that ***diagonalizable*** means that there exists $A \in GL(n,\mathbb{C})$ such that

$$
AXA^{-1} = \text{Diag}(\lambda_1,\lambda_2,\dots,\lambda_n)
$$

where $\text{Diag}(\lambda_1,\lambda_2,\dots,\lambda_n)$ is a diagonal matrix with λ_i in the ith row and column, i.e.

$$\text{Diag}(\lambda_1, \lambda_2, \ldots, \lambda_n) \equiv \begin{pmatrix} \lambda_1 & & & \\ & \lambda_2 & & \\ & & \cdots & \\ & & & \lambda_n \end{pmatrix}. \tag{4.60}$$

In this case, we have

$$\begin{aligned} \det e^X &= \det(Ae^X A^{-1}) \\ &= \det e^{AXA^{-1}} \\ &= \det e^{\text{Diag}(\lambda_1, \lambda_2, \ldots, \lambda_n)} \quad \text{check!} \\ &= \det \text{Diag}(e^{\lambda_1}, e^{\lambda_2}, \ldots, e^{\lambda_n}) \quad \text{(see exercise below)} \\ &= e^{\lambda_1} e^{\lambda_2} \cdots e^{\lambda_n} \\ &= e^{\lambda_1 + \lambda_2 + \cdots + \lambda_n} \\ &= e^{\text{Tr}\, X} \end{aligned} \tag{4.61}$$

and so the formula holds. □

Exercise 4.24. Prove (4.61). That is, use the definition of the matrix exponential as a power series to show that

$$e^{\text{Diag}(\lambda_1, \lambda_2, \ldots, \lambda_n)} = \text{Diag}(e^{\lambda_1}, e^{\lambda_2}, \ldots, e^{\lambda_n}).$$

In addition to being a useful formula, Proposition 4.6 also provides a nice geometric interpretation of the Tr functional. Consider an arbitrary one-parameter subgroup $\{e^{tX}\} \subset GL(n, \mathbb{C})$. We have

$$\det e^{tX} = e^{\text{Tr}\, tX} = e^{t\, \text{Tr}\, X},$$

so taking the derivative with respect to t and evaluating at $t = 0$ gives

$$\left. \frac{d}{dt} \det e^{tX} \right|_{t=0} = \text{Tr}\, X.$$

Since the determinant measures how an operator or matrix changes volumes (cf. Example 3.28), this tells us that the trace of the generator X gives the *rate at which volumes change* under the action of the corresponding one-parameter subgroup e^{tX}.

Example 4.29. $\mathfrak{so}(n)$ *and* $\mathfrak{su}(n)$*, the Lie algebras of* $SO(n)$ *and* $SU(n)$

Recall that $SO(n)$ and $SU(n)$ are the subgroups of $O(n)$ and $U(n)$ consisting of matrices with unit determinant. What additional condition must we impose on the generators X to ensure that $\det e^{tX} = 1\ \forall\, t$? From (4.59), it's clear that $\det e^{tX} = 1\ \forall\, t$ if and only if

$$\operatorname{Tr} X = 0.$$

In the case of $\mathfrak{o}(n)$, this is actually already satisfied, since the anti-symmetry condition implies that all the diagonal entries are zero. Thus, $\mathfrak{so}(n) = \mathfrak{o}(n)$, and both Lie algebras will henceforth be denoted as $\mathfrak{so}(n)$. We could have guessed that they would be equal; as discussed above, generators X are in one-to-one correspondence with "infinitesimal" transformations $I + \epsilon X$, so the Lie algebra just tells us about transformations that are close to the identity. However, the discussion from Example 4.15 generalizes easily to show that $O(n)$ is a group with two components, and that the component which contains the identity is just $SO(n)$. Thus the set of "infinitesimal" transformations of both groups should be equal, and so should their Lie algebras. The same argument applies to $SO(3,1)_o$ and $O(3,1)$; their Lie algebras are thus identical, and will both be denoted as $\mathfrak{so}(3,1)$.

For $\mathfrak{su}(n)$ the story is a little different. Here, the anti-hermiticity condition only guarantees that the trace of an anti-Hermitian matrix is pure imaginary, so demanding that it actually be zero is a bona fide constraint. Thus, $\mathfrak{su}(n)$ can without redundancy be described as the set of traceless, anti-Hermitian $n \times n$ matrices. The tracelessness condition provides one additional constraint beyond anti-hermiticity, so that

$$\dim \mathfrak{su}(n) = \dim \mathfrak{u}(n) - 1 = n^2 - 1.$$

Can you find a nice basis for $\mathfrak{su}(n)$? □

Now that we are acquainted with the Lie algebras of our favorite matrix Lie groups (viewed as isometry groups), it is time to point out some common features that they share. First off, they are all real vector spaces, as you can easily check.[28] Secondly, they are *closed under commutators*, in the sense that if X and Y are elements of the Lie algebra, then so is

$$[X, Y] \equiv XY - YX.$$

You will check this in Exercise 4.25 below. Thirdly, these Lie algebras (and, in fact, all sets of matrices) satisfy the ***Jacobi Identity***,

$$[[X, Y], Z] + [[Y, Z], X] + [[Z, X], Y] = 0 \ \ \forall\, X, Y, Z \in \mathfrak{g}. \tag{4.62}$$

This can be verified directly by expanding the commutators, and doesn't depend on any special properties of $\mathfrak{g}$, only on the definition of the commutator. We will have more to say about the Jacobi identity later.

[28] $\mathfrak{gl}(n, \mathbb{C})$ can also be considered a complex vector space, but we won't consider it as such in this text.

Exercise 4.25. (a) Show directly from the defining conditions (4.56), (4.57), and (4.58) that $\mathfrak{so}(n)$, $\mathfrak{u}(n)$, and $\mathfrak{so}(n-1,1)$ are vector spaces, i.e. closed under scalar multiplication and addition. Impose the additional tracelessness condition and show that $\mathfrak{su}(n)$ is a vector space as well.

(b) Similarly, show directly from the defining conditions that $\mathfrak{so}(n)$, $\mathfrak{u}(n)$, and $\mathfrak{so}(n-1,1)$ are closed under commutators.

(c) Prove the *cyclic property* of the Trace functional,

$$\mathrm{Tr}(A_1 A_2 \cdots A_n) = \mathrm{Tr}(A_2 \cdots A_n A_1), \quad A_i \in M_n(\mathbb{C})$$

and use this to show directly that $\mathfrak{su}(n)$ is closed under commutators.

The reason we've singled out these peculiar-seeming properties is that they will turn out to be important, and it turns out that *all* Lie algebras of matrix Lie groups enjoy them:

Proposition 4.2. *Let $\mathfrak{g}$ be the Lie algebra of a matrix Lie group G. Then $\mathfrak{g}$ satisfies the following:*

1. *$\mathfrak{g}$ is a real vector space*
2. *$\mathfrak{g}$ is closed under commutators*
3. *All elements of $\mathfrak{g}$ obey the Jacobi identity.*

Proof sketch. Proving this turns out to be somewhat technical, so we'll just sketch a proof here and refer you to Hall [11] for the details. Let $\mathfrak{g}$ be the Lie algebra of a matrix Lie group G, and let $X, Y \in \mathfrak{g}$. We'd first like to show that $X + Y \in \mathfrak{g}$. Since $\mathfrak{g}$ is closed under real scalar multiplication (why?), proving $X + Y \in \mathfrak{g}$ will be enough to show that $\mathfrak{g}$ is a real vector space. The proof of this hinges on the following identity, known as the ***Lie Product Formula***, which we state but don't prove:

$$e^{X+Y} = \lim_{m\to\infty} \left(e^{\frac{X}{m}} e^{\frac{Y}{m}}\right)^m .$$

This formula should be thought of as expressing the addition operation in $\mathfrak{g}$ in terms of the product operation in G. With this in hand, we note that $A_m \equiv (e^{\frac{X}{m}} e^{\frac{Y}{m}})^m$ is a convergent sequence, and that every term in the sequence is in G since it is a product of elements in G. Furthermore, the limit matrix $A = e^{X+Y}$ is in $GL(n, \mathbb{C})$ by (4.54). By the definition of a matrix Lie group, then, $A = e^{X+Y} \in G$, and thus $\mathfrak{g}$ is a real vector space.

The second task is to show that $\mathfrak{g}$ is closed under commutators. First, we claim that for any $X \in \mathfrak{gl}(n, \mathbb{C})$ and $A \in GL(n, \mathbb{C})$,

$$e^{AXA^{-1}} = Ae^X A^{-1}. \tag{4.63}$$

You can easily verify this by expanding the power series on both sides. This implies that if $X \in \mathfrak{g}$ and $A \in G$, then $AXA^{-1} \in \mathfrak{g}$ as well, since

$$e^{tAXA^{-1}} = Ae^{tX} A^{-1} \in G \ \forall t \in \mathbb{R}.$$

Now let $A = e^{tY}$, $Y \in \mathfrak{g}$. Then $e^{tY}Xe^{-tY} \in \mathfrak{g}\ \forall t$, and we can compute the derivative of this expression at $t = 0$:

$$\begin{aligned} \frac{d}{dt}e^{tY}Xe^{-tY}\bigg|_{t=0} &= Ye^{tY}Xe^{-tY}\big|_{t=0} - e^{tY}Xe^{-tY}Y\big|_{t=0} \\ &= YX - XY. \end{aligned} \tag{4.64}$$

Since we also have

$$\frac{d}{dt}e^{tY}Xe^{-tY}\bigg|_{t=0} = \lim_{h\to 0}\frac{e^{hY}Xe^{-hY} - X}{h}$$

and the right side is always in $\mathfrak{g}$ since $\mathfrak{g}$ is a vector space,[29] this shows that $YX - XY = [Y, X]$ is in $\mathfrak{g}$.

The third and final task would be to verify the Jacobi identity, but as we pointed out above this holds for *any* set of matrices, and can be easily verified by direct computation. This completes the proof sketch. □

Before moving on to the next section, we should discuss the significance of the commutator. We proved above that all Lie algebras are closed under commutators, but so what? Why is this property worth singling out? Well, it turns out that the algebraic structure of the commutator on $\mathfrak{g}$ is closely related to the algebraic structure of the product on G. This is most clearly manifested in the ***Baker–Campbell–Hausdorff*** (BCH) formula, which for X and Y sufficiently small[30] expresses $e^X e^Y$ as a single exponential:

$$e^X e^Y = e^{X+Y+\frac{1}{2}[X,Y]+\frac{1}{12}[X,[X,Y]]-\frac{1}{12}[Y,[X,Y]]+\cdots}. \tag{4.66}$$

It can be shown[31] that the series in the exponent converges for such X and Y, and that the series consists *entirely of iterated commutators*, so that the exponent really is an element of $\mathfrak{g}$ (if the exponent had a term like XY in it, this would not be the case since matrix Lie algebras are in general not closed under ordinary matrix multiplication). Thus, the BCH formula plays a role analagous to that of the Lie product formula, but in the other direction: while the Lie product formula

[29] To be rigorous, we also need to note that $\mathfrak{g}$ is closed in the topological sense, but this can be regarded as a technicality.

[30] The size of a matrix $X \in M_n(\mathbb{C})$ is usually expressed by the ***Hilbert–Schmidt norm***, defined as

$$||X|| \equiv \sum_{i,j=1}^{n} |X_{ij}|^2. \tag{4.65}$$

If we introduce the basis $\{E_{ij}\}$ of $M_n(\mathbb{C})$ from Example 2.9, then we can identify $M_n(\mathbb{C})$ with $\mathbb{C}^{n^2}$, and then (4.65) is just the standard Hermitian inner product.

[31] See Hall [11] for a nice discussion and Varadarajan [20] for a complete proof.

expresses Lie algebra addition in terms of group multiplication, the BCH formula *expresses group multiplication in terms of the commutator on the Lie algebra*. The BCH formula thus tells us that

Much of the group structure of G is encoded in the commutator on $\mathfrak{g}$.

You will calculate the first few terms of the exponential in the BCH formula in Problem 4-11.

4.7 The Lie Algebras of Classical and Quantum Physics

We are now ready for a detailed look at the Lie algebras of the matrix Lie groups we discussed in Sect. 4.3.

Example 4.30. $\mathfrak{so}(2)$

As discussed in Example 4.26, $\mathfrak{so}(2)$ consists of all antisymmetric 2×2 matrices. All such matrices are of the form

$$\begin{pmatrix} 0 & -a \\ a & 0 \end{pmatrix}$$

and so $\mathfrak{so}(2)$ is one-dimensional and we may take as a basis

$$X = \begin{pmatrix} 0 & -1 \\ 1 & 0 \end{pmatrix}.$$

You will explicitly compute that

$$e^{\theta X} = \begin{pmatrix} \cos\theta & -\sin\theta \\ \sin\theta & \cos\theta \end{pmatrix}, \quad \theta \in \mathbb{R} \tag{4.67}$$

so that X really does generate counterclockwise rotations in the $x - y$ plane. Recall from Sect. 4.1 that X also generates rotations in the sense that for any position vector $\mathbf{r}$, $X\mathbf{r}$ is just the direction in which $\mathbf{r}$ would change under rotation. This means that X induces a vector *field* $X^\sharp$, given at an arbitrary point $\mathbf{r}$ by

$$X^\sharp(\mathbf{r}) \equiv X\mathbf{r}.$$

This vector field is depicted in Fig. 4.8.

Exercise 4.26. Verify (4.67) by explicitly summing the power series for $e^{\theta X}$.

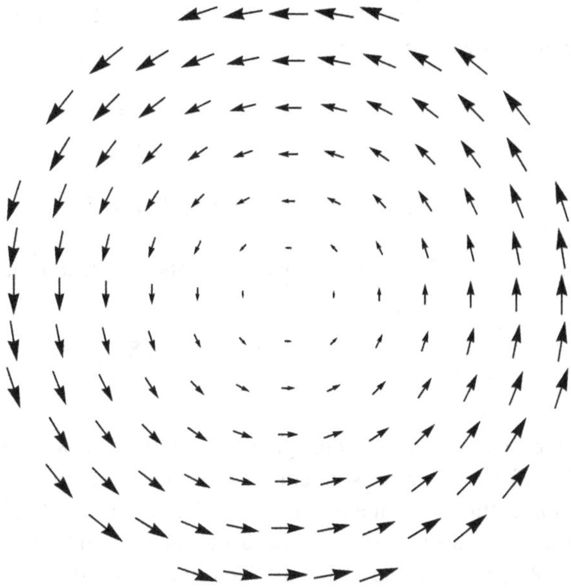

Fig. 4.8 The vector field $X^\sharp(\mathbf{r}) \equiv X\mathbf{r}$ induced by the $\mathfrak{so}(2)$ element X. This is one sense in which X "generates" rotations

Example 4.31. $\mathfrak{so}(3)$

As remarked in Example 4.26, the matrices A_{ij} form a basis for $\mathfrak{o}(n)$. We now specialize to $n = 3$ and rename the A_{ij} as

$$L_x \equiv \begin{pmatrix} 0 & 0 & 0 \\ 0 & 0 & -1 \\ 0 & 1 & 0 \end{pmatrix}, \quad L_y \equiv \begin{pmatrix} 0 & 0 & 1 \\ 0 & 0 & 0 \\ -1 & 0 & 0 \end{pmatrix}, \quad L_z = \begin{pmatrix} 0 & -1 & 0 \\ 1 & 0 & 0 \\ 0 & 0 & 0 \end{pmatrix} \tag{4.68}$$

which you may recognize from Sects. 4.1 and 4.5. Note that we can write the components of *all* of these matrices in one equation as

$$\boxed{(L_i)_{jk} = -\epsilon_{ijk},} \tag{4.69}$$

an expression which will prove useful later. An arbitrary element $X \in \mathfrak{so}(3)$ looks like

$$X = \begin{pmatrix} 0 & -z & y \\ z & 0 & -x \\ -y & x & 0 \end{pmatrix} = xL_x + yL_y + zL_z. \tag{4.70}$$

Note that $\dim \mathfrak{so}(3) = 3$, which is also the number of parameters in $SO(3)$.

You can check, as we did in Exercise 4.2, that the commutators of the basis elements work out to be

$$[L_x, L_y] = L_z$$
$$[L_y, L_z] = L_x$$
$$[L_z, L_x] = L_y$$

or, if we label the generators with numbers instead of letters,

$$[L_i, L_j] = \sum_{k=1}^{3} \epsilon_{ijk} L_k. \tag{4.71}$$

(We will frequently change notation in this way when it proves convenient; hopefully, this causes no confusion). These are, of course, the well-known angular momentum commutation relations of quantum mechanics. The relation between the Lie algebra of $\mathfrak{so}(3)$ and the usual angular momentum operators (2.17) will be explained in the next chapter. Note that if we defined Hermitian $\mathfrak{so}(3)$ generators by $L'_i \equiv iL_i$, the commutation relations would become

$$[L'_i, L'_j] = \sum_{k=1}^{3} i\epsilon_{ijk} L'_k,$$

which is the form found in most quantum mechanics texts.

We can think of the generator X in (4.70) as generating a rotation about the axis $[X]_{\{L_i\}} = (x, y, z)$, as follows: for any $v = (v_x, v_y, v_z) \in \mathbb{R}^3$, you can check that

$$Xv = \begin{pmatrix} 0 & -z & y \\ z & 0 & -x \\ -y & x & 0 \end{pmatrix} \begin{pmatrix} v_x \\ v_y \\ v_z \end{pmatrix} = (x, y, z) \times (v_x, v_y, v_z) = [X] \times v$$

so that if v lies along the axis $[X]$, then $Xv = 0$. This then implies that

$$e^{tX}[X] = [X]$$

(why?), so e^{tX} must be a rotation about $[X]$ (since it leaves $[X]$ fixed). In fact, if we take $[X]$ to be a unit vector and rename it $\hat{\mathbf{n}}$, and also rename t as θ, you will show below that

$$e^{\theta X} = \begin{pmatrix} n_x^2(1-\cos\theta)+\cos\theta & n_x n_y(1-\cos\theta)-n_z\sin\theta & n_x n_z(1-\cos\theta)+n_y\sin\theta \\ n_y n_x(1-\cos\theta)+n_z\sin\theta & n_y^2(1-\cos\theta)+\cos\theta & n_y n_z(1-\cos\theta)-n_x\sin\theta \\ n_z n_x(1-\cos\theta)-n_y\sin\theta & n_z n_y(1-\cos\theta)+n_x\sin\theta & n_z^2(1-\cos\theta)+\cos\theta \end{pmatrix}, \tag{4.72}$$

proving our claim from Example 4.14 that this matrix represents a rotation about $\hat{\mathbf{n}}$ by an angle θ.

The astute reader may recall that we already made a connection between antisymmetric 3×3 matrices and rotations back in Example 3.30. There we saw that if $A(t)$ was a time-dependent orthogonal matrix representing the rotation of a rigid body, then the associated angular velocity bivector (in the space frame) was $[\tilde{\omega}] = \frac{dA}{dt}A^{-1}$. If we let $A(t)$ be a one-parameter subgroup $A(t) = e^{tX}$ generated by some $X \in \mathfrak{so}(3)$, then the associated angular velocity bivector is

$$\begin{aligned} [\tilde{\omega}] &= \frac{dA(t)}{dt}A^{-1} \\ &= Xe^{tx}e^{-tx} \\ &= X. \end{aligned} \tag{4.73}$$

Thus:

The angular velocity bivector is just a rotation generator.

(Recall that we gave a more heuristic demonstration of this fact back in Example 4.3 at the beginning of this chapter, which may be worth re-reading now that you've seen a more precise treatment.) Furthermore, applying the J map from Example 3.10 to both sides of (4.73) gives

$$[\boldsymbol{\omega}] = [X]$$

and so the pseudovector $\boldsymbol{\omega}$ is just the rotation generator expressed in coordinates $[X]$. Note that this agrees with our discussion above, where we found that $[X]$ gave the axis of rotation.

Exercise 4.27. Let $X \in \mathfrak{so}(3)$ be given by

$$X = \begin{pmatrix} 0 & -n_z & n_y \\ n_z & 0 & -n_x \\ -n_y & n_x & 0 \end{pmatrix}$$

where $\hat{\mathbf{n}} = (n_x, n_y, n_z)$ is a unit vector. Verify (4.72) by explicitly summing the power series for $e^{\theta X}$.

Exercise 4.28. Using the basis $\mathcal{B} = \{L_x, L_y, L_z\}$ for $\mathfrak{so}(3)$, show that

$$[[X, Y]]_{\mathcal{B}} = [X]_{\mathcal{B}} \times [Y]_{\mathcal{B}},$$

which shows that in components, the commutator is given by the usual cross product on $\mathbb{R}^3$.

Note: On the left-hand side of the above equation the two sets of brackets have entirely different meanings. The inner bracket is the Lie bracket, while the outer bracket denotes the component representation of an element of the Lie algebra.

Example 4.32. $\mathfrak{su}(2)$

We already met $\mathfrak{su}(2)$, the set of 2×2 traceless anti-Hermitian matrices, in Example 4.22. We took as a basis

$$S_x \equiv \frac{1}{2}\begin{pmatrix} 0 & -i \\ -i & 0 \end{pmatrix}, \quad S_y \equiv \frac{1}{2}\begin{pmatrix} 0 & -1 \\ 1 & 0 \end{pmatrix}, \quad S_z \equiv \frac{1}{2}\begin{pmatrix} -i & 0 \\ 0 & i \end{pmatrix}.$$

Notice again that the number of parameters in the matrix Lie group $SU(2)$ is equal to the dimension of its Lie algebra $\mathfrak{su}(2)$. You can check that the commutation relation for these basis vectors is

$$[S_i, S_j] = \sum_{k=1}^{3} \epsilon_{ijk} S_k$$

which is the same as the $\mathfrak{so}(3)$ commutation relations! Does that mean that $\mathfrak{su}(2)$ and $\mathfrak{so}(3)$ are, in some sense, the same? And is this in some way related to the homomorphism from $SU(2)$ to $SO(3)$ that we discussed in Example 4.22? The answer to both these questions is yes, as we will discuss in the next few sections. We will also see that, just as $X = xL_x + yL_y + zL_z \in \mathfrak{so}(3)$ can be interpreted as generating a rotation about $(x, y, z) \in \mathbb{R}^3$, so can

$$Y = xS_1 + yS_2 + zS_3 = \tfrac{1}{2}\begin{pmatrix} -iz & -i(x - iy) \\ -i(x + iy) & iz \end{pmatrix} \in \mathfrak{su}(2).$$

In fact, you can show that

$$e^{\theta n^i S_i} = \begin{pmatrix} \cos(\theta/2) - i n_z \sin(\theta/2) & (-in_x - n_y)\sin(\theta/2) \\ (-in_x + n_y)\sin(\theta/2) & \cos(\theta/2) + i n_z \sin(\theta/2) \end{pmatrix}, \qquad (4.74)$$

which we claimed earlier implements a rotation by angle θ around axis $\hat{\mathbf{n}} \equiv (n_x, n_y, n_z)$. To prove this, see Exercise 4.29 and Problem 4-7.

Exercise 4.29. Let $\hat{\mathbf{n}} = (n^1, n^2, n^3)$ be a unit vector. Prove (4.74) by direct calculation, and use that to show that

$$e^{\theta n^i S_i} = \cos(\theta/2)\, I + 2\sin(\theta/2)\, n^i S_i.$$

You will use this formula in Problem 4-7.

Example 4.33. $\mathfrak{so}(3,1)$

From Example 4.28, we know that an arbitrary $X \in \mathfrak{so}(3,1)$ can be written as

$$X = \begin{pmatrix} X' & \mathbf{a} \\ \mathbf{a} & 0 \end{pmatrix}$$

with $X' \in \mathfrak{so}(3)$ and $\mathbf{a} \in \mathbb{R}^3$. Embedding the L_i of $\mathfrak{so}(3)$ into $\mathfrak{so}(3,1)$ as

$$\tilde{L}_i \equiv \begin{pmatrix} L_i & \mathbf{0} \\ \mathbf{0} & 0 \end{pmatrix}$$

and defining new generators

$$K_1 \equiv \begin{pmatrix} 0&0&0&1 \\ 0&0&0&0 \\ 0&0&0&0 \\ 1&0&0&0 \end{pmatrix}, \quad K_2 \equiv \begin{pmatrix} 0&0&0&0 \\ 0&0&0&1 \\ 0&0&0&0 \\ 0&1&0&0 \end{pmatrix}, \quad K_3 \equiv \begin{pmatrix} 0&0&0&0 \\ 0&0&0&0 \\ 0&0&0&1 \\ 0&0&1&0 \end{pmatrix}$$

we have the following commutation relations (Exercise 4.30):

$$[\tilde{L}_i, \tilde{L}_j] = \sum_{k=1}^{3} \epsilon_{ijk} \tilde{L}_k \tag{4.75a}$$

$$[\tilde{L}_i, K_j] = \sum_{k=1}^{3} \epsilon_{ijk} K_k \tag{4.75b}$$

$$[K_i, K_j] = -\sum_{k=1}^{3} \epsilon_{ijk} \tilde{L}_k. \tag{4.75c}$$

As we mentioned in Example 4.28, the K_i can be interpreted as generating boosts along their corresponding axes. In fact, you will show in Exercise 4.30 that for $u \in \mathbb{R}$,

$$e^{uK_3} = \begin{pmatrix} 1&0&0&0 \\ 0&1&0&0 \\ 0&0&\cosh u & -\sinh u \\ 0&0&-\sinh u & \cosh u \end{pmatrix}, \tag{4.76}$$

which we know from Example 4.17 represents a boost of speed $\beta = v/c = \tanh u$ in the z-direction. We will comment on the appearance of the rapidity u here in Box 4.5 below.

The commutation relations in (4.75) bear interpretation. Equation (4.75a) is of course just the usual $\mathfrak{so}(3)$ commutation relations. Equation (4.75b) says that the K_i transform like a vector under rotations [note the similarity with (3.57)]. Finally, (4.75c) says that if we perform two successive boosts in different directions, the *order* matters. Furthermore, the difference between performing them in one order and the other is actually a rotation! This then implies, by the Baker–Campbell–Hausdorff formula (4.66), that the product of two finite boosts is not necessarily another boost, as mentioned in Example 4.17.

Exercise 4.30. Verify the commutation relations (4.75), and verify (4.76) by explicitly summing the exponential power series. For a further challenge, sum the exponential power series for $e^{u^i K_i}$ to get (4.33), which represents a boost in the direction $\mathbf{u}$.

Box 4.5 *The Addition of Rapidities and the Einstein Velocity Addition Law*
The fact that the rapidity u appears in the exponent in (4.76) shows that u is a kind of "boost angle" which provides a natural measure for the magnitude of a boost. In fact, just as the angles of rotation about a single axis add, so do boosts in a given direction: letting $K \equiv K_3$ and invoking the addition property of exponents[32] we have

$$e^{u_1 K} e^{u_2 K} = e^{(u_1+u_2)K}, \tag{4.77}$$

which says that a boost by rapidity u_2 followed by a boost by u_1 (all in the z-direction) is equivalent to a boost by $u_1 + u_2$. Of course, the same is *not* true for boost *velocities*. For these, we invoke the addition law for hyperbolic tangents,

$$\tanh(u_1 + u_2) = \frac{\tanh u_1 + \tanh u_2}{1 + \tanh u_1 \tanh u_2}, \tag{4.78}$$

which combined with the substitutions $\beta_{1+2} \equiv \tanh(u_1 + u_2)$ and $\beta_i = \tanh u_i$ becomes

$$\beta_{1+2} = \frac{\beta_1 + \beta_2}{1 + \beta_1 \beta_2}.$$

This, of course, is just Einstein's relativistic velocity addition law; we see here that it follows from the additivity of rapidities, and that it is essentially just the addition law for hyperbolic tangents!

Exercise 4.31.
Derive (4.78) by substituting (4.76) into (4.77).

Example 4.34. $\mathfrak{sl}(2, \mathbb{C})_{\mathbb{R}}$

$\mathfrak{sl}(2, \mathbb{C})_{\mathbb{R}}$ is defined to be the Lie algebra of $SL(2, \mathbb{C})$, viewed as a *real* vector space. Since $SL(2, \mathbb{C})$ is just the set of all 2×2 complex matrices with unit determinant, $\mathfrak{sl}(2, \mathbb{C})_{\mathbb{R}}$ is just the set of all *traceless* 2×2 complex matrices, and thus could be

[32] Which, of course, only holds for matrices when the matrices being added commute.

viewed as a *complex* vector space, though we won't take that point of view here.[33] A (real) basis for $\mathfrak{sl}(2,\mathbb{C})_{\mathbb{R}}$ is

$$S_1 = \frac{1}{2}\begin{pmatrix} 0 & -i \\ -i & 0 \end{pmatrix}, \quad S_2 = \frac{1}{2}\begin{pmatrix} 0 & -1 \\ 1 & 0 \end{pmatrix}, \quad S_3 = \frac{1}{2}\begin{pmatrix} -i & 0 \\ 0 & i \end{pmatrix}$$

$$\tilde{K}_1 \equiv \frac{1}{2}\begin{pmatrix} 0 & 1 \\ 1 & 0 \end{pmatrix}, \quad \tilde{K}_2 \equiv \frac{1}{2}\begin{pmatrix} 0 & -i \\ i & 0 \end{pmatrix}, \quad \tilde{K}_3 \equiv \frac{1}{2}\begin{pmatrix} 1 & 0 \\ 0 & -1 \end{pmatrix}.$$

Note that $\tilde{K}_i = iS_i$. This fact simplifies certain computations, such as the ones you will perform in checking that these generators satisfy the following commutation relations:

$$[S_i, S_j] = \sum_{k=1}^{3} \epsilon_{ijk} S_k$$

$$\left[S_i, \tilde{K}_j\right] = \sum_{k=1}^{3} \epsilon_{ijk} \tilde{K}_k$$

$$\left[\tilde{K}_i, \tilde{K}_j\right] = -\sum_{k=1}^{3} \epsilon_{ijk} S_k.$$

These are identical to the $\mathfrak{so}(3,1)$ commutation relations! As in the case of $\mathfrak{su}(2)$ and $\mathfrak{so}(3)$, this is intimately related to the homomorphism from $SL(2,\mathbb{C})$ to $SO(3,1)_o$ that we described in Example 4.23. This will be discussed in Sect. 4.9.

Exercise 4.32. Check that the $\tilde{K}_i$ generate boosts, as you would expect, by explicitly calculating $e^{u^i \tilde{K}_i}$ to get (4.36).

4.8 Abstract Lie Algebras

So far we have considered Lie algebras associated with matrix Lie groups, and we sketched proofs that these sets are real vector spaces which are closed under commutators. As in the case of abstract vector spaces and groups, however, we can

[33]The space of all traceless 2×2 complex matrices viewed as a *complex* vector space is denoted $\mathfrak{sl}(2,\mathbb{C})$ without the $\mathbb{R}$ subscript. In this case, the S_i suffice to form a basis. In this text, however, we will usually take Lie algebras to be real vector spaces, even if they naturally form complex vector spaces as well. You should be aware, though, that $\mathfrak{sl}(2,\mathbb{C})$ is fundamental in the theory of *complex* Lie algebras, and so in the math literature the Lie algebra of $SL(2,\mathbb{C})$ is almost always considered to be the complex vector space $\mathfrak{sl}(2,\mathbb{C})$ rather than the real Lie algebra $\mathfrak{sl}(2,\mathbb{C})_{\mathbb{R}}$. We will have more to say about $\mathfrak{sl}(2,\mathbb{C})$ in the appendix.

now turn around and use these properties to *define* abstract Lie algebras. This will clarify the nature of the Lie algebras we've already met, as well as permit discussion of other examples relevant for physics.

That said, a ***(real, abstract) Lie algebra*** is defined to be a real vector space $\mathfrak{g}$ equipped with a bilinear map $[\cdot,\cdot] : \mathfrak{g} \times \mathfrak{g} \to \mathfrak{g}$ called the ***Lie bracket*** which satisfies

1. $[X,Y] = -[Y,X] \;\forall\, X,Y \in \mathfrak{g}$ (Antisymmetry)
2. $[[X,Y],Z] + [[Y,Z],X] + [[Z,X],Y] = 0 \;\forall\, X,Y,Z \in \mathfrak{g}$ (Jacobi identity)

By construction, all Lie algebras of matrix Lie groups satisfy this definition (when we take the bracket to be the commutator), and we will see that it is precisely the above properties of the commutator that make those Lie algebras useful in applications. Furthermore, there are some (abstract) Lie algebras that arise in physics for which the bracket is *not* a commutator, and which are not usually associated with a matrix Lie group; this definition allows us to include those algebras in our discussion. We'll meet a few of these algebras below, but first we consider two basic examples.

Example 4.35. $\mathfrak{gl}(V)$ *The Lie algebra of linear operators on a vector space*

Let V be a (possibly infinite-dimensional) vector space. We can turn $\mathcal{L}(V)$, the set of all linear operators on V, into a Lie algebra by taking the Lie bracket to be the commutator, i.e.

$$[T,U] \equiv TU - UT \quad T,U \in \mathcal{L}(V).$$

Note that this is a commutator of *operators*, not matrices, though of course there is a nice correspondence between the two when V is finite-dimensional and we introduce a basis. This Lie bracket is obviously anti-symmetric and can be seen to obey the Jacobi identity, so it turns $\mathcal{L}(V)$ into a Lie algebra which we'll denote by $\mathfrak{gl}(V)$.[34] We'll have more to say about $\mathfrak{gl}(V)$ as we progress.

Example 4.36. $\mathfrak{isom}(V)$ *The Lie algebra of anti-Hermitian operators*

Consider the setup of the previous example, except now let V be an inner product space. For any $T \in \mathcal{L}(V)$, the inner product on V allows us [via (4.13)] to define its adjoint $T^\dagger$, and we can then define $\mathfrak{isom}(V) \subset \mathfrak{gl}(V)$ to be the set of all ***anti-Hermitian*** operators, i.e. those which satisfy

$$\boxed{T^\dagger = -T.} \tag{4.79}$$

[34] There is a subtlety here: the vector space underlying $\mathfrak{gl}(V)$ is of course just $\mathcal{L}(V)$, so the difference between the two is just that one comes equipped with a Lie bracket, and the other is considered as a vector space with no additional structure.

You can easily verify that $\mathfrak{isom}(V)$ is a Lie subalgebra of $\mathfrak{gl}(V)$. (A ***Lie subalgebra*** of a Lie algebra $\mathfrak{g}$ is a vector subspace $\mathfrak{h} \subset \mathfrak{g}$ that is also closed under the Lie bracket, and hence forms a Lie algebra itself).

This definition is very reminiscent of the characterization of $\mathfrak{u}(n)$ that we gave in Example 4.27; in fact, if V is complex n-dimensional and we introduce an orthonormal basis, then $\mathfrak{isom}(V) = \mathfrak{u}(n)$! The reason we introduce $\mathfrak{isom}(V)$ here as an *abstract* Lie algebra is that in the infinite-dimensional case we cannot view $\mathfrak{isom}(V)$ as the Lie algebra associated to a matrix Lie group, because we haven't developed any theory for infinite-dimensional matrices. Nonetheless, $\mathfrak{isom}(V)$ should be thought of as a coordinate-free, infinite-dimensional analog of $\mathfrak{u}(n)$, and we'll find that it plays a central role in quantum mechanics, as we'll see in the next example.

Example 4.37. *The Poisson bracket on phase space*

Consider a physical system with a $2n$ dimensional phase space P parametrized by n generalized coordinates q_i and the n conjugate momenta p_i. The set of all complex-valued, infinitely differentiable[35] functions on P is a real vector space which we'll denote by $\mathcal{C}(P)$. We can turn $\mathcal{C}(P)$ into a Lie algebra using the ***Poisson bracket*** as our Lie bracket, where the Poisson bracket is defined by

$$\{f, g\} \equiv \sum_i \frac{\partial f}{\partial q_i}\frac{\partial g}{\partial p_i} - \frac{\partial g}{\partial q_i}\frac{\partial f}{\partial p_i}, \qquad f, g \in \mathcal{C}(P).$$

The anti-symmetry of the Poisson bracket is clear, and the Jacobi identity can be verified directly by a brute-force calculation.

The functions in $\mathcal{C}(P)$ are known as ***observables***, and the Poisson bracket thus turns the set of observables into one huge[36] Lie algebra. The standard (or ***canonical***) quantization prescription, as developed by Dirac, Heisenberg, and the other founders of quantum mechanics, is to then interpret this Lie algebra of observables as a Lie subalgebra of $\mathfrak{isom}(\mathcal{H})$ for some Hilbert space $\mathcal{H}$ (this identification is known as a Lie algebra ***representation***, which is the subject of the next chapter). The commutator of the observables in $\mathfrak{isom}(\mathcal{H})$ is then just given by the Poisson bracket of the corresponding functions in $\mathcal{C}(P)$. Thus the set of all observables in quantum mechanics forms a Lie algebra, which is one of our main reasons for studying Lie algebras here.

Though $\mathcal{C}(P)$ is in general infinite-dimensional, it often has interesting finite-dimensional Lie subalgebras. For instance, if $P = \mathbb{R}^6$ and the q_i are just the usual

[35]A function is "infinitely differentiable" if it can be differentiated an arbitrary number of times. Besides the step function and its derivative, the Dirac delta "function", most functions that one meets in classical physics and quantum mechanics are infinitely differentiable. This includes the exponential and trigonometric functions, as well as any other function that permits a power series expansion.

[36]By this we mean infinite-dimensional, and usually requiring a basis that *cannot* be indexed by the integers but rather must be labeled by elements of $\mathbb{R}$ or some other continuous set. You should recall from Sect. 3.7 that $L^2(\mathbb{R})$ was another such "huge" vector space.

cartesian coordinates for $\mathbb{R}^3$, we can consider the three components of the angular momentum,

$$\begin{aligned} J_1 &= q_2 p_3 - q_3 p_2 \\ J_2 &= q_3 p_1 - q_1 p_3 \\ J_3 &= q_1 p_2 - q_2 p_1, \end{aligned} \tag{4.80}$$

all of which are in $\mathcal{C}(\mathbb{R}^6)$. You will check below that the Poisson bracket of these functions turns out to be

$$\{J_i, J_j\} = \sum_{k=1}^{3} \epsilon_{ijk} J_k, \tag{4.81}$$

which are of course the familiar angular momentum ($\mathfrak{so}(3)$) commutation relations; as mentioned above, this is in fact where the angular momentum commutation relations come from! This then implies that $\mathfrak{so}(3)$ is a Lie subalgebra of $\mathcal{C}(\mathbb{R}^6)$. You may be wondering, however, why the angular momentum commutation relations are the same as the $\mathfrak{so}(3)$ commutation relations. What do the functions J_i have to do with generators of rotations? The answer has to do with a general relationship between symmetries and conserved quantities, which we summarize as follows (note: the following discussion is rather dense, and can be omitted on a first reading).

Consider a classical system (i.e., a phase space P together with a Hamiltonian $H \in \mathcal{C}(P)$) which has a matrix Lie group G of canonical transformations[37] acting on it. If H is invariant[38] under the action of G, then G is said to be a group of ***symmetries*** of the system. In this case, one can then show[39] that for every $X \in \mathfrak{g}$ there is a function $f_X \in \mathcal{C}(P)$ which is *constant* along particle trajectories in P (where trajectories, of course, are given by solutions to Hamilton's equations). This fact is known as ***Noether's Theorem***, and it tells us that every element X of the Lie algebra gives a conserved quantity f_X. Furthermore, the Poisson bracket between two such functions f_X and f_Y is given just by the function associated with the Lie bracket of the corresponding elements of $\mathfrak{g}$, i.e.

$$\{f_X, f_Y\} = f_{[X,Y]}. \tag{4.82}$$

If $G = SO(3)$ acting on P by rotations, then it turns out[40] that

---

[37] That is, one-to-one and onto transformations which preserve the form of Hamilton's equations. See Goldstein [8].

[38] Let $\phi_g : P \to P$ be the transformation of P corresponding to the group element $g \in G$. Then H is *invariant* under G if $H(\phi_g(p)) = H(p)\ \forall\ p \in P,\ g \in G$.

[39] Under some mild assumptions. See Cannas [4] or Arnold [1], for example.

[40] See Arnold [1] for a discussion of Noether's theorem and a derivation of an equivalent formula.

$$f_X(q_i, p_i) = (\mathbf{J}(q_i, p_i) \,|\, [X]), \quad X \in \mathfrak{so}(3)$$

where $\mathbf{J} = (J_1, J_2, J_3)$ is the usual angular momentum vector with components given by (4.80) and $(\cdot\,|\,\cdot)$ is the usual inner product on $\mathbb{R}^3$. In particular, for $X = L_i \in \mathfrak{so}(3)$ we have

$$f_{L_i} = (\mathbf{J} \,|\, [L_i]) = (\mathbf{J} \,|\, e_i) = J_i.$$

Thus, *the conserved quantity associated with rotations about the* i*th axis is just the* i*th component of angular momentum.* This is the connection between angular momentum and rotations, and from (4.82) we see that the J_i *must* have the same commutation relations as the L_i, which is of course what we found in (4.81). □

Exercise 4.33. Verify (4.81). Also show that for a function $F \in C(\mathbb{R}^6)$ that depends *only* on the coordinates q_i,

$$\{p_j, F(q_i)\} = -\frac{\partial F}{\partial q_j}$$

and

$$\{J_3, F(q_i)\} = -q_1 \frac{\partial F}{\partial q_2} + q_2 \frac{\partial F}{\partial q_1} = -\frac{\partial F}{\partial \phi}$$

where ϕ is the azimuthal angle. We will interpret these results in the next chapter.

If we have a one-dimensional system with position coordinate q and conjugate momentum p, then $P = \mathbb{R}^2$ and $C(P)$ contains another well-known Lie algebra: the Heisenberg algebra.

Example 4.38. *The Heisenberg algebra*

Define the ***Heisenberg algebra*** H to be the span of $\{q, p, 1\} \subset C(\mathbb{R}^2)$, where $1 \in H$ is just the constant function with value 1. The only nontrivial Poisson bracket is between p and q, which you can check is just

$$\{q, p\} = 1.$$

Apart from the factors of $\hbar$ (which we've dropped throughout the text) and i (which is just an artifact of the physicist's definition of a Lie algebra), this is the familiar commutation relation from quantum mechanics. H is clearly closed under the Lie (Poisson) bracket, and is thus a Lie subalgebra of $C(\mathbb{R}^2)$.

Can p and q be thought of as generators of specific transformations? Well, one of the most basic representations of p and q as operators is on the vector space $L^2(\mathbb{R})$, where

$$\begin{aligned} \hat{q} f(x) &= x f(x) \\ \hat{p} f(x) &= -\frac{df}{dx} \end{aligned} \tag{4.83}$$

(again we drop the factor of i, in disagreement with the physicist's convention). Note that $[\hat{q}, \hat{p}] = \{q, p\} = 1$. If we exponentiate $\hat{p}$, we find (see Exercise 4.34 below)

$$e^{t\hat{p}} f(x) = f(x - t)$$

so that $\hat{p} = -\frac{d}{dx}$ generates translations along the x-axis! It follows that $\{e^{t\hat{p}} \mid t \in \mathbb{R}\}$ is the one-parameter subgroup of all translations along the x-axis. What about $\hat{q}$? Well, if we work in the momentum representation of Example 3.17 so that we're dealing with the Fourier transform $\phi(p)$ of $f(x)$, you know from Exercise 3.19 that $\hat{q}$ is represented by $i\frac{d}{dp}$. Multiplying by i gives $i\hat{q} = -\frac{d}{dp}$ and thus

$$e^{it\hat{q}}\phi(p) = \phi(p - t),$$

which you can think of as a "translation in momentum space". We will explain the extra factor of i in the next chapter.

If we treat H as a complex vector space, then we can consider another common basis for it, which is $\{Q, P, 1\} \subset C(\mathbb{R}^2)$ where

$$Q \equiv \frac{p + iq}{\sqrt{2}}$$

$$P \equiv \frac{p - iq}{\sqrt{2}i}.$$

You can check that

$$\{Q, P\} = 1 \tag{4.84}$$

which you may recognize from classical mechanics as the condition that Q and P be ***canonical variables*** (See Goldstein [8]). Q and P are well suited to the solution of the one-dimensional harmonic oscillator problem, and you may recognize their formal similarity to the raising and lowering operators a and $a^\dagger$ employed in the quantum-mechanical version of the same problem.

Q and P are not easily interpreted as generators of specific transformations, and our discussion of them helps explain why we defined *abstract* Lie algebras—so that we could work with spaces that behave *like* the Lie algebras of matrix Lie groups (in that they are vector spaces with a Lie bracket), but aren't necessarily Lie algebras of matrix Lie groups themselves.

Exercise 4.34. Show by exponentiating $\hat{p} = -\frac{d}{dx}$ that $e^{t\hat{p}} f(x)$ is just the power series expansion for $f(x - t)$.

Exercise 4.35. Verify (4.84).

4.9 Homomorphism and Isomorphism Revisited

In Sect. 4.4 we used the notion of a group homomorphism to make precise the relationship between $SU(2)$ and $SO(3)$, as well as $SL(2,\mathbb{C})$ and $SO(3,1)_o$. Now we will define the corresponding notion for Lie algebras to make precise the relationship between $\mathfrak{su}(2)$ and $\mathfrak{so}(3)$, as well as $\mathfrak{sl}(2,\mathbb{C})_{\mathbb{R}}$ and $\mathfrak{so}(3,1)$. We will also show how these relationships between Lie algebras arise as a consequence of the relationships between the corresponding groups.

That said, we define a ***Lie algebra homomorphism*** from a Lie algebra $\mathfrak{g}$ to a Lie algebra $\mathfrak{h}$ to be a linear map $\phi : \mathfrak{g} \to \mathfrak{h}$ that preserves the Lie bracket, in the sense that

$$\boxed{[\phi(X), \phi(Y)] = \phi([X,Y]) \quad \forall\, X, Y \in \mathfrak{g}.} \tag{4.85}$$

If ϕ is a vector space isomorphism (which implies that $\mathfrak{g}$ and $\mathfrak{h}$ have the same dimension), then ϕ is said to be a ***Lie algebra isomorphism***. In this case, there is a one-to-one correspondence between $\mathfrak{g}$ and $\mathfrak{h}$ that preserves the bracket, so just as with group isomorphisms we consider $\mathfrak{g}$ and $\mathfrak{h}$ to be equivalent and write $\mathfrak{g} \simeq \mathfrak{h}$.

Sometimes the easiest way to prove that two Lie algebras $\mathfrak{g}$ and $\mathfrak{h}$ are isomorphic is with an astute choice of bases. Let $\{X_i\}_{i=1\ldots n}$ and $\{Y_i\}_{i=1\ldots n}$ be bases for $\mathfrak{g}$ and $\mathfrak{h}$ respectively. Then the commutation relations take the form

$$[X_i, X_j] = \sum_{k=1}^{n} c_{ij}{}^{k} X_k$$

$$[Y_i, Y_j] = \sum_{k=1}^{n} d_{ij}{}^{k} Y_k$$

where the numbers $c_{ij}{}^{k}$ and $d_{ij}{}^{k}$ are known as the ***structure constants*** of $\mathfrak{g}$ and $\mathfrak{h}$. (This is a bit of a misnomer, though, since the structure constants depend on a choice of basis, and are not as inherent a feature of the algebra as the name implies.) If one can exhibit bases such that $c_{ij}{}^{k} = d_{ij}{}^{k} \;\; \forall\, i, j, k$, then it's easy to check that the map

$$\begin{aligned} \phi : \quad & \mathfrak{g} \to \mathfrak{h} \\ & v^i X_i \mapsto v^i Y_i \end{aligned}$$

is a Lie algebra isomorphism. We will use this below to show that $\mathfrak{so}(3) \simeq \mathfrak{su}(2)$ and $\mathfrak{so}(3,1) \simeq \mathfrak{sl}(2,\mathbb{C})_{\mathbb{R}}$.

Box 4.6 *Structure Constants as Tensor Components*
The notation $c_{ij}{}^{k}$ suggests that the structure constants are the components of a tensor, and indeed this is the case. Define a (2,1) tensor T on a Lie algebra $\mathfrak{g}$ by

$$T(X, Y, f) \equiv f([X, Y]) \quad \forall\, X, Y \in \mathfrak{g},\ f \in \mathfrak{g}^*.$$

Then, letting $\{f^k\}_{i=1\ldots n}$ be the basis dual to $\{X_i\}_{i=1\ldots n}$, T has components

$$T_{ij}{}^k = f^k([X_i, X_j]) = f^k(c_{ij}{}^l X_l) = c_{ij}{}^k$$

as expected. Note that T is essentially just the Lie bracket, except we have to feed the vector that the bracket produces to a dual vector to get a number.

Example 4.39. $\mathfrak{gl}(V)$ *and* $\mathfrak{gl}(n, C)$

Let V be an n-dimensional vector space over a set of scalars C. If we choose a basis for V, we can define a map from the Lie algebra $\mathfrak{gl}(V)$ of linear operators on V to the matrix Lie algebra $\mathfrak{gl}(n, C)$ by $T \mapsto [T]$. You can easily check that this is a Lie algebra isomorphism, so $\mathfrak{gl}(V) \simeq \mathfrak{gl}(n, C)$. If V is an inner product space, we can restrict this isomorphism to $\mathfrak{isom}(V) \subset \mathfrak{gl}(V)$ to get

$$\mathfrak{isom}(V) \simeq \mathfrak{o}(n) \quad \text{if } C = \mathbb{R}$$

$$\mathfrak{isom}(V) \simeq \mathfrak{u}(n) \quad \text{if } C = \mathbb{C}.$$

Example 4.40. *The* ad *homomorphism*

Let $\mathfrak{g}$ be a Lie algebra. Recall from Example 2.15 that we can use the bracket to turn $X \in \mathfrak{g}$ into a linear operator by sticking X in one argument of the bracket and leaving the other open, as in $[X, \cdot]$. One can easily check that $[X, \cdot]$ is a linear map from $\mathfrak{g}$ to $\mathfrak{g}$, and we denote this linear operator by ad_X. We thus have

$$\boxed{\mathrm{ad}_X(Y) \equiv [X, Y] \quad X, Y \in \mathfrak{g}.}$$

Note that we have turned X into a linear operator on the very space in which it lives. Furthermore, in the case where $\mathfrak{g}$ is a Lie algebra of linear operators (or matrices), the operator ad_X is then a *linear operator on a space of linear operators*! This idea was already introduced in the context of linear operators back in Example 2.15. Our reason for introducing this construction here is that it actually defines a linear map (check!)

$$\begin{aligned} \mathrm{ad} : \mathfrak{g} &\longrightarrow \mathfrak{gl}(\mathfrak{g}) \\ X &\mapsto \mathrm{ad}_X \end{aligned}$$

between two Lie algebras. Is this map a Lie algebra homomorphism? It is if

$$\mathrm{ad}_{[X,Y]} = [\mathrm{ad}_X, \mathrm{ad}_Y] \tag{4.86}$$

[notice that the bracket on the left is taken in $\mathfrak{g}$ and on the right in $\mathfrak{gl}(\mathfrak{g})$]. You will verify (4.86) in Exercise 4.36 below, where you will find that it is equivalent to the Jacobi identity! In fact, this is one way of interpreting the Jacobi identity—it guarantees that ad is a Lie algebra homomorphism for any Lie algebra $\mathfrak{g}$.

We will see in the next chapter that the ad homomorphism occurs frequently in physics, and we'll find in Chap. 6 that it's also crucial to a proper understanding of spherical tensors. More mathematically, the ad homomorphism is also fundamental in the beautiful structure theory of Lie algebras; see Hall [11]. □

Exercise 4.36. Verify that (4.86) is equivalent to the Jacobi identity.

Before we get to more physical examples, we need to explain how a continuous homomorphism from a matrix Lie group G to a matrix Lie group H leads to a Lie algebra homomorphism from $\mathfrak{g}$ to $\mathfrak{h}$. This is accomplished by the following proposition:

Proposition 4.3. *Let $\Phi : G \to H$ be a continuous homomorphism from a matrix Lie group G to a matrix Lie group H. Then this induces a Lie algebra homomorphism $\phi : \mathfrak{g} \to \mathfrak{h}$ given by*

$$\boxed{\phi(X) = \frac{d}{dt}\, \Phi(e^{tX})\Big|_{t=0}.} \tag{4.87}$$

Proof heuristics: Before diving into the proof, let's get a geometric sense of what this proposition is saying and why it should be true. Recall from Sect. 4.5 that we can think of $\mathfrak{g}$ as the tangent plane to G at the identity e; Proposition 4.3 then says a Lie group homomorphism $\Phi : G \to H$ induces a map from the tangent space at the identity in G to that of H. Furthermore, if we recall that a tangent vector to the identity in G (i.e., $X \in \mathfrak{g}$) is associated with one-parameter subgroup $\gamma(t) = e^{tX}$, then the corresponding one-parameter subgroup in H is just $(\Phi \circ \gamma)(t) = \Phi(e^{tX})$. Taking the derivative of this curve yields $\phi(X)$, just as stated in (4.87), and we then have the essential relation

$$\boxed{\Phi(e^{tX}) = e^{t\phi(X)}} \tag{4.88}$$

which we will prove below. These heuristics are illustrated in Fig. 4.9.

Proof of Proposition 4.3. This proof is a little long, but is a nice application of all that we've been discussing. The idea is first to check that (4.87) defines an element of $\mathfrak{h}$, and then check that ϕ really is a Lie algebra homomorphism by checking that it's linear and preserves the Lie bracket.

Let $\Phi : G \to H$ be a homomorphism satisfying the hypotheses of the proposition, and let $\{e^{tX}\}$ be a one-parameter subgroup in G. It's easy to see (check!)

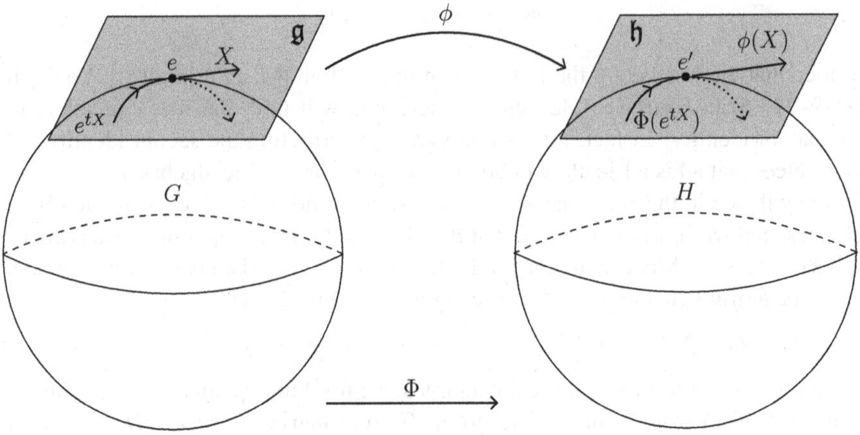

Fig. 4.9 Schematic representation of the Lie algebra homomorphism $\phi : \mathfrak{g} \to \mathfrak{h}$ induced by a Lie group homomorphism $\Phi : G \to H$. The linear map ϕ should be pictured as a map between the tangent planes at the identity of G and H. The vector $\phi(X)$ is just given by the tangent vector to $\Phi(e^{tX})$ at e'

that $\{\Phi(e^{tX})\}$ is a one-parameter subgroup of H, and hence by the discussion in Sect. 4.5 there must be a $Z \in \mathfrak{h}$ such that $\Phi(e^{tX}) = e^{tZ}$. We can thus define our map ϕ from $\mathfrak{g}$ to $\mathfrak{h}$ by $\phi(X) \equiv Z$. We then have

$$\frac{d}{dt}\,\Phi(e^{tX})\big|_{t=0} = Z$$

or equivalently

$$\Phi(e^{tX}) = e^{t\phi(X)}, \tag{4.89}$$

as suggested above. In addition to thinking of ϕ as a map between tangent planes, we can also roughly think of ϕ as the "infinitesimal" version of Φ, with ϕ taking "infinitesimal" transformations in $\mathfrak{g}$ to "infinitesimal" transformations in $\mathfrak{h}$.

Is ϕ a Lie algebra homomorphism? There are several things to check. First, we must check that ϕ is linear, which we can prove by checking that $\phi(sX) = s\phi(X)$ and $\phi(X + Y) = \phi(X) + \phi(Y)$. To check that $\phi(sX) = s\phi(X)$, we can just differentiate using the chain rule:

$$\begin{aligned}
\phi(sX) &= \frac{d}{dt}\Phi(e^{tsX})\bigg|_{t=0} \\
&= \frac{d(st)}{dt}\,\frac{d}{d(st)}\Phi(e^{tsX})\bigg|_{t=st=0} \\
&= sZ \\
&= s\phi(X).
\end{aligned}$$

Checking that $\phi(X+Y) = \phi(X)+\phi(Y)$ is a little more involved and involves a bit of calculation. The idea behind this calculation is to use the Lie product formula to express the addition in $\mathfrak{g}$ in terms of the group product in G, then use the fact that Φ is a homomorphism to express this in terms of the product in H, and then use the Lie product formula in reverse to express this in terms of addition in $\mathfrak{h}$. We have:

$$
\begin{aligned}
\phi(X+Y) &= \frac{d}{dt}\Phi(e^{t(X+Y)})\Big|_{t=0} \\
&= \frac{d}{dt}\Phi\left(\lim_{m\to\infty}(e^{\frac{tX}{m}}e^{\frac{tY}{m}})^m\right)\Big|_{t=0} && \text{by Lie Product formula} \\
&= \frac{d}{dt}\lim_{m\to\infty}\Phi\left((e^{\frac{tX}{m}}e^{\frac{tY}{m}})^m\right)\Big|_{t=0} && \text{since } \Phi \text{ is continuous} \\
&= \frac{d}{dt}\lim_{m\to\infty}\left(\Phi(e^{\frac{tX}{m}})\Phi(e^{\frac{tY}{m}})\right)^m\Big|_{t=0} && \text{since } \Phi \text{ is a homomorphism} \\
&= \frac{d}{dt}\lim_{m\to\infty}\left(e^{\frac{t\phi(X)}{m}}e^{\frac{t\phi(Y)}{m}}\right)^m\Big|_{t=0} && \text{by (4.88)} \\
&= \frac{d}{dt}e^{t(\phi(X)+\phi(Y))}\Big|_{t=0} && \text{by Lie product formula} \\
&= \phi(X)+\phi(Y).
\end{aligned}
$$

So we've established that ϕ is a linear map from $\mathfrak{g}$ to $\mathfrak{h}$. Does it preserve the bracket? Yes, but proving this also requires a bit of calculation. The idea behind this calculation is just to express everything in terms of one-parameter subgroups and then use the fact that Φ is a homomorphism. You can skip this calculation on a first reading, if desired. We have

$$
\begin{aligned}
[\phi(X),\phi(Y)] &= \frac{d}{dt}\, e^{t\phi(X)}\phi(Y)e^{-t\phi(X)} && \text{by \~{}(4.64)} \\
&= \frac{d}{dt}\,\Phi(e^{tX})\phi(Y)\Phi(e^{-tX}) && \text{by (4.88)} \\
&= \frac{d}{dt}\,\Phi(e^{tX})\left(\frac{d}{ds}e^{s\phi(Y)}\right)\Phi(e^{-tX}) \\
&= \frac{d}{dt}\frac{d}{ds}\,\Phi(e^{tX})\Phi(e^{sY})\Phi(e^{-tX}) && \text{by (4.88)} \\
&= \frac{d}{dt}\frac{d}{ds}\,\Phi(e^{tX}e^{sY}e^{-tX}) && \text{since } \Phi \text{ is a homomorphism} \\
&= \frac{d}{dt}\frac{d}{ds}\,\Phi\left(e^{(s\,e^{tX}Ye^{-tX})}\right) && \text{by (4.63)}
\end{aligned}
$$

$$= \frac{d}{dt} \phi \left(e^{tX} Y e^{-tX} \right) \qquad \text{by definition of } \phi$$

$$= \phi \left(\frac{d}{dt} e^{tX} Y e^{-tX} \right) \qquad \text{by Exercise 4.37 below}$$

$$= \phi \left([X, Y] \right) \qquad \text{by (4.64)}$$

and so ϕ is a Lie algebra homomorphism, and the proof is complete. □

Exercise 4.37. Let ϕ be a linear map from a finite-dimensional vector space V to a finite-dimensional vector space W, and let $\gamma(t) : \mathbb{R} \to V$ be a differentiable V-valued function of t (you can think of this as a path in V parametrized by t). Show from the definition of a derivative that

$$\frac{d}{dt}\phi(\gamma(t)) = \phi\left(\frac{d}{dt}\gamma(t)\right) \quad \forall\, t. \tag{4.90}$$

Now let's make all this concrete by considering some examples.

Example 4.41. *$SU(2)$ and $SO(3)$ revisited*

Recall from Example 4.22 that we have a homomorphism $\rho : SU(2) \to SO(3)$ defined by the equation

$$[AXA^\dagger]_{\mathcal{B}} = \rho(A)[X]_{\mathcal{B}} \tag{4.91}$$

where $X \in \mathfrak{su}(2)$, $A \in SU(2)$, and $\mathcal{B} = \{S_x, S_y, S_z\}$. The induced Lie algebra homomorphism ϕ is given by

$$\phi(Y) = \frac{d}{dt}\rho(e^{tY})|_{t=0},$$

and you will show below that this gives

$$\phi(S_i) = L_i \quad i = 1, 2, 3. \tag{4.92}$$

This means that ϕ is one-to-one and onto, and since the commutation relations (i.e., structure constants) are the same for the S_i and L_i, we can then conclude from our earlier discussion that ϕ is a Lie algebra isomorphism, and thus $\mathfrak{su}(2) \simeq \mathfrak{so}(3)$. You may have already known or guessed this, but the important thing to keep in mind is that this Lie *algebra* isomorphism actually stems from the *group* homomorphism between $SU(2)$ and $SO(3)$.

Exercise 4.38. Calculate $\frac{d}{dt}\rho(e^{tS_i})|_{t=0}$ and verify (4.92). **Hint**: First, heed the warning of Box 4.7 below. Then, use the definition of ρ given in (4.91), let X be arbitrary, and plug in $A = e^{tS_i}$. You'll have to do the calculation separately for $i = 1, 2, 3$. Also, it may save time to compute $\frac{d}{dt}(e^{tS_i} X e^{-tS_i})$ symbolically (using the product rule) *before* plugging in coordinate expressions for the various matrices.

Box 4.7 *A Word About Calculating Induced Lie Algebra Homomorphisms*
A word of warning is required here. In calculating induced Lie algebra homomorphisms via the definition (4.87), one may be tempted to move the derivative $\frac{d}{dt}$ "through" the group homomorphism Φ, and thus calculate $\phi(X)$ as $\Phi(\frac{d}{dt}e^{tX})$. This may even seem justified by (4.90). Note, however, that the expression $\Phi(\frac{d}{dt}e^{tX})$ is nonsensical, since the argument (when evaluated at $t = 0$) lives in the Lie algebra $\mathfrak{g}$, whereas the domain of Φ is the Lie *group* G. This suggests that (4.90) may not generally apply to Lie group homomorphisms; indeed, a key assumption in its derivation was that the map in question be a *linear map between vector spaces*, rather than a more general Lie group homomorphism.

Example 4.42. *$SL(2,\mathbb{C})$ and $SO(3,1)_o$ revisited*

Just as in the last example, we'll now examine the Lie algebra isomorphism between $\mathfrak{sl}(2,\mathbb{C})_{\mathbb{R}}$ and $\mathfrak{so}(3,1)$ that arises from the homomorphism $\rho : SL(2,\mathbb{C}) \to SO(3,1)_o$. Recall that ρ was defined by

$$[AXA^\dagger]_{\mathcal{B}} = \rho(A)[X]_{\mathcal{B}}$$

where $A \in SL(2,\mathbb{C})$, $X \in H_2(\mathbb{C})$, and $\mathcal{B} = \{\sigma_x, \sigma_y, \sigma_z, I\}$. The induced Lie algebra homomorphism is given again by

$$\phi(Y) = \frac{d}{dt}\rho(e^{tY})|_{t=0},$$

which, as you will again show, yields

$$\begin{aligned} \phi(S_i) &= \tilde{L}_i \\ \phi(\tilde{K}_i) &= K_i . \end{aligned} \tag{4.93}$$

Thus ϕ is one-to-one, onto, and preserves the bracket (since the $\tilde{L}_i$ and K_i have the same structure constants as the S_i and $\tilde{K}_i$); thus, $\mathfrak{sl}(2,\mathbb{C})_{\mathbb{R}} \simeq \mathfrak{so}(3,1)$. □

Exercise 4.39. Verify (4.93).

What is the moral of the story from the previous two examples? How should one think about these groups and their relationships? Well, the homomorphisms ρ allows us to interpret any $A \in SU(2)$ as a rotation and any $A \in SL(2,\mathbb{C})$ as a restricted Lorentz transformation. As we mentioned before, though, ρ is two-to-one, and in fact A and $-A$ in $SU(2)$ correspond to the same rotation in $SO(3)$, and likewise for $SL(2,\mathbb{C})$. However, A and $-A$ are not "close" to each other; for instance, if we consider an infinitesimal transformation $A = I + \epsilon X$, we have $-A = -I - \epsilon X$, which is *not* close to the identity (though it is close to $-I$). Thus, the fact that ρ is not one-to-one cannot be discerned by examining the neighborhood around a given matrix; one has to look at the *global* structure of the group for that. So one might say that *locally*, $SU(2)$ and $SO(3)$ are identical, but *globally* they differ. In particular, they are identical when one looks at elements near the identity, which is why their Lie algebras are isomorphic. The same comments hold for $SL(2,\mathbb{C})$ and $SO(3,1)_o$.

One important fact to take away from this is that **the correspondence between matrix Lie groups and Lie algebras is not one-to-one**; two different matrix Lie groups might have isomorphic Lie algebras. Thus, if we start with a Lie algebra, there is no way to associate to it a unique matrix Lie group. This fact will have important implications in the next chapter.

Example 4.43. *The* Ad *and* ad *homomorphisms*

You may have found it curious that $\mathfrak{su}(2)$ was involved in the group homomorphism between $SU(2)$ and $SO(3)$. This is no accident, and Example 4.41 is actually an instance of a much more general construction which we now describe. Consider a matrix Lie group G and its Lie algebra $\mathfrak{g}$. We know that for any $A \in G$ and $X \in \mathfrak{g}$, AXA^{-1} is also in $\mathfrak{g}$, so we can actually define a linear operator Ad_A on $\mathfrak{g}$ by

$$\boxed{\mathrm{Ad}_A(X) = AXA^{-1}, \quad X \in \mathfrak{g}.}$$

We can think of Ad_A as the linear operator which takes a matrix X and applies the similarity transformation corresponding to A, as if A was implementing a change of basis. This actually allows us to define a group homomorphism

$$\begin{aligned} \mathrm{Ad} : G &\to GL(\mathfrak{g}) \\ A &\mapsto \mathrm{Ad}_A, \end{aligned}$$

where you should quickly verify that $\mathrm{Ad}_A \mathrm{Ad}_B = \mathrm{Ad}_{AB}$. Since Ad is a homomorphism between the two matrix Lie groups G and $GL(\mathfrak{g})$, we can consider the induced Lie algebra homomorphism $\phi : \mathfrak{g} \to \mathfrak{gl}(\mathfrak{g})$. What does ϕ look like? Well, if $X \in \mathfrak{g}$, then $\phi(X) \in \mathfrak{gl}(\mathfrak{g})$ is the linear operator given by

$$\phi(X) = \frac{d}{dt} \mathrm{Ad}_{e^{tX}} \bigg|_{t=0}.$$

To figure out what this is, exactly, we evaluate the right-hand side on $Y \in \mathfrak{g}$:

$$\begin{aligned} \phi(X)(Y) &= \frac{d}{dt} \left(\mathrm{Ad}_{e^{tX}}(Y) \right) \bigg|_{t=0} \\ &= \frac{d}{dt} \left(e^{tX} Y e^{-tX} \right) \bigg|_{t=0} \\ &= [X, Y] \\ &= \mathrm{ad}_X(Y) \end{aligned}$$

so ϕ is nothing but the ad homomorphism of Example 4.40! Thus ad is the "infinitesimal" version of Ad, and the **commutator is the infinitesimal version of the similarity transformation.**

Note also that since Ad is a homomorphism, $\mathrm{Ad}_{e^{tX}}$ is a one-parameter subgroup in $GL(\mathfrak{g})$, and its derivative at $t = 0$ is ad_X. From (4.88) we then have

$$\mathrm{Ad}_{e^{tX}} = e^{t\,\mathrm{ad}_X} \tag{4.94}$$

as an equality in $GL(\mathfrak{g})$. In other words,

$$e^{tX}Ye^{-tX} = Y + t[X,Y] + \frac{t^2}{2}[X,[X,Y]] + \frac{t^3}{3!}[X,[X,[X,Y]]] + \ldots. \tag{4.95}$$

It is a nice exercise to expand the left-hand side of this equation as a power series and verify the equality; this is Problem 4-13. □

Exercise 4.40. Let $X, H \in M_n(C)$. Use (4.94) to show that

$$[X,H] = 0 \iff e^{tX}He^{-tX} = H \quad \forall t \in \mathbb{R}.$$

If we think of H as a quantum-mechanical Hamiltonian, this shows how the invariance properties of the Hamiltonian (like invariance under rotations R) can be formulated in terms of commutators with the corresponding generators.

To make the connection between all this and Example 4.41, we note that usually Ad_A will preserve a metric on $\mathfrak{g}$ (known as the ***Killing Form*** K; see Problem 4-12), and thus

$$\mathrm{Ad} : G \to \mathrm{Isom}(\mathfrak{g}).$$

In the case of $G = SU(2)$ above, $\mathfrak{g} = \mathfrak{su}(2)$ is three-dimensional and K is positive-definite, so[41]

$$\mathrm{Isom}(\mathfrak{su}(2)) \simeq O(3),$$

and thus $\mathrm{Ad} : SU(2) \to O(3)$. You can check that this map is identical to the homomorphism ρ described in Example 4.22, and so we actually have $\mathrm{Ad} : SU(2) \to SO(3)$! Thus the homomorphism between $SU(2)$ and $SO(3)$ is nothing but the Adjoint map of $SU(2)$, where $\mathrm{Ad}(g)$, $g \in SU(2)$ is orthogonal with respect to the Killing form on $\mathfrak{su}(2)$. We also have the corresponding Lie algebra homomorphism $\mathrm{ad} : \mathfrak{su}(2) \to \mathfrak{so}(3)$, and we know that this must be equal to ϕ from Example 4.41; thus, $[\mathrm{ad}_{S_i}]_{\mathcal{B}} = \phi(S_i) = L_i$.

---

[41] Of course, this identification depends on a choice of basis, which was made when we chose to work with $\mathcal{B} = \{S_1, S_2, S_3\}$.

Exercise 4.41. Using $[S_i, S_j] = \sum_{k=1}^{3} \epsilon_{ijk} S_k$, compute the matrix representation of ad_{S_i} in the basis $\mathcal{B}$ and verify explicitly that $[\text{ad}_{S_i}]_{\mathcal{B}} = L_i$.

Example 4.44. *The relationship between determinant and trace*

We saw in Exercise 4.16 that $\det : GL(n, \mathbb{C}) \to \mathbb{C}^*$ is a group homomorphism, and in fact it is a continuous homomorphism between matrix Lie groups (recall that $\mathbb{C}^* \simeq GL(1, \mathbb{C})$). Proposition 4.3 then tells us that det induces a Lie algebra homomorphism $\phi : \mathfrak{gl}(n, \mathbb{C}) \to \mathbb{C}$. What is ϕ? In Problem 4-14 you will prove (by direct calculation) the remarkable fact that ϕ is nothing but the trace functional. We can then apply (4.88) with $t = 1$ to immediately obtain

$$\det e^X = e^{\text{Tr}X} \quad \forall\, X \in \mathfrak{gl}(n, \mathbb{C}),$$

a result which we proved only for diagonalizable matrices back in Proposition 4.6.

Chapter 4 Problems

Note: Problems marked with an "*" tend to be longer, and/or more difficult, and/or more geared towards the completion of proofs and the tying up of loose ends. Though these problems are still worthwhile, they can be skipped on a first reading.

4-1. (*) In this problem we show that $SO(n)$ can be characterized as the set of all linear operators which take orthonormal (ON) bases into orthonormal bases and can be obtained continuously from the identity.

(a) The easy part. Show that if a linear operator R takes ON bases into ON bases and is continuously obtainable from the identity, then $R \in SO(n)$. It's immediate that $R \in O(n)$; the trick here is showing that $\det R = 1$.

(b) The converse. If $R \in SO(n)$, then it's immediate that R takes ON bases into ON bases. The slightly nontrivial part is showing that R is continuously obtainable from the identity. Prove this using induction, as follows: First, show that the claim is trivially true for $SO(1)$. Then suppose that the claim is true for $n - 1$. Take $R \in SO(n)$ and show that it can be continuously connected (via orthogonal similarity transformations) to a matrix of the form

$$\begin{pmatrix} 1 & \\ & R' \end{pmatrix}, \quad R' \in SO(n-1).$$

The claim then follows since by hypotheses R' can be continuously connected to the identity. (Hint: You'll need the 2-D rotation which takes e_1 into Re_1.)

4-2. In this problem we prove Euler's theorem that any $R \in SO(3)$ has an eigenvector with eigenvalue 1. This means that all vectors v proportional to this eigenvector are invariant under R, i.e. $Rv = v$, and so R fixes a line in space, known as the ***axis of rotation***.

(a) Show that λ being an eigenvalue of R is equivalent to $\det(R - \lambda I) = 0$. Refer to Problem 3-5 if necessary.

(b) Prove Euler's theorem by showing that

$$\det(R - I) = 0.$$

Do this using the orthogonality condition and properties of the determinant. You should not have to work in components.

4-3. Show that the matrix (4.24) is just the component form (in the standard basis) of the linear operator

$$R(\hat{\mathbf{n}}, \theta) = L(\hat{\mathbf{n}}) \otimes \hat{\mathbf{n}} + \cos\theta\, (I - L(\hat{\mathbf{n}}) \otimes \hat{\mathbf{n}}) + \sin\theta\, \hat{\mathbf{n}} \times$$

where (as you should recall) $L(v)(w) = (v|w)$, and the last term eats a vector v and spits out $\sin\theta\, \hat{\mathbf{n}} \times v$. Show that the first term is just the projection onto the axis of rotation $\hat{\mathbf{n}}$, and that the second and third terms just give a counterclockwise rotation by θ in the plane perpendicular to $\hat{\mathbf{n}}$. Convince yourself that this is exactly what a rotation about $\hat{\mathbf{n}}$ should do.

4-4. (∗) Show that $SO(3,1)_o$ is a subgroup of $O(3,1)$. Remember that $SO(3,1)_o$ is defined by 3 conditions: $|A| = 1$, $A_{44} > 1$, and (4.18). Proceed as follows:

(a) Show that $I \in SO(3,1)_o$.

(b) Show that if $A \in SO(3.1)_o$, then $A^{-1} \in SO(3,1)_o$. Do this as follows:

(i) Verify that $|A^{-1}| = 1$

(ii) Show that A^{-1} satisfies (4.18). Use this to deduce that A^T does also.

(iii) Write out the 44 component of (4.18) for both A and A^{-1}. You should get equations of the form

$$a_0^2 = 1 + \mathbf{a}^2$$
$$b_0^2 = 1 + \mathbf{b}^2. \quad (4.96)$$

where $b_0 = (A^{-1})_{44}$. Clearly this implies $b_0 < -1$ or $b_0 > 1$. Now, write out the 44 component of the equation $AA^{-1} = I$. You should find

$$a_0 b_0 = 1 - \mathbf{a} \cdot \mathbf{b}.$$

If we let $a \equiv |\mathbf{a}|$, $b \equiv |\mathbf{b}|$ then the last equation implies

$$1 - ab < a_0 b_0 < 1 + ab. \tag{4.97}$$

Assume $b_0 < -1$ and use (4.96) to derive a contradiction to (4.97), hence showing that $b_0 = (A^{-1})_{44} > 1$, and that $A^{-1} \in SO(3,1)_o$.

(c) Show that if $A, B \in SO(3,1)_o$, then $AB \in SO(3,1)_o$. You may have to do some inequality manipulating to show that $(AB)_{44} > 0$.

4-5. (∗) In this problem we prove that any $A \in SO(3,1)_o$ can be written as a product of a rotation and a boost.

(a) If A is not a pure rotation, then there is some relative velocity $\boldsymbol{\beta}$ between the standard reference frame and the new one described by A. Use the method of Exercise 4.13 to find $\boldsymbol{\beta}$ in terms of the components of A.

(b) Let L be the pure boost of the form (4.31) corresponding to the $\boldsymbol{\beta}$ you found above. Show that $L^{-1}A$ is a rotation, by calculating that

$$(L^{-1}A)_{44} = 1$$
$$(L^{-1}A)_{i4} = (L^{-1}A)_{4i} = 0, \quad i = 1, 2, 3.$$

Conclude that $A = L(L^{-1}A)$ is the desired decomposition.

4-6. (∗) In this problem we show that any $A \in SL(2,\mathbb{C})$ can be decomposed as $A = \tilde{L}U$ where $U \in SU(2)$ and $\tilde{L}$ is of the form (4.36). Unfortunately, it's a little too much work to prove this from scratch, so we'll start with the ***polar decomposition theorem***, which states that any $A \in SL(2,\mathbb{C})$ can be decomposed as $A = HU$ where $U \in SU(2)$ and H is Hermitian and ***positive***, which means that $(v|Hv) > 0$ for all nonzero $v \in \mathbb{C}^2$ (here $(\cdot\,|\,\cdot)$ is the standard Hermitian inner product on $\mathbb{C}^2$). The polar decomposition theorem can be thought of as a higher-dimensional analog of the polar form of a complex number, $z = re^{i\theta}$. You'll show below that the set of positive, Hermitian $H \in SL(2,\mathbb{C})$ is exactly the set of matrices of the form (4.36). This, combined with the polar decomposition theorem, yields the desired result. For more on the polar decomposition theorem itself, including a proof, see Hall [11].

(a) Show that an arbitrary Hermitian $H \in SL(2,\mathbb{C})$ can be written as

$$H = \begin{pmatrix} a+b & z \\ \bar{z} & a-b \end{pmatrix}$$

where

$$a, b \in \mathbb{R},\ z \in \mathbb{C},\ a^2 - b^2 - |z|^2 = 1. \tag{4.98}$$

(b) Show that any numbers a, b, z satisfying (4.98) can be written in the form

$$a = \pm \cosh u$$
$$b = v_z \sinh u$$
$$z = (v_x - i v_y) \sinh u$$

for some $u \in \mathbb{R}$ and unit vector $v \in \mathbb{R}^3$. Then show that positivity requires that $a = +\cosh u$. This puts H in the form (4.36).

(c) It remains to be shown that an H of the form (4.36) is actually positive. To do this, we employ a theorem (see Hoffman and Kunze [13]) which states that a matrix B is positive if and only if it is Hermitian and all its ***principal minors*** are positive, where its principal minors $\Delta_k(B)$, $k \leq n$ are the $k \times k$ partial determinants defined by

$$\Delta_k(B) \equiv \begin{vmatrix} B_{11} & \cdots & B_{1k} \\ \cdots & \cdots & \cdots \\ B_{k1} & \cdots & B_{kk} \end{vmatrix}.$$

Show that both the principal minors $\Delta_1(H)$ and $\Delta_2(H)$ are positive, so that H is positive.

4-7. (∗) In this problem we find an explicit formula for the map $\rho : SU(2) \to O(3)$ of Example 4.22 and use it to prove that ρ maps $SU(2)$ onto $SO(3)$ and has kernel $\pm I$.

(a) Take an arbitrary $SU(2)$ matrix A of the form (4.25) and calculate $AXA^\dagger$ for an arbitrary $X \in \mathfrak{su}(2)$ of the form (4.40).
(b) Decomposing α and β into their real and imaginary parts, use (a) to compute the column vector $[AXA^\dagger]$.
(c) Use b) to compute the 3×3 matrix $\rho(A)$. Recall that $\rho(A)$ is defined by the equation $\rho(A)[X] = [AXA^\dagger]$.
(d) Parametrize α and β as

$$\alpha = e^{i(\psi+\phi)/2} \cos\frac{\theta}{2}$$
$$\beta = i e^{i(\psi-\phi)/2} \sin\frac{\theta}{2}$$

as in (4.26). Substitute this into your expression for $\rho(A)$ and show that this is the transpose (or inverse, by the orthogonality relation) of (4.23).
(e) Parametrize α and β as in (4.74) by

$$\alpha = \cos(\theta/2) - i n_z \sin(\theta/2)$$

$$\beta = (-in_x - n_y)\sin(\theta/2)$$

and substitute this into your expression for $\rho(A)$ to get (4.24).

(f) Conclude that $\det \rho(A) = 1$ and that ρ maps $SU(2)$ onto $SO(3)$. Also use your expression from (c) to show that the kernel of ρ is $\pm I$. (It's obvious that $\pm I \in K$. What takes a little calculation is showing that $\pm I$ is *all* of K.)

4-8. (∗) In this problem we find an explicit formula for the map $\rho : SL(2,\mathbb{C}) \to O(3,1)$ of Example 4.23, and use it to prove that ρ maps $SL(2,\mathbb{C})$ onto $SO(3,1)_o$ and has kernel $\pm I$. Note that since any $A \in SL(2,\mathbb{C})$ can be decomposed into $A = \tilde{L}\tilde{R}$ with $\tilde{R} \in SU(2)$ and $\tilde{L}$ of the form (4.36), and any $B \in SO(3,1)_o$ can be decomposed as $B = LR'$ where L is of the form (4.33) and $R' \in SO(3) \subset SO(3,1)_o$, our task will be complete if we can show that $\rho(\tilde{L}) = L$. We'll do this as follows:

(a) To simplify computation, write the matrix (4.36) as

$$\tilde{L} = \begin{pmatrix} a+b & z \\ \bar{z} & a-b \end{pmatrix} \quad a,b \in \mathbb{R},\ z \in \mathbb{C}$$

and calculate $\tilde{L}X\tilde{L}^\dagger$ for $X \in H_2(\mathbb{C})$ of the form (4.44).

(b) Decomposing z into its real and imaginary parts, compute the column vector $[\tilde{L}X\tilde{L}^\dagger]$.

(c) Use (b) to compute the 4×4 matrix $\rho(\tilde{L})$.

(d) Substitute back in the original expressions for a, b, and z

$$\begin{aligned} a &= \cosh u/2 \\ b &= \frac{u_z}{u}\sinh u/2 \\ z &= \frac{1}{u}(u_x - iu_y)\sinh u/2 \end{aligned}$$

into $\rho(\tilde{L})$ to obtain (4.33).

4-9. (∗) In this problem we'll prove the claim from Example 4.25 that, for a given permutation $\sigma \in S_n$, the number of transpositions in any decomposition of σ is either always odd or always even.

(a) Consider the polynomial

$$p(x_1, \cdots, x_n) = \prod_{i<j}(x_i - x_j).$$

For example, for $n = 3$ and $n = 4$ this gives

$$p(x_1, x_2, x_3) = (x_1 - x_2)(x_1 - x_3)(x_2 - x_3)$$
$$p(x_1, x_2, x_3, x_4) = (x_1 - x_2)(x_1 - x_3)(x_1 - x_4) \times (x_2 - x_3)(x_2 - x_4)(x_3 - x_4).$$

Define an action of $\sigma \in S_n$ on p by

$$(\sigma p)(x_1, \cdots, x_n) \equiv p(x_{\sigma(1)}, \cdots, x_{\sigma(n)}) = \prod_{i<j}(x_{\sigma(i)} - x_{\sigma(j)}).$$

Convince yourself that $\sigma p = \pm p$.

(b) Let $\tau \in S_n$ be a transposition. Prove (or at least convince yourself) that $\tau p = -p$.

(c) Now assume that σ has a decomposition into an even number of transpositions. Use p to prove that σ can then never have a decomposition into an *odd* number of transpositions. Use the same logic to show that if σ has a decomposition into an odd number of transpositions, then all of its decompositions must have an odd number of transpositions.

4-10. (*) Consider the subset $H \subset GL(n, \mathbb{C})$ consisting of those matrices whose entries are real and rational. Show that H is in fact a subgroup, and construct a sequence of matrices in H that converge to an invertible matrix with *irrational* entries (there are many ways to do this!). This shows that H is a subgroup of $GL(n, \mathbb{C})$ which is *not* a matrix Lie group.

4-11. In this problem we'll calculate the first few terms in the Baker–Campbell–Hausdorff formula (4.66).

Let G be a Lie group and let $X, Y \in \mathfrak{g}$. Suppose that X and Y are small, so that $e^X e^Y$ is close to the identity and hence has a logarithm computable by the power series for ln,

$$\ln(X) = -\sum_{k=1}^{\infty} \frac{(I - X)^k}{k}.$$

By explicitly expanding out the relevant power series, show that up to third order in X and Y,

$$\ln(e^X e^Y) = X + Y + \frac{1}{2}[X, Y] + \frac{1}{12}[X, [X, Y]] - \frac{1}{12}[Y, [X, Y]] + \cdots .$$

Note that one would actually have to compute higher order terms to verify that they can be written as commutators, and this gets very tedious. A more sophisticated proof is needed to show that every term in the series is an iterated commutator and hence an element of $\mathfrak{g}$. See Varadarajan [20] for such a proof.

4-12. Let G be a matrix Lie group. Its Lie algebra $\mathfrak{g}$ comes equipped with a symmetric $(2, 0)$ tensor known as its ***Killing Form***, denoted K and defined by

$$K(X, Y) \equiv -\mathrm{Tr}(\mathrm{ad}_X \mathrm{ad}_Y).$$

(a) Show that K is Ad-invariant, in the sense that

$$K(\mathrm{Ad}_A(X), \mathrm{Ad}_A(Y)) = K(X, Y) \quad \forall X, Y \in \mathfrak{g}, \ \ A \in G.$$

(b) You know from Exercise 4.41 that $[\mathrm{ad}_{S_i}] = L_i$. Use this to compute the components of K in the $\{S_i\}$ basis and prove that

$$[K] = 2I.$$

Thus, K is positive definite. This means that K is an inner product on $\mathfrak{su}(2)$, and so from part (a) we conclude that $\mathrm{Ad}_A \in \mathrm{Isom}(\mathfrak{su}(2)) \simeq O(3)$.

4-13. Prove directly that

$$\mathrm{Ad}_{e^{tX}} = e^{t\,\mathrm{ad}_X} \tag{4.99}$$

by induction, as follows: first verify that the terms first order in t on either side are equal. Then, assume that the nth order terms are equal (where n is an arbitrary integer), and use this to prove that the $n+1$th order terms are equal. Induction then shows that the terms of every order are equal, and so (4.99) is proven.

4-14. In Example 4.44 we claimed that the trace functional Tr is the Lie algebra homomorphism ϕ induced by the determinant function, when the latter is considered as a homomorphism

$$\begin{aligned} \det : GL(n, \mathbb{C}) &\to \mathbb{C}^* \\ A &\mapsto \det A. \end{aligned}$$

This problem asks you to prove this claim.
Begin with the definition (4.87) of ϕ, which in this case says

$$\phi(X) = \frac{d}{dt}\det(e^{tX})|_{t=0}.$$

To show that $\phi(X) = \mathrm{Tr}X$, expand the exponential above to first order in t, plug into the determinant using the formula (3.72), and expand this to first order in t using properties of the determinant. This should yield the desired result.

Chapter 5
Basic Representation Theory

Now that we are familiar with groups and, in particular, the various transformation groups (i.e., matrix Lie groups) that arise in physics, we are ready to look at objects that "transform" in specific ways under the action of these groups. The notion of an object "transforming" in a specific way is made precise by the mathematical notion of a ***representation***, which is essentially just a way of representing the elements of a group or Lie algebra as operators on a vector space; the objects which "transform" are then just elements of the vector space.

Representations are important in both classical and quantum physics. In classical physics, they clarify what we mean by a particular object's "transformation properties." In quantum mechanics representations actually provide the basic mathematical framework, since the Lie algebra of observables $\mathfrak{g}$ acts on the Hilbert space at hand, making it a representation of $\mathfrak{g}$. Furthermore, representation theory clarifies the notion of "vector" and "tensor" operators, which are usually introduced in a somewhat ad-hoc way (much as we did in Sect. 3.7!). Finally, representation theory allows us to easily and generally prove many of the quantum-mechanical "selection rules" that are so handy in computation.

As is now our custom, we begin this chapter with some heuristics which hopefully motivate the basic definitions of representation theory, as well as the questions it seeks to answer.

5.1 Invitation: Symmetry Groups and Quantum Mechanics

The basic notions of representation theory are very natural, perhaps even obvious. Nonetheless, they can be very helpful in organizing our thinking about the plethora of linear operators and corresponding vector spaces that arise in physics, particularly in quantum theory. In this section we'll show how the notion of a representation arises very naturally in quantum mechanics, and is in fact the essence

N. Jeevanjee, *An Introduction to Tensors and Group Theory for Physicists*,
DOI 10.1007/978-3-319-14794-9_5

of the canonical quantization prescription. We'll also consider the plurality of representations of a given symmetry, how that plurality depends on the symmetry under consideration, and in particular we'll revisit various forms of the angular momentum operators and see how representation theory might help us make sense of them.

The definition of a representation arises very naturally if we consider symmetry transformations in quantum mechanics. As we saw in the last chapter, the basic symmetries of space and spacetime (translations, rotations, Lorentz transformations, parity, etc.) are mathematically embodied in groups acting on $\mathbb{R}^3$ or $\mathbb{R}^4$. For a quantum-mechanical system with Hilbert space $\mathcal{H}$, there should then be some corresponding action of these groups on $\mathcal{H}$. Furthermore, this action should be via unitary operators (i.e., isometries; cf. Sect. 4.2). This is because quantum-mechanical states are represented by unit vectors, and so any symmetry transformation which takes a state into another state must preserve norms. This implies (as can be shown via an argument similar to the one used in Exercise 4.21) that such a transformation must be unitary.

Let G be our symmetry group. Then there should be a map

$$\Pi : G \to \text{Isom}(\mathcal{H}),$$

where for any $g \in G$, $\Pi(g)$ implements the transformation g on our quantum-mechanical system. It is only natural to require that the implementation of g_1 followed by the implementation of g_2 should be the same as implementing $g_2 g_1$. This means that we must have

$$\Pi(g_2)\Pi(g_1) = \Pi(g_2 g_1).$$

This, of course, just says that Π must be a group homomorphism! Note that Π is of a restricted class of homomorphisms, in that it maps G into a group of linear operators. This motivates the definition of a ***representation*** of a group G as a group homomorphism

$$\Pi : G \to GL(V) \quad \text{for some vector space } V.$$

If, as is often the case, V is an inner product space and $\Pi : G \to \text{Isom}(V)$, then the representation Π is said to be ***unitary***.

If G is a matrix Lie group, then Proposition 4.3 tells us that any[1] representation $\Pi : G \to GL(V)$ induces a Lie algebra homomorphism $\pi : \mathfrak{g} \to \mathfrak{gl}(V)$. Such homomorphisms are similarly known as ***Lie algebra representations***. If V is an inner product space and Π is unitary, then by the argument leading to (4.55) π will map $\mathfrak{g}$ into the space $\mathfrak{isom}(V)$ of anti-Hermitian operators (cf. Example 4.36). In this case π is also said to be ***unitary***. With this terminology we can say that *any quantum-mechanical Hilbert space* $\mathcal{H}$ *should carry a unitary representation* $\pi : \mathfrak{g} \to \mathfrak{isom}(\mathcal{H})$ *of any physically relevant Lie algebra* $\mathfrak{g}$.

[1]We ignore here the (very mild) requirement of continuity of Π. For more on this see Hall [11], Sect. 1.6.

This raises some obvious and important questions. For a given $\mathfrak{g}$, do any such representations exist? Are they unique? If not, how many are there? These are some of the questions representation theory seeks to answer.

Unsurprisingly, the answer depends very much on $\mathfrak{g}$. Consider first the Lie algebra $\mathcal{C}(P)$ of observables for a classical system with phase space P, discussed in Example 4.37. We mentioned there that the canonical prescription for quantizing a classical system is to interpret the elements of $\mathcal{C}(P)$ as anti-Hermitian operators on some Hilbert space $\mathcal{H}$, where the commutation relation between operators is just given by the Poisson bracket of the corresponding functions in $\mathcal{C}(P)$. This, however, is just the definition of a unitary Lie algebra representation! We can thus reformulate the standard quantization prescription as:

> **The canonical quantization of a classical system with phase space P consists of finding a unitary representation of its Lie algebra of observables $\mathcal{C}(P)$.**

The canonical quantization prescription is thus just a statement about representations.

The question remains, though, about the existence and uniqueness of representations of $\mathcal{C}(P)$. To answer this, let's consider the simplest case where our physical system has just one-dimension with coordinate q, along with conjugate momentum p, so that $P = \mathbb{R}^2$. Let's further restrict ourselves to the Heisenberg algebra $H \subset \mathcal{C}(\mathbb{R}^2)$, introduced in Example 4.38. Any representation of $\mathcal{C}(\mathbb{R}^2)$ will induce a representation of H, so we content ourselves with asking whether H has any unitary representations.[2] This question is answered by the celebrated *Stone-von Neumann theorem*, which says that (up to a change of basis and under certain technical assumptions), there is only *one* unitary representation π of H. This, then, must be given by the familiar $\hat{q}$ and $\hat{p}$ operators acting on $V = L^2(\mathbb{R})$, as in (4.83). We can thus rest easy because the one representation of H that we know is essentially the only game in town!

Of course, the Heisenberg algebra H is not the only Lie algebra of physical interest. While $\hat{q}$ and $\hat{p}$ in H generate translations in momentum and position space (as discussed in Example 4.38), there are other symmetries, such as rotations, that should also be represented on $\mathcal{H}$. To a certain degree this happens automatically: you may recall from Example 4.37 that in three spatial dimensions $\mathfrak{so}(3) \subset \mathcal{C}(P)$, so that the usual representation of $\mathcal{C}(P)$ on $L^2(\mathbb{R}^3)$ induces an $\mathfrak{so}(3)$ representation on the same space. This corresponds to *orbital* angular momentum, and the form of the representation is given by (5.3) below. But, we also know that particles carry an additional "spin" degree of freedom, unrelated to their spatial degrees of freedom, which form additional representations of $\mathfrak{so}(3)$ that must be "tensor-producted" with $L^2(\mathbb{R}^3)$. The "theory of angular momentum," which is really just $\mathfrak{so}(3)$ representation theory in disguise, says that the only relevant possibilities are

---

[2]Extending such representations to all of $\mathcal{C}(P)$ is possible in some senses, leading to ***deformation quantization*** and ***geometric quantization***, but such topics are far outside the scope of this text. A standard reference for geometric quantization is Woodhouse [22], and an introduction to deformation quantization can be found in Zachos [23].

the spin s representations, described in Example 3.20.[3] We have, however, already met several other incarnations of the $\mathfrak{so}(3)$ generators L_i. Are these related to the spin s representations, and if so, how?

Let us revisit these various versions of the $\mathfrak{so}(3)$ generators L_i, viewing them properly as representations $\pi : \mathfrak{so}(3) \to \mathfrak{gl}(V)$. First, consider the $\mathfrak{so}(3) \simeq \mathfrak{su}(2)$ representation given by the $\mathfrak{su}(2)$ matrices S_i from Example 4.32. These matrices obey the $\mathfrak{so}(3)$ commutation relations, so we can take $V = \mathbb{C}^2$ and define

$$\pi(L_x) \equiv \frac{1}{2}\begin{pmatrix} 0 & -i \\ -i & 0 \end{pmatrix}, \quad \pi(L_y) \equiv \frac{1}{2}\begin{pmatrix} 0 & -1 \\ 1 & 0 \end{pmatrix}, \quad \pi(L_z) \equiv \frac{1}{2}\begin{pmatrix} -i & 0 \\ 0 & i \end{pmatrix}. \tag{5.1}$$

This is, of course, the familiar $s = 1/2$ representation.

A slightly less straightforward example is given by taking $V = \mathbb{R}^3$ and $\pi : \mathfrak{so}(3) \to \mathfrak{gl}(\mathbb{R}^3)$ to be the identity map, so that

$$\pi(L_x) = L_x = \begin{pmatrix} 0 & 0 & 0 \\ 0 & 0 & -1 \\ 0 & 1 & 0 \end{pmatrix}, \quad \pi(L_y) = L_y = \begin{pmatrix} 0 & 0 & 1 \\ 0 & 0 & 0 \\ -1 & 0 & 0 \end{pmatrix},$$

$$\pi(L_z) = L_z = \begin{pmatrix} 0 & -1 & 0 \\ 1 & 0 & 0 \\ 0 & 0 & 0 \end{pmatrix}. \tag{5.2}$$

These matrices act on three-dimensional vectors, which are sometimes said to be "spin-one," suggesting that this representation is related to the $s = 1$ representation on $\mathbb{C}^3$. That representation is complex, however, and usually features a diagonalized L_z. Nonetheless, it is straightforward to extend real representations to complex ones (as we'll do in Sect. 5.10), at which point we can make a (complex) change of basis to diagonalize L_z. This turns the matrices in (5.2) into the familiar $s = 1$ representation, reproduced in (5.4) below. This change of basis should be familiar, actually; you already used it in Exercise 3.9 to diagonalize essentially the same version of L_z as above! This suggests that representations which differ by only a change of basis should be considered "equivalent"; we will take this up in Sect. 5.6.

As yet another example, recall from Sect. 2.4 that on the vector space of polynomials on $\mathbb{R}^3$, we can represent the L_i as

$$\begin{aligned} \pi(L_x) &= i\left(z\frac{\partial}{\partial y} - y\frac{\partial}{\partial z}\right) \\ \pi(L_y) &= i\left(x\frac{\partial}{\partial z} - z\frac{\partial}{\partial x}\right) \\ \pi(L_z) &= i\left(y\frac{\partial}{\partial x} - x\frac{\partial}{\partial y}\right). \end{aligned} \tag{5.3}$$

[3]We will prove this result ourselves in Sect. 5.9. Also, note the contrast between $\mathfrak{so}(3)$ and $C(P)$; $C(P)$ has essentially only one unitary representation, whereas $\mathfrak{so}(3)$ has an infinite number!

It is straightforward to check that these satisfy the required commutation relations, and so this gives yet another representation of $\mathfrak{so}(3)$, in fact an infinite-dimensional one. How does this relate to the spin s representations?

To see the connection, let's restrict the differential operators in (5.3) to just the degree 1 polynomials and represent them in the $l = 1$ spherical harmonic basis (cf. Example 2.16 and Exercise 2.11). This yields

$$[\pi(L_x)] = \frac{i}{\sqrt{2}}\begin{pmatrix} 0 & 1 & 0 \\ 1 & 0 & -1 \\ 0 & -1 & 0 \end{pmatrix}, \quad [\pi(L_y)] = \frac{1}{\sqrt{2}}\begin{pmatrix} 0 & 1 & 0 \\ -1 & 0 & -1 \\ 0 & 1 & 0 \end{pmatrix},$$

$$[\pi(L_z)] = \begin{pmatrix} -i & 0 & 0 \\ 0 & 0 & 0 \\ 0 & 0 & i \end{pmatrix} \tag{5.4}$$

which is just the familiar $s = 1$ representation. Thus the $l = 1$ spherical harmonic representation of $\mathfrak{so}(3)$ must be equivalent to the standard spin-one representation! It turns out, unsurprisingly, that this is in fact true for all $l \in \mathbb{N}$; we'll prove this in Sect. 5.9. Furthermore, this suggests that infinite-dimensional representations like the space of polynomials on $\mathbb{R}^3$ may "decompose" into simpler representations which are easier to describe. Section 5.7 describes this decomposition process, as well as the notion of ***irreducibility*** which defines what we mean by a "simple" representation. It will turn out that the irreducible representations of $\mathfrak{so}(3)$ are (up to equivalence) the spin s representations. Furthermore, *any* finite-dimensional representation (and some infinite-dimensional representations too!) of $\mathfrak{so}(3)$ can be decomposed into a collection of irreducible representations. (Such decomposable collections often arise as tensor products of simpler representations; we'll study this in Sect. 5.4.) Thus, virtually any $\mathfrak{so}(3)$ representation can, with perhaps a little work decomposing and changing bases, be viewed as a collection of spin s representations. Analogous results hold for $\mathfrak{so}(3, 1)$, which we'll prove in Sect. 5.11.

With this motivation we now proceed to the precise definitions, as well as a wealth of examples.

Box 5.1 *Internal Symmetries*
Before moving on, it's worth pointing out that the spin degree of freedom discussed above is just one example of what are known as "internal" degrees of freedom. These are unrelated to spatial degrees of freedom, and manifest as additional Hilbert spaces $\mathcal{H}_{\text{internal}}$ which one must tensor product with $L^2(\mathbb{R}^n)$ (in n spatial dimensions). Since $\mathcal{H}_{\text{internal}}$ is unrelated to physical space, the symmetry groups that act on it need not be related to the symmetries of space-time. Nature indeed takes this liberty, leading to the $SU(2)$ "isospin," $SU(3)$ "flavor", and $SU(3)$ "color" symmetries, with corresponding representations on $\mathcal{H}_{\text{internal}}$. These symmetries and their representations are much more relevant

for particle physics and quantum field theory than for the single-particle quantum mechanics we focus on, but the basic representation theory we present here is a necessary prelude to those more advanced topics. An excellent reference for $SU(3)$ representation theory is Hall [11].

5.2 Representations: Definitions and Basic Examples

In the previous section we gave a preliminary definition of a Lie algebra representation as simply a Lie algebra homomorphism where the target space (or range) is the Lie algebra $\mathfrak{gl}(V)$ of linear operators on a vector space V. Similarly, a group representation was simply a group homomorphism where the target space was a group of linear operators on V. In both cases, we should think of the resulting operators as "representing" the elements of our group or Lie algebra. With these basic ideas in mind, we now give the precise definitions.

A ***representation*** of a group G is a vector space V together with a group homomorphism $\Pi : G \to GL(V)$. Sometimes they are written as a pair (Π, V), though occasionally when the homomorphism Π is understood we'll just talk about V, which is known as the ***representation space***. If V is a real vector space, then we say that (Π, V) is a ***real*** representation, and similarly if V is a complex vector space. If G is a matrix Lie group, and V is finite-dimensional, and the group homomorphism $\Pi : G \to GL(V)$ is continuous, then Π induces a Lie algebra homomorphism $\pi : \mathfrak{g} \to \mathfrak{gl}(V)$ by (4.87). Any homomorphism from $\mathfrak{g}$ to $\mathfrak{gl}(V)$ for some V is known as a ***Lie algebra representation***, so every finite-dimensional representation of a Lie group G induces a representation of the corresponding Lie algebra $\mathfrak{g}$. The converse is not true, however; not every representation of $\mathfrak{g}$ comes from a corresponding representation of G. This is intimately connected with the fact that for a given Lie algebra $\mathfrak{g}$, there is no unique matrix Lie group G that one can associate with it. We'll discuss this in detail in the case of $\mathfrak{su}(2)$ and $\mathfrak{so}(3, 1)$ later.

In many of our physical applications the vector space V will come equipped with an inner product $(\cdot | \cdot)$ which is preserved by the operators Π_g, or in other words $\Pi : G \to \text{Isom}(V)$. In this case, we say that Π is a ***unitary*** representation, since each $\Pi(g)$ will be a unitary operator (cf. Example 4.8). The induced Lie algebra representation π then maps $\mathfrak{g}$ into $\mathfrak{isom}(V)$ (by the argument leading to (4.55), in which case π is also referred to as ***unitary***; this terminology applies to any such $\pi : \mathfrak{g} \to \mathfrak{isom}(V)$, regardless of whether it is induced by a unitary group representation.

It is thus very natural to require symmetry generators and observables in quantum mechanics to be anti-Hermitian. There is, however, another reason for insisting on anti-Hermitian operators, which is that division by i then yields Hermitian operators, which are diagonalizable with real eigenvalues (cf. Box 4.4). These two requirements are logically independent, and there is no reason a priori to suppose that they can be met simultaneously. Thus, it is very convenient that anti-Hermitian operators fulfill both!

Box 5.2 *Application to Infinite-Dimensions*
As with our discussion of tensors, we will treat mainly finite-dimensional vector spaces here but will occasionally be interested in infinite-dimensional applications. Rigorous treatment of these applications can be subtle and technical, though, so as before we will extend our results to certain infinite-dimensional cases without addressing the issues related to infinite-dimensionality. Again, you should be assured that all such applications are legitimate and can, in theory, be justified.

Example 5.1. *The trivial representation*

As the name suggests, this example will be somewhat trivial, though we will end up referring to it later. For any group G (matrix or discrete) and vector space V, define the ***trivial representation*** of G on V by

$$\Pi(g) = I \quad \forall\, g \in G.$$

You will verify below that this is a representation. Suppose in addition that G is a matrix Lie group. What is the Lie algebra representation induced by Π? For all $X \in \mathfrak{g}$ we have

$$\begin{aligned}\pi(X) &= \frac{d}{dt}\Pi(e^{tX}) \\ &= \frac{d}{dt} I \\ &= 0\end{aligned}$$

so the trivial representation of a Lie algebra is given by $\pi(X) = 0\ \forall\, X \in \mathfrak{g}$.

Exercise 5.1. Let our representation space be $V = \mathbb{C}$. Show that $GL(\mathbb{C}) \simeq GL(1, \mathbb{C}) \simeq \mathbb{C}^*$ where $\mathbb{C}^*$ is the group of nonzero complex numbers. Then verify that $\Pi : G \to \mathbb{C}^*$ as defined above is a group homomorphism, hence a representation. Also verify that $\pi : \mathfrak{g} \to \mathfrak{gl}(\mathbb{C}) \simeq \mathbb{C}$ given by $\pi(X) = 0\ \ \forall\, X$ is a Lie algebra representation.

Example 5.2. *The fundamental representation* $\Pi = \mathrm{Id}$

Let G be a matrix Lie group. By definition, G is a subset of $GL(n, C) = GL(C^n)$ for some n, so we can simply interpret the elements of G as operators (acting by matrix multiplication) on $V = C^n$. This yields the ***fundamental*** (or ***standard***) representation of G, in which case the group homomorphism Π is just the identity map Id.

If $G = O(3)$ or $SO(3)$, then $V = \mathbb{R}^3$ and the fundamental representation is known as the ***vector*** representation. If $G = SU(2)$, then $V = \mathbb{C}^2$ and the

Table 5.1 Summary of the fundamental representation for various matrix Lie groups G, including the representation space V and common nomenclature

Fundamental representations		
Group G	V	Name
$SU(2)$	$\mathbb{C}^2$	Spinor
$SO(3)$	$\mathbb{R}^3$	Vector
$SO(3,1)$	$\mathbb{R}^4$	Four-vector
$SL(2,\mathbb{C})$	$\mathbb{C}^2$	Spinor (relativistic)

fundamental representation is known as the ***spinor*** representation. If $G = SO(3,1)_o$ or $O(3,1)$, then $V = \mathbb{R}^4$, and the fundamental representation is also known as the ***vector*** (or sometimes ***four-vector***) representation. If $G = SL(2,\mathbb{C})$,then $V = \mathbb{C}^2$, and the fundamental representation is also known as the ***spinor*** representation. Vectors in this last representation are sometimes referred to more specifically as ***left-handed spinors***, and are used to describe massless relativistic spin 1/2 particles.[4] This is summarized in Table 5.1.

Each of these group representations induces a representation of the corresponding Lie algebra which then goes by the same name, and which is also given just by interpreting the elements of $\mathfrak{g} \subset \mathfrak{gl}(n, C)$ as linear operators. Since Lie algebras are vector spaces and a representation π is a linear map, we can describe any Lie algebra representation completely just by giving the image of the basis vectors under π (this is one of the nice features of Lie algebra representations; they are much easier to concretely visualize). Thus, the vector representation of $\mathfrak{so}(3)$ is given by

$$\pi(L_x) = \begin{pmatrix} 0 & 0 & 0 \\ 0 & 0 & -1 \\ 0 & 1 & 0 \end{pmatrix}$$

$$\pi(L_y) = \begin{pmatrix} 0 & 0 & 1 \\ 0 & 0 & 0 \\ -1 & 0 & 0 \end{pmatrix}$$

$$\pi(L_z) = \begin{pmatrix} 0 & -1 & 0 \\ 1 & 0 & 0 \\ 0 & 0 & 0 \end{pmatrix},$$

where, again, π is just the identity. Likewise, the spinor representation of $\mathfrak{su}(2)$ is given by

[4]There is, of course, such a thing as a ***right-handed spinor*** as well, which we'll meet in the next section and which is also used to describe massless spin 1/2 particles. The right- and left-handed spinors are known collectively as ***Weyl spinors***, in contrast to the ***Dirac spinors***, which are used to describe *massive* spin 1/2 particles. We shall discuss Dirac spinors towards the end of this chapter.

$$\pi(S_x) = \frac{1}{2}\begin{pmatrix} 0 & -i \\ -i & 0 \end{pmatrix}$$

$$\pi(S_y) = \frac{1}{2}\begin{pmatrix} 0 & -1 \\ 1 & 0 \end{pmatrix}$$

$$\pi(S_z) = \frac{1}{2}\begin{pmatrix} -i & 0 \\ 0 & i \end{pmatrix}$$

and similarly for the vector representation of $\mathfrak{so}(3,1)$ and the spinor representation of $\mathfrak{sl}(2,\mathbb{C})_{\mathbb{R}}$.

Exercise 5.2. Show that the fundamental representations of $SO(3)$, $O(3)$, and $SU(2)$ are unitary. (The fundamental representations of $SO(3,1)_o$, $O(3,1)$, and $SL(2,\mathbb{C})$ are *not* unitary, which can be guessed from the fact that the matrices in these groups are not unitary matrices. This stems from the fact that these groups preserve the Minkowski metric, which is not an inner product.)

Example 5.3. *The adjoint representation*

A less trivial class of examples is given by the Ad homomorphism of Example 4.43. Recall that Ad is a map from G to $GL(\mathfrak{g})$, where the operator Ad_A (for $A \in G$) is defined by

$$\mathrm{Ad}_A(X) = AXA^{-1} \quad X \in \mathfrak{g}.$$

In the context of representation theory, the Ad homomorphism is known as the ***adjoint*** representation $(\mathrm{Ad}, \mathfrak{g})$. Note that the vector space of the adjoint representation is just the Lie algebra of G! The adjoint representation is thus quite a natural construction, and is ubiquitous in representation theory (and elsewhere!) for that reason. To get a handle on what the adjoint representation looks like for some of the groups we've been working with, we consider the corresponding Lie algebra representation $(\mathrm{ad}, \mathfrak{g})$, which you will recall acts as

$$\mathrm{ad}_X(Y) = [X, Y] \quad X, Y \in \mathfrak{g}.$$

For $\mathfrak{so}(3)$ with basis $\mathcal{B} = \{L_i\}_{i=1-3}$, you have already calculated in Exercise 4.41 (using the isomorphic $\mathfrak{su}(2)$ with basis $\{S_i\}_{i=1-3}$) that

$$[\mathrm{ad}_{L_i}]_{\mathcal{B}} = L_i$$

so for $\mathfrak{so}(3)$ the adjoint representation and fundamental representation are identical! (Note that we didn't have to choose a basis when describing the fundamental representation because the use of the standard basis there is implicit.) Does this mean that the adjoint representations of the corresponding groups $SO(3)$ and $O(3)$ are also identical to the vector representation? Not quite. The adjoint representation of $SO(3)$ is identical to the vector representation (as we'll show), but that does not carry over to $O(3)$; for $O(3)$, the inversion transformation $-I$ acts as minus the identity in the vector representation, but in the adjoint representation acts as

$$\mathrm{Ad}_{-I}(X) = (-I)X(-I) = X \tag{5.5}$$

so Ad_{-I} is the identity! Thus the vector and adjoint representations of $O(3)$, though similar, are not identical, and so the adjoint representation is known as the ***pseudovector*** representation. This will be discussed further in the next section.

What about the adjoint representations of $SU(2)$ and $\mathfrak{su}(2)$? Well, we already met these representations in Examples 4.22 and 4.41, and since $\mathfrak{su}(2) \simeq \mathfrak{so}(3)$ and the adjoint representation of $\mathfrak{so}(3)$ is the vector representation, the adjoint representations of both $SU(2)$ and $\mathfrak{su}(2)$ are also known as their vector representations.

As for the adjoint representations of $SO(3,1)_o$ and $O(3,1)$, it is again useful to consider first the adjoint representation of their common Lie algebra, $\mathfrak{so}(3,1)$. The vector space here is $\mathfrak{so}(3,1)$ itself, which is six-dimensional and spanned by the basis $\mathcal{B} = \{\tilde{L}_i, K_j\}_{i,j=1-3}$. You will compute in Exercise 5.3 below that the matrix forms of $\mathrm{ad}_{\tilde{L}_i}$ and ad_{K_i} are (in 3×3 block matrix form)

$$\begin{aligned}[\mathrm{ad}_{\tilde{L}_i}] &= \begin{pmatrix} L_i & 0 \\ 0 & L_i \end{pmatrix} \\ [\mathrm{ad}_{K_i}] &= \begin{pmatrix} 0 & -L_i \\ L_i & 0 \end{pmatrix}.\end{aligned} \tag{5.6}$$

From this we see that the $\tilde{L}_i$ and K_i both transform like vectors under rotations ($\mathrm{ad}_{\tilde{L}_i}$), but are mixed under boosts (ad_{K_i}). This is reminiscent of the behavior of the electric and magnetic field vectors, and it turns out (as we'll see in Sect. 5.5) that the action of $\mathfrak{so}(3,1)$ acting on itself via the adjoint representation is identical to the action of Lorentz transformation generators on the antisymmetric field tensor $F_{\mu\nu}$ from Example 3.16. The adjoint representation for $\mathfrak{so}(3,1)$ is thus also known as the ***antisymmetric 2nd rank tensor*** representation, as is the adjoint representation of the corresponding groups $SO(3,1)_o$ and $O(3,1)$. We omit a discussion of the adjoint representation of $SL(2,\mathbb{C})$ for technical reasons.[5] □

Exercise 5.3. Verify Eq. (5.6).

Table 5.2 below summarizes the representations we just discussed, along with the fundamental representations from Table 5.1.

[5] Namely, that the vector space in question, $\mathfrak{sl}(2,\mathbb{C})_{\mathbb{R}}$, is usually regarded as a three-dimensional complex vector space in the literature, not as a six-dimensional real vector space (which is the viewpoint of interest for us), so to avoid confusion we omit this topic. This won't affect any discussions of physical applications.

Table 5.2 Summary of the adjoint and fundamental representations for various matrix Lie groups G, including the representation spaces V and common nomenclature

	Adjoint		Fundamental	
Group	$V = \mathfrak{g}$	Name	V	Name
$SO(3)$	$\mathfrak{so}(3)$	Vector	$\mathbb{R}^3$	Vector
$O(3)$	$\mathfrak{so}(3)$	Pseudovector	$\mathbb{R}^3$	Vector
$SU(2)$	$\mathfrak{su}(2)$	Vector	$\mathbb{C}^2$	Spinor
$SO(3,1)_o$	$\mathfrak{so}(3,1)$	Antisymmetric 2nd rank tensor	$\mathbb{R}^4$	Four-vector
$O(3,1)$	$\mathfrak{so}(3,1)$	Antisymmetric 2nd rank tensor	$\mathbb{R}^4$	Four-vector
$SL(2,\mathbb{C})$	—		$\mathbb{C}^2$	Spinor (relativistic)

5.3 Further Examples

The fundamental and adjoint representations of a matrix Lie group are the most basic examples of representations and are the ones out of which most others can be built, as we'll see in the next section. There are, however, a few other representations of matrix Lie groups that you are probably already familiar with, as well as a few representations of abstract Lie algebras and discrete groups that are worth discussing. We'll discuss these in this section, and return in the next section to developing the general theory.

Example 5.4. *Representations of* $\mathbb{Z}_2$

The notion of representations is useful not just for matrix Lie groups, but for more general groups as well. Consider the finite group $\mathbb{Z}_2 = \{1, -1\}$. For any vector space V, we can define the ***alternating*** representation (Π_{alt}, V) of $\mathbb{Z}_2$ by

$$\begin{aligned}\Pi_{\text{alt}}(1) &= I\\ \Pi_{\text{alt}}(-1) &= -I.\end{aligned}$$

This, along with the trivial representation, allows us to succinctly distinguish between the vector and pseudovector representations of $O(3)$[6]; when restricted to $\mathbb{Z}_2 \simeq \{I, -I\} \subset O(3)$, the fundamental (vector) representation of $O(3)$ becomes Π_{alt}, whereas the adjoint (pseudovector) representation becomes Π_{trivial}.

Another place where these representations of $\mathbb{Z}_2$ crop up is the theory of identical particles. Recall from Example 4.25 that there is a homomorphism

$$\text{sgn} : S_n \to \mathbb{Z}_2$$

[6]And, as we'll see, between the vector and pseudovector representations of $O(3,1)$.

which tells us whether a given permutation is even or odd. If we then compose this map with either of the two representations introduced above, we get two representations of S_n: $\Pi_{\text{alt}} \circ \text{sgn}$, which is known as the ***sgn*** (read: "sign") representation of S_n, and $\Pi_{\text{trivial}} \circ \text{sgn}$, which is just the trivial representation of S_n. If we consider an n-particle system with Hilbert space $\mathcal{T}_n^0(\mathcal{H})$, then from Example 4.25 we know that $S^n(\mathcal{H}) \subset \mathcal{T}_p^0(\mathcal{H})$ furnishes the trivial representation of S_n, whereas $\Lambda^n\mathcal{H} \subset \mathcal{T}_p^0(\mathcal{H})$ furnishes the sgn representation of S_n. This allows us to restate the symmetrization postulate once again, in its arguably most succinct form:

> **Symmetrization Postulate III**: For a system composed of n identical particles, any state of the system lives either in the trivial representation of S_n (in which case the particles are known as ***bosons***) or in the sgn representation of S_n (in which case the particles are known as ***fermions***).

□

Box 5.3 *Proving the Symmetrization Postulate*

Though the symmetrization postulate, by name, implies that it is to be treated as an assumption, it is in fact deducible from the physically motivated requirement that any n-particle state be invariant (up to a phase) under particle interchange. More formally, we give the following proposition and proof:

Proposition 5.1. *Let $\mathcal{H}_{\text{tot}} \equiv \mathcal{T}_n^0(\mathcal{H})$ be the Hilbert space of a system with n identical particles, and let $\Pi : S_n \to GL(\mathcal{H}_{\text{tot}})$ be a representation of S_n on $\mathcal{H}_{\text{tot}}$ given by*

$$(\Pi(\sigma))(v_1 \otimes v_2 \otimes \cdots \otimes v_n) = v_{\sigma(1)} \otimes v_{\sigma(2)} \otimes \cdots \otimes v_{\sigma(n)} \quad \forall \sigma \in S_n,\ v_i \in \mathcal{H}.$$

Then if $\psi \in \mathcal{H}_{\text{tot}}$ satisfies

$$\Pi(\sigma)(\psi) = c\psi \quad \forall \sigma \in S_n, \text{ with } |c| = 1 \tag{5.7}$$

(where c may depend on σ and ψ), then ψ is an element of either $S^n(\mathcal{H})$ or $\Lambda^n\mathcal{H}$.

Proof. Equation (5.7) tells us that we can restrict Π to $\text{Span}\{\psi\}$ to get a complex one-dimensional representation

$$\Pi_\psi : S_n \to GL(\text{Span}\{\psi\}) \simeq \mathbb{C}^*.$$

Furthermore, if we consider a transposition $\tau \in S_n$, we have $\tau^2 = I$ which means

$$\psi = \Pi_\psi(\tau^2)\psi = c^2\psi \implies c = \pm 1 \tag{5.8}$$

and so $\Pi_\psi : S_n \to \mathbb{Z}_2$. All we need to prove, then, is that the trivial and sgn representations are the only such representations of S_n.

We proceed by contradiction. Assume that $\Pi_\psi : S_n \to \mathbb{Z}_2$ is neither the trivial nor the sgn representation. Then there must exist transpositions τ_+, τ_- such that $\Pi_\psi(\tau_\pm) = \pm 1$. Assume, without loss of generality, that $\tau_+ = (12)$, where the notation (12) denotes the transposition that switches indices 1 and 2. Note that $(12) = (21) = (12)^{-1}$. We write τ_- more generally as (ij), $i, j \in \{1, \ldots, n\}$. Then, letting $\rho \equiv (1i)(2j) \in S_n$, consider the permutation

$$\rho\tau_+\rho^{-1} = (1i)(2j)(12)(2j)(1i).$$

By carefully tracing through the action of this permutation on the indices 1,2,i,j, you should convince yourself that, in fact, $\rho\tau_+\rho^{-1} = \tau_-$. But this implies

$$-1 = \Pi_\psi(\tau_-) = \Pi_\psi(\rho\tau_+\rho^{-1}) = \Pi_\psi(\rho)\Pi_\psi(\tau_+)\Pi_\psi(\rho^{-1}) = \Pi_\psi(\tau_+) = +1.$$

This contradiction completes the proof. □

Example 5.5. *The four-vector representation of $SL(2,\mathbb{C})$*

Recall from Example 4.23 that $SL(2,\mathbb{C})$ acts on $H_2(\mathbb{C})$ by sending $X \to AXA^\dagger$, where $X \in H_2(\mathbb{C})$ and $A \in SL(2,\mathbb{C})$. It's easy to see that this actually defines a representation $(\Pi, H_2(\mathbb{C}))$ given by

$$\Pi(A)(X) \equiv AXA^\dagger.$$

We already saw that if we take $\mathcal{B} = \{\sigma_x, \sigma_y, \sigma_z, I\}$ as a basis for $H_2(\mathbb{C})$, then $[\Pi(A)] \in SO(3,1)$, so in this basis the action of $\Pi(A)$ looks like the action of restricted Lorentz transformations on four-vectors. Hence $(\Pi, H_2(\mathbb{C}))$ is also known as the ***four-vector*** representation of $SL(2,\mathbb{C})$.

Example 5.6. *The right-handed spinor representation of $SL(2,\mathbb{C})$*

As one might expect, the ***right-handed spinor*** representation $(\bar{\Pi}, \mathbb{C}^2)$ of $SL(2,\mathbb{C})$ is closely related to the fundamental (left-handed spinor) representation. It is defined simply by taking the adjoint inverse of the fundamental representation, that is

$$\bar{\Pi}(A)\, v \equiv A^{\dagger -1}\, v, \qquad A \in SL(2,\mathbb{C}),\ v \in \mathbb{C}^2.$$

You can easily check that this defines a bona fide representation of $SL(2,\mathbb{C})$. The usual four component Dirac spinor can be thought of as a kind of "sum" of a left-handed spinor and a right-handed one, as we'll discuss later on. □

The next few examples are instances of a general class of representations that is worth describing briefly. Say we are given a finite-dimensional vector space V, a representation Π of G on V, and a possibly infinite-dimensional vector space $C(V)$ of functions on V. (This could be, for instance, the set $P_l(V)$ of all polynomial functions of a fixed degree l, or the set of all infinitely differentiable complex-valued functions $C(V)$, or the set of all square-integrable functions $L^2(V)$.) Then the representation Π on V induces a representation $\tilde{\Pi}$ on $C(V)$ as follows: if $f \in C(V)$, then the function $\tilde{\Pi}_g f \in C(V)$ is just given by $f \circ \Pi_g^{-1}$, or

$$\boxed{(\tilde{\Pi}_g f)(v) \equiv f(\Pi_g^{-1} v) \quad g \in G,\ f \in C(V),\ v \in V.} \tag{5.9}$$

There are a couple things to check. First, it must be verified that $\tilde{\Pi}_g f$ is actually an element of $C(V)$; this, of course, depends on the exact nature of $C(V)$ and of Π and must be checked independently for each example. Assuming this is true, we also need to verify that $\tilde{\Pi}(g)$ is a linear operator and that it satisfies $\tilde{\Pi}(gh) = \tilde{\Pi}(g)\tilde{\Pi}(h)$. This computation is the same for all such examples, and you will perform it in Exercise 5.4 below.

Exercise 5.4. Confirm that if (Π, V) is a representation of G, then $\tilde{\Pi}_g$ is a *linear* operator on $C(V)$, and that $\tilde{\Pi}(g_1)\tilde{\Pi}(g_2) = \tilde{\Pi}(g_1 g_2)$. What happens if you try to define $\tilde{\Pi}$ using g instead of g^{-1}?

Example 5.7. *The spin s representation of $\mathfrak{su}(2)$ and polynomials on $\mathbb{C}^2$*

We already know that the Hilbert space corresponding to a spin s particle fixed in space is $\mathbb{C}^{2s+1}$. Since the spin angular momentum $\mathbf{S}$ is an observable for this system and the S_i have the $\mathfrak{su}(2)$ structure constants, $\mathbb{C}^{2s+1}$ must be a representation of $\mathfrak{su}(2)$. How is this representation defined? Usually one answers this question by considering the eigenvalues of S_z and showing that for any finite-dimensional representation V, the eigenvalues of S_z lie between $-s$ and s for some half-integral s. V is then defined to be the span of the eigenvectors of S_z (from which we conclude that $\dim V = 2s + 1$), and the action of S_x and S_y is determined by the $\mathfrak{su}(2)$ commutation relations. This construction is important and will be presented in Sect. 5.9, but it is also rather abstract; for the time being, we present an alternate construction which is more concrete.

Consider the set of all degree l polynomials on $\mathbb{C}^2$, i.e. the set of all degree l polynomials in the two complex variables z_1 and z_2. This is a complex vector space, denoted $P_l(\mathbb{C}^2)$, and has basis

$$\mathcal{B}_l = \{z_1^{l-k} z_2^k \mid 0 \le k \le l\} = \{z_1^l,\ z_1^{l-1} z_2,\ \ldots,\ z_1 z_2^{l-1},\ z_2^l\}$$

and hence dimension $l + 1$. The fundamental representation of $SU(2)$ on $\mathbb{C}^2$ then induces an $SU(2)$ representation $(\Pi_l, P_l(\mathbb{C}^2))$ as described above: given a polynomial function $p \in P_l(\mathbb{C}^2)$ and $A \in SU(2)$, $\Pi_l(A)(p)$ is the degree l polynomial given by

$$(\Pi_l(A)(p))(v) \equiv p(A^{-1}v) \quad v \in \mathbb{C}^2.$$

To make this representation concrete, consider the degree one polynomial $p(z_1, z_2) = z_1$. This polynomial function just picks off the first coordinate of any $v \in \mathbb{C}^2$. Let

$$A = \begin{pmatrix} \alpha & \beta \\ -\bar{\beta} & \bar{\alpha} \end{pmatrix}$$

so that

$$A^{-1} = \begin{pmatrix} \bar{\alpha} & -\beta \\ \bar{\beta} & \alpha \end{pmatrix}.$$

Then

$$(\Pi_1(A)p)(v) = p(A^{-1}v)$$

but

$$A^{-1}v = \begin{pmatrix} \bar{\alpha} & -\beta \\ \bar{\beta} & \alpha \end{pmatrix} \begin{pmatrix} z_1 \\ z_2 \end{pmatrix} = \begin{pmatrix} \bar{\alpha} z_1 - \beta z_2 \\ \bar{\beta} z_1 + \alpha z_2 \end{pmatrix} \tag{5.10}$$

so then

$$\Pi_1(A) z_1 = \bar{\alpha} z_1 - \beta z_2.$$

Likewise, (5.10) tells us that

$$\Pi_1(A) z_2 = \bar{\beta} z_1 + \alpha z_2.$$

From this, the action of $\Pi_l(A)$ on higher order polynomials can be determined since $\Pi_l(A)(z_1^{l-k} z_2^k) = (\Pi_1(A) z_1)^{l-k} (\Pi_1(A) z_2)^k$.

We can then consider the induced Lie algebra representation $(\pi_l, P_l(\mathbb{C}^2))$, in which (as you will show) $z_1^{l-k} z_2^k$ is an eigenvector of S_z with eigenvalue $i(\frac{l}{2} - k)$, so that

$$[S_z]_{\mathcal{B}_l} = i \begin{pmatrix} l/2 & & & \\ & l/2 - 1 & & \\ & & \ldots & \\ & & & -l/2 \end{pmatrix}. \tag{5.11}$$

If we let $s \equiv l/2$, then we recognize this as the usual form (up to that pesky factor of i) of S_z acting on the Hilbert space of a spin s particle. □

Exercise 5.5. Use the definition of induced Lie algebra representations and the explicit form of e^{tS_i} in the fundamental representation [which can be deduced from (4.74)] to compute the action of the operators $\pi_1(S_i)$ on the functions z_1 and z_2 in $P_1(\mathbb{C}^2)$. Show that we can write these operators in differential form as

$$\begin{aligned} \pi_1(S_1) &= \frac{i}{2}\left(z_2\frac{\partial}{\partial z_1} + z_1\frac{\partial}{\partial z_2}\right) \\ \pi_1(S_2) &= \frac{1}{2}\left(z_2\frac{\partial}{\partial z_1} - z_1\frac{\partial}{\partial z_2}\right) \\ \pi_1(S_3) &= \frac{i}{2}\left(z_1\frac{\partial}{\partial z_1} - z_2\frac{\partial}{\partial z_2}\right). \end{aligned} \tag{5.12}$$

Prove that these expressions also hold for the operators $\pi_l(S_i)$ on $P_l(\mathbb{C}^2)$. Verify the $\mathfrak{su}(2)$ commutation relations directly from these expressions. Finally, use (5.12) to show that

$$(\pi_l(S_z))(z_1^{l-k}z_2^k) = i(l/2-k)z_1^{l-k}z_2^k.$$

Warning: This is a challenging exercise, but worthwhile. See the next example for an analogous calculation for $SO(3)$. Also, don't forget that for a Lie group representation Π, induced Lie algebra representation π, and Lie algebra element X,

$$\pi(X) = \frac{d}{dt}\Pi(e^{tX}) \neq \Pi\left(\frac{d}{dt}e^{tX}\right),$$

as discussed in Box 4.7.

Example 5.8. *$L^2(\mathbb{R}^3)$ as a representation of $SO(3)$*

Recall from Example 3.18 that $L^2(\mathbb{R}^3)$ is the set of all complex-valued square-integrable functions on $\mathbb{R}^3$, and is physically interpreted as the Hilbert space of a spinless particle moving in three-dimensions. As in the previous example, the fundamental representation of $SO(3)$ on $\mathbb{R}^3$ induces an $SO(3)$ representation $(\Pi, L^2(\mathbb{R}^3))$ by

$$(\Pi_R(f))(\mathbf{x}) \equiv f(R^{-1}\mathbf{x}) \qquad f \in L^2(\mathbb{R}^3),\ R \in SO(3),\ \mathbf{x} \in \mathbb{R}^3.$$

One can think of $\Pi(R)f$ as just a "rotated" version of the function f. This representation is unitary, as we'll now digress for a moment to show.

Let $(\cdot\,|\,\cdot)$ denote the Hibert space inner product on $L^2(\mathbb{R}^3)$. To show that Π is unitary, we need to show that

$$(\Pi(R)f\,|\,\Pi(R)g) = (f\,|\,g) \ \ \forall\, f, g \in L^2(\mathbb{R}^3).$$

First off, we have

$$(\Pi(R)f|\Pi(R)g) = \int d^3x\,(\Pi_R\bar{f})(\mathbf{x})\,(\Pi_R g)(\mathbf{x})$$
$$= \int d^3x\;\bar{f}(R^{-1}\mathbf{x})\,g(R^{-1}\mathbf{x}).$$

If we now change variables to $\mathbf{x}' = R^{-1}\mathbf{x}$ and remember to include the Jacobian, we get

$$(\Pi(R)f|\Pi(R)g) = \int d^3x' \left|\frac{\partial(x,y,z)}{\partial(x',y',z')}\right| \bar{f}(\mathbf{x}')\,g(\mathbf{x}')$$
$$= \int d^3x'\,\bar{f}(\mathbf{x}')\,g(\mathbf{x}')$$
$$= (f|g),$$

where you will verify below that the Jacobian determinant $\left|\frac{\partial(x,y,z)}{\partial(x',y',z')}\right|$ is equal to one. This should be no surprise; the Jacobian tells us how volumes change under a change of variables, but since in this case the change of variables is given just by a rotation (which we know preserves volumes), we should expect the Jacobian to be one.

Though the definition of Π might look strange, it is actually the action of rotations on position kets that we're familiar with; if we act on the basis "vector" $|\mathbf{x}_0\rangle = \delta(\mathbf{x}-\mathbf{x}_0)$, we find that (for arbitrary $\psi \in L^2(\mathbb{R}^3)$)

$$\langle\psi|\Pi_R|\mathbf{x}_0\rangle = \int d^3x\,\bar{\psi}(\mathbf{x})\delta(R^{-1}\mathbf{x}-\mathbf{x}_0)$$
$$= \int d^3x'\,\bar{\psi}(R\mathbf{x}')\delta(\mathbf{x}'-\mathbf{x}_0) \quad \text{where we let } \mathbf{x}' \equiv R^{-1}\mathbf{x}$$
$$= \bar{\psi}(R\mathbf{x_0})$$

hence we must have

$$\Pi_R|\mathbf{x}_0\rangle = |R\mathbf{x}_0\rangle$$

which is the familiar action of rotations on position kets.

What does the corresponding representation of $\mathfrak{so}(3)$ look like?[7] As mentioned above, we can get a handle on that just by computing $\pi(L_i)$, since π is a linear map and any $X \in \mathfrak{so}(3)$ can just be expressed as a linear combination of the L_i. Hence we compute:

$$\begin{aligned}
(\pi_{L_i} f)(\mathbf{x}) &= \frac{d}{dt}\,(\Pi_{e^{tL_i}} f)(\mathbf{x})|_{t=0} && \text{by the definition of } \pi \\
&= \frac{d}{dt}\, f(e^{-tL_i}\mathbf{x})\Big|_{t=0} && \text{by the definition of } \Pi \\
&= \sum_{j=1}^{3} \frac{\partial f}{\partial x^j}(\mathbf{x})\,\frac{d}{dt}\,(e^{-tL_i}\mathbf{x})^j\Big|_{t=0} && \text{by the multivariable chain rule} \\
&= \sum_{j=1}^{3} \frac{\partial f}{\partial x^j}(\mathbf{x})(-L_i\mathbf{x})^j \\
&= -\sum_{j,k=1}^{3} \frac{\partial f}{\partial x^j}(\mathbf{x})(L_i)_{jk}x^k \\
&= \sum_{j,k=1}^{3} \epsilon_{ijk}\, x^k\,\frac{\partial f}{\partial x^j}(\mathbf{x}) && \text{by (4.69)}
\end{aligned}$$

so that, relabeling dummy indices,

$$\pi(L_i) = -\sum_{j,k=1}^{3} \epsilon_{ijk}\, x^j\,\frac{\partial}{\partial x^k}\,.$$

More concretely, we have

$$\begin{aligned}
\pi(L_x) &= z\frac{\partial}{\partial y} - y\frac{\partial}{\partial z} \\
\pi(L_y) &= x\frac{\partial}{\partial z} - z\frac{\partial}{\partial x} \\
\pi(L_z) &= y\frac{\partial}{\partial x} - x\frac{\partial}{\partial y}
\end{aligned} \tag{5.13}$$

[7]We should mention here that the infinite-dimensionality of $L^2(\mathbb{R}^3)$ makes a proper treatment of the induced Lie algebra representation quite subtle; for instance, we calculate in this example that the elements of $\mathfrak{so}(3)$ are to be represented by differential operators, yet not all functions in $L^2(\mathbb{R}^3)$ are differentiable! (Just think of a step function which is equal to 1 inside the unit sphere and 0 outside the unit sphere; this function is not differentiable at $r = 1$.) In this example and elsewhere, we ignore such subtleties.

which, up to our usual factor of i, is just (2.17)! □

Exercise 5.6. Verify that if $\mathbf{x}' = R^{-1}\mathbf{x}$ for some rotation $R \in SO(3)$, then

$$\left|\frac{\partial(x,y,z)}{\partial(x',y',z')}\right| = 1. \tag{5.14}$$

Example 5.9. *$\mathcal{H}_l(\mathbb{R}^3)$, $\tilde{\mathcal{H}}_l$ and $L^2(S^2)$ as representations of $SO(3)$*

Recall from Chap. 2 that $\mathcal{H}_l(\mathbb{R}^3)$ is the vector space of all harmonic complex-valued degree l polynomials on $\mathbb{R}^3$. Since $\mathcal{H}_l(\mathbb{R}^3)$ is a space of functions on $\mathbb{R}^3$, we get an $SO(3)$ representation of the same form as in the last example, namely

$$(\Pi_l(R)(f))(\mathbf{x}) \equiv f(R^{-1}\mathbf{x}) \qquad f \in \mathcal{H}_l(\mathbb{R}^3),\ R \in SO(3),\ \mathbf{x} \in \mathbb{R}^3.$$

You will check below that if f is a harmonic polynomial of degree l, then $\Pi_l(R)f$ is too, so that Π_l really is a representation on $\mathcal{H}_l(\mathbb{R}^3)$. The induced $\mathfrak{so}(3)$ representation also has the same form as in the previous example.

If we now restrict all the functions in $\mathcal{H}_l(\mathbb{R}^3)$ to the unit sphere, we get a representation of $SO(3)$ on $\tilde{\mathcal{H}}_l$, the space of spherical harmonics of degree l. Concretely, we can describe this representation by writing $Y(\theta,\phi)$ as $Y(\hat{\mathbf{n}})$, where $\hat{\mathbf{n}}$ is a unit vector giving the point on the sphere which corresponds to (θ,ϕ). Then the $SO(3)$ representation $(\tilde{\Pi}, \tilde{\mathcal{H}}_l)$ is given simply by

$$(\tilde{\Pi}_l(R)Y)(\hat{\mathbf{n}}) \equiv Y(R^{-1}\hat{\mathbf{n}}). \tag{5.15}$$

The interesting thing about this representation is that it turns out to be unitary! The inner product in this case is just given by integration over the sphere with the usual area form, i.e.

$$(Y_1(\theta,\phi)|Y_2(\theta,\phi)) \equiv \int_0^{\pi}\int_0^{2\pi} \bar{Y}_1(\theta,\phi)Y_2(\theta,\phi)\,\sin\theta\,d\phi\,d\theta.$$

Proving that (5.15) is unitary with respect to this inner product is straightforward but tedious, so we omit the calculation.[8]

One nice thing about this inner product, though, is that we can use it to define a notion of square-integrability just as we did for $\mathbb{R}$ and $\mathbb{R}^3$: we say that a $\mathbb{C}$-valued function $Y(\theta,\phi)$ on the sphere is ***square-integrable*** if

$$(Y(\theta,\phi)|Y(\theta,\phi)) = \int_0^{\pi}\int_0^{2\pi} |Y(\theta,\phi)|^2\,\sin\theta\,d\phi\,d\theta < \infty.$$

[8] Just as in the previous example, however, one can define transformed coordinates θ' and ϕ' and then unitarity hinges on the Jacobian determinant $\left|\frac{\partial(\theta,\phi)}{\partial(\theta',\phi')}\right|$ being equal to one, which it is because rotations preserve area on the sphere.

Just as with square-integrable functions on $\mathbb{R}$, the set of all such functions forms a Hilbert space, usually denoted as $L^2(S^2)$, where S^2 denotes the (two-dimensional) unit sphere in $\mathbb{R}^3$. It's easy to see[9] that each $\tilde{\mathcal{H}}_l \subset L^2(S^2)$, and in fact it turns out that all the $\tilde{\mathcal{H}}_l$ taken together are actually equal to $L^2(S^2)$! We'll discuss this further in Sect. 5.7 , but for now we note that this implies that the set $\{Y^l_m \mid 0 \le l < \infty,\ -l \le m \le l\}$ of *all* the spherical harmonics form an (orthonormal) basis for $L^2(S^2)$. This can be thought of as a consequence of the *spectral theorems* of functional analysis, crucial to quantum mechanics, which tell us that, under suitable hypotheses, the eigenfunctions of a self-adjoint linear operator (in this case the spherical laplacian Δ_{S^2}) form an orthonormal basis for the Hilbert space on which it acts. □

Exercise 5.7. Let $f \in \mathcal{H}_l(\mathbb{R}^3)$. Convince yourself that $\Pi_R f$ is also a degree l polynomial, and then use the chain rule to show that it's harmonic, hence an element of $\mathcal{H}_l(\mathbb{R}^3)$. The orthogonality of R should be crucial in your calculation!

Exercise 5.8. If you've never done so, find the induced $\mathfrak{so}(3)$ representation $\tilde{\pi}_l$ on $\tilde{\mathcal{H}}_l$ by expressing (5.13) in spherical coordinates. You should get

$$\begin{aligned}
\tilde{\pi}_l(L_x) &= \sin\phi\frac{\partial}{\partial\theta} + \cot\theta\cos\phi\frac{\partial}{\partial\phi} \\
\tilde{\pi}_l(L_y) &= -\cos\phi\frac{\partial}{\partial\theta} + \cot\theta\sin\phi\frac{\partial}{\partial\phi} \\
\tilde{\pi}_l(L_z) &= -\frac{\partial}{\partial\phi}.
\end{aligned}$$

Check directly that these satisfy the $\mathfrak{so}(3)$ commutation relations, as they should.

The representations we've discussed so far have primarily been representations of matrix Lie groups and their associated Lie algebras. As we mentioned earlier, though, there are abstract Lie algebras which have physically relevant representations too. We'll meet a couple of those now.

Example 5.10. *The Heisenberg algebra acting on* $L^2(\mathbb{R})$

We've mentioned this representation a few times already in this text, but we discuss it here to formalize it and place it in its proper context. The Heisenberg algebra $H = \text{Span}\{q, p, 1\} \subset \mathcal{C}(\mathbb{R}^2)$ has a unitary Lie algebra representation π on $L^2(\mathbb{R})$ given by

[9] Each $Y \in \tilde{\mathcal{H}}_l$ is the restriction of a polynomial to S^2 and is hence continuous, hence $|Y|^2$ must have a finite maximum $M \in \mathbb{R}$. This implies

$$(Y|Y) < 4\pi M < \infty.$$

$$\begin{aligned}(\pi_q(f))(x) &= ix\,f(x)\\ (\pi_p(f))(x) &= -\frac{df}{dx}(x)\\ (\pi_1(f))(x) &= if(x)\end{aligned} \qquad (5.16)$$

To verify that π is indeed a Lie algebra representation, one needs only to verify that the one nontrivial bracket is preserved, i.e. that $[\pi(q), \pi(p)] = \pi([q, p])$. This should be a familiar fact by now, and is readily verified if not. Showing that π is unitary requires a little bit of calculation, which you will perform below. The factors of i appearing above, especially the one in (5.16), may look funny; you should keep in mind, though, that the absence of these factors of i in the physics literature is again an artifact of the physicist's convention in defining Lie algebras (cf. Box 4.4), and that these factors of i are crucial for ensuring that the above operators are anti-Hermitian, as will be seen below. Note that the usual physics notation for these operators is $\hat{q} \equiv \pi_q = \pi(q)$ and $\hat{p} \equiv \pi_p = \pi(p)$.

Exercise 5.9. Verify that $\pi(q), \pi(p)$, and $\pi(1)$ are all anti-Hermitian operators with respect to the usual inner product on $L^2(\mathbb{R})$. Exponentiate these operators to find, for $t, a, \theta \in \mathbb{R}$,

$$\begin{aligned}(e^{t\pi(q)} f)(x) &= e^{itx} f(x)\\ (e^{a\pi(p)} f)(x) &= f(x-a)\\ (e^{\theta\pi(1)} f)(x) &= e^{i\theta} f(x)\end{aligned}$$

and conclude, as we saw (in part) in Example 4.38, that $\pi(q)$ generates translations in momentum space, $\pi(p)$ generates translations in x, and $\pi(1)$ generates multiplication by a phase factor.

Example 5.11. *The adjoint representation of $C(P)$*

Recall from Example 4.40 that for any Lie algebra $\mathfrak{g}$, regardless of whether or not it is the Lie algebra of a matrix Lie group, there is a Lie algebra homomorphism

$$\begin{aligned}\mathrm{ad} : \mathfrak{g} &\to \mathfrak{gl}(\mathfrak{g})\\ X &\mapsto \mathrm{ad}_X\end{aligned}$$

and hence a representation of $\mathfrak{g}$ on itself. Suppose that $\mathfrak{g} = C(\mathbb{R}^6)$, the Lie algebra of observables on the phase space $\mathbb{R}^6$ which corresponds to a single particle living in three-dimensional space. What does the adjoint representation of $C(\mathbb{R}^6)$ look like? Since $C(\mathbb{R}^6)$ is infinite-dimensional, computing the matrix representations of basis elements is not really feasible. Instead, we pick a few important elements of $C(\mathbb{R}^6)$ and determine how they act on $C(\mathbb{R}^6)$ as linear differential operators. First, consider $\mathrm{ad}_{q_i} \in \mathfrak{gl}(C(\mathbb{R}^6))$. We have, for arbitrary $f \in C(\mathbb{R}^6)$,

$$\mathrm{ad}_{q_i} f = \{q_i, f\} = \frac{\partial f}{\partial p_i}$$

and so just as $-\frac{\partial}{\partial x}$ generates translation in the x-direction, $\mathrm{ad}_{q_i} = \frac{\partial}{\partial p_i}$ generates translation in the p_i direction in phase space. Similarly, you can calculate that

$$\mathrm{ad}_{p_i} = -\frac{\partial}{\partial q_i} \tag{5.17}$$

so that ad_{p_i} generates translation in the q_i direction. If $f \in C(\mathbb{R}^6)$ depends only on the q_i and not the p_i, then one can show that

$$\mathrm{ad}_{L_3} f = -\frac{\partial f}{\partial \phi}, \tag{5.18}$$

where $\phi \equiv \tan^{-1}(q_2/q_1)$ is the azimuthal angle, so that L_3 generates rotations around the z-axis. Finally, for arbitrary $f \in C(\mathbb{R}^6)$, one can show using Hamilton's equations that

$$\mathrm{ad}_H = -\frac{d}{dt} \tag{5.19}$$

so that the Hamiltonian generates time translations. These facts are all part of the ***Poisson Bracket formulation*** of classical mechanics, and it is from this formalism that quantum mechanics gets the notion that the various "symmetry generators" (which are of course just elements of the Lie algebra of the symmetry group G in question) that act on a Hilbert space should correspond to physical observables.

Exercise 5.10. Verify (5.17)–(5.19). If you did Exercise 4.33, you only need to verify (5.19), for which you'll need Hamilton's equations $-\frac{\partial H}{\partial q_i} = \frac{dp_i}{dt}$, $\frac{\partial H}{\partial p_i} = \frac{dq_i}{dt}$.

5.4 Tensor Product Representations

The next step in our study of representations is to learn how to take tensor products of representations. This is important for several reasons: First, as we will soon see, almost all representations of interest can be viewed as tensor products of other, more basic representations. Second, tensor products are ubiquitous in quantum mechanics (since they represent the addition of degrees of freedom), so we better know how they interact with representations. Finally, tensors can be understood as elements of tensor product spaces (cf. Sect. 3.5) and so understanding the tensor product of representations will allow us to understand more fully what is meant by the statement that a particular object "transforms like a tensor."

Suppose, then, that we have two representations (Π_1, V_1) and (Π_2, V_2) of a group G. Then we can define their ***tensor product representation*** $(\Pi_1 \otimes \Pi_2, V_1 \otimes V_2)$ by

$$\boxed{(\Pi_1 \otimes \Pi_2)(g) \equiv \Pi_1(g) \otimes \Pi_2(g) \in \mathcal{L}(V_1 \otimes V_2),} \tag{5.20}$$

where you should recall from (3.56) how $\Pi_1(g) \otimes \Pi_2(g)$ acts on $V_1 \otimes V_2$. It is straightforward to check that this really does define a representation of G. If G is a matrix Lie group, we can then calculate the corresponding Lie algebra representation:

$$\begin{aligned}
&(\pi_1 \otimes \pi_2)(X) \\
&\equiv \frac{d}{dt} \left. \Pi_1(e^{tX}) \otimes \Pi_2(e^{tX})\right|_{t=0} \\
&= \lim_{h\to 0} \left[\frac{\Pi_1(e^{hX}) \otimes \Pi_2(e^{hX}) - \left.\Pi_1(e^{tX}) \otimes \Pi_2(e^{tX})\right|_{t=0}}{h} \right] \\
&= \lim_{h\to 0} \left[\frac{\Pi_1(e^{hX}) \otimes \Pi_2(e^{hX}) - I \otimes I}{h} \right] \\
&= \lim_{h\to 0} \left[\frac{\Pi_1(e^{hX}) \otimes \Pi_2(e^{hX}) - I \otimes \Pi_2(e^{hX}) + I \otimes \Pi_2(e^{hX}) - I \otimes I}{h} \right] \\
&= \lim_{h\to 0} \left[\frac{(\Pi_1(e^{hX}) - I)}{h} \otimes \Pi_2(e^{hX}) \right] + \lim_{h\to 0} \left[I \otimes \frac{(\Pi_2(e^{hX}) - I)}{h} \right] \\
&= \boxed{\pi_1(X) \otimes I + I \otimes \pi_2(X),}
\end{aligned} \tag{5.21}$$

where in the second-to-last line we used the bilinearity of the tensor product. If we think of π_i as a sort of "derivative" of Π_i, then one can think of (5.21) as a kind of product rule. In fact, the above calculation is totally analogous to the proof of the product rule from single-variable calculus! It is a nice exercise to directly verify that $\pi_1 \otimes \pi_2$ is a Lie algebra representation; this is Exercise 5.11 below.

Exercise 5.11. Verify that (5.21) defines a Lie algebra representation. Mainly, this consists of verifying that

$$[(\pi_1 \otimes \pi_2)(X), (\pi_1 \otimes \pi_2)(Y)] = (\pi_1 \otimes \pi_2)([X, Y]).$$

You may find the form of (5.21) familiar from the discussion below (3.56), as well as from other quantum mechanics texts; we are now in a position to explain this connection, as well as clarify what is meant by the terms "additive" and "multiplicative" quantum numbers.

Example 5.12. *Quantum mechanics, tensor product representations, and additive and multiplicative quantum numbers*

We have already discussed how in quantum mechanics one adds degrees of freedom by taking a tensor product of Hilbert spaces. We have also discussed (in Example 4.37) how a matrix Lie group of symmetries of a physical system (i.e., a matrix Lie group G that acts on the phase space P and preserves the Hamiltonian H) gives rise to a Lie algebra of observables isomorphic to its own

Lie algebra $\mathfrak{g}$, and how the quantum-mechanical Hilbert space associated with that system should be a representation of $\mathfrak{g}$. Thus, if we have a composite physical system represented by a Hilbert space $\mathcal{H} = \mathcal{H}_1 \otimes \mathcal{H}_2$, and if the $\mathcal{H}_i$ carry representations π_i of some matrix Lie group of symmetries G, then it's natural to take as an additional axiom that G is represented on $\mathcal{H}$ by the tensor product representation (5.20), which induces the representation (5.21) of $\mathfrak{g}$ on $\mathcal{H}$.

For example, let G be the group of rotations $SO(3)$, and let $\mathcal{H}_1 = L^2(\mathbb{R}^3)$ correspond to the spatial degrees of freedom of a particle of spin s and $\mathcal{H}_2 = \mathbb{C}^{2s+1}$, $2s \in \mathbb{N}$ correspond to the internal spin degree of freedom. Then $\mathfrak{so}(3)$ is represented on the total space $\mathcal{H} = L^2(\mathbb{R}^3) \otimes \mathbb{C}^{2s+1}$ by

$$(\pi \otimes \pi_s)(L_i) = \pi(L_i) \otimes I + I \otimes \pi_s(L_i), \tag{5.22}$$

where π is the representation of Example 5.8 and π_s is the spin s representation from Example 5.7. If we identify $(\pi \otimes \pi_s)(L_i)$ with $\mathbf{J}_i$, the ith component of the total angular momentum operator, and $\pi(L_i)$ with $\mathbf{L}_i$, the ith component of the orbital angular momentum operator, and $\pi_s(L_i)$ with $\mathbf{S}_i$, the ith component of the spin angular momentum operator, then (5.22) is just the component form of

$$\mathbf{J} = \mathbf{L} \otimes I + I \otimes \mathbf{S},$$

the familiar equation expressing the total angular momentum as the sum of the spin and orbital angular momentum. We thus see that the form of this equation, which we weren't in a position to understand (mathematically) when we first discussed it in Example 3.20, is dictated by representation theory, and in particular *by the form (5.21) of the induced representation of a Lie algebra on a tensor product space*. The same is true for other symmetry generators, like the translation generator p in the Heisenberg algebra. If we have two particles in one-dimension with corresponding Hilbert spaces $\mathcal{H}_i$, $i = 1, 2$, along with representations π_i of the Heisenberg algebra, then the representation of p on the total space $\mathcal{H} = \mathcal{H}_1 \otimes \mathcal{H}_2$ is just

$$(\pi_1 \otimes \pi_2)(p) = \pi_1(p) \otimes I + I \otimes \pi_2(p) \equiv \hat{p}_1 \otimes I + I \otimes \hat{p}_2,$$

where $\hat{p}_i \equiv \pi_i(p)$. This expresses the fact that the total momentum is just the sum of the momenta of the individual particles!

More generally, (5.21) can be seen as the mathematical expression of the fact that **physical observables corresponding to generators in the Lie algebra are additive**. More precisely, we have the following: let $v_i \in \mathcal{H}_i$, $i = 1, 2$ be eigenvectors of operators $\pi_i(A)$ with eigenvalues a_i, where A is an element of the Lie algebra of a symmetry group G. Then $v_1 \otimes v_2$ is an eigenvector of $(\pi_1 \otimes \pi_2)(A)$ with eigenvalue $a_1 + a_2$. In other words, the eigenvalue a of A is an ***additive quantum number***. Most familiar quantum numbers, such as energy, momentum, and angular momentum, are additive quantum numbers, but there are exceptions. One such exception is the parity operator P. If we have two three-dimensional physical systems with corresponding Hilbert spaces $\mathcal{H}_i$, then the $\mathcal{H}_i$ should furnish representations Π_i of $O(3)$. Now, $P = -I \in O(3)$, and if $v_i \in \mathcal{H}_i$ are eigenvectors of $\Pi_i(P)$ with eigenvalues λ_i, then $v_1 \otimes v_2 \in \mathcal{H}_1 \otimes \mathcal{H}_2$ has eigenvalue

$$\begin{aligned}((\Pi_1 \otimes \Pi_2)(P))(v_1 \otimes v_2) &= \Pi_1(P)v_1 \otimes \Pi_2(P)v_2 \\ &= (\lambda_1\lambda_2)v_1 \otimes v_2,\end{aligned}$$

where we used the property (3.37c) of the tensor product in the last line. Thus parity is known as a ***multiplicative quantum number***. This is due to the fact that the parity operator P is an element of the symmetry *group*, whereas most other observables are elements of the symmetry *algebra* (usually the Lie algebra corresponding to the symmetry group). □

Our next order of business is to clarify what it means for an object to "transform like a tensor." We already addressed this to a certain degree in Sect. 3.2 when we discussed change of bases. There, however, we looked at how a change of basis affects the component representation of tensors, which was the passive point of view. Here we will take the active point of view, where instead of changing bases we will be considering a group G acting on a vector space V via some representation Π. Taking the active point of view should nonetheless give the same transformation laws, and we'll indeed see that considering tensor product representations of G in components reproduces the formulae from Sect. 3.2, but in the active form.

Recall from Sect. 3.5 that the set of tensors of rank (r, s) on a vector space V is just

$$\mathcal{T}_s^r(V) = \underbrace{V^* \otimes \ldots . \otimes V^*}_{r \text{ times}} \otimes \underbrace{V \otimes \cdots \otimes V}_{s \text{ times}}.$$

Given a representation Π of G on V, we'd like to extend this representation to the vector space $\mathcal{T}_s^r(V)$. To do this, we need to specify a representation of G on V^*. This is easily done: V^* is a vector space of functions on V (just the linear functions, in fact), so we can use (5.9) to obtain the ***dual representation*** (Π^*, V^*), defined as

$$\boxed{(\Pi_g^* f)(v) \equiv f(\Pi_g^{-1} v) \quad g \in G,\ f \in V^*,\ v \in V.} \tag{5.23}$$

This representation has the nice property that if $\{e_i\}$ and $\{e^i\}$ are dual bases for V and V^*, then the bases $\{\Pi(g)e_i\}$ and $\{\Pi^*(g)e^i\}$ are also dual to each other for any $g \in G$. You will check this in Exercise 5.12 below.

With the dual representation in hand, we can then consider the tensor product representation

$$\Pi_s^r \equiv \underbrace{\Pi^* \otimes \ldots . \otimes \Pi^*}_{r \text{ times}} \otimes \underbrace{\Pi \otimes \cdots \otimes \Pi}_{s \text{ times}}.$$

This is just given by applying by the appropriate operator $\Pi(g)$ or $\Pi^*(g)$ to each factor in the tensor product, i.e.

$$\boxed{(\Pi^r_s(g))(f_1\otimes\cdots\otimes f_r\otimes v_1\otimes\cdots\otimes v_s)=\Pi^*(g)f_1\otimes\cdots\otimes\Pi^*(g)f_r\otimes\Pi(g)v_1\otimes\cdots\otimes\Pi(g)v_s.}\tag{5.24}$$

If G is a matrix Lie group, then the corresponding Lie algebra representation π^r_s is given by repeated application of the product rule (5.21), which produces $r+s$ terms with either a $\pi(g)$ or a $\pi^*(g)$ acting on one of the factors in each term. That is,

$$\begin{aligned}(\pi^r_s(g))&(f_1\otimes\cdots\otimes f_r\otimes v_1\otimes\cdots\otimes v_s)\\ &=\pi^*(g)f_1\otimes\cdots\otimes f_r\otimes v_1\otimes\cdots\otimes v_s\\ &\quad+f_1\otimes\pi^*(g)f_2\otimes\cdots\otimes f_r\otimes v_1\otimes\cdots\otimes v_s+\cdots\\ &\quad+f_1\otimes\cdots\otimes\pi^*(g)f_r\otimes v_1\otimes\cdots\otimes v_s\\ &\quad+f_1\otimes\cdots\otimes f_r\otimes\pi(g)v_1\otimes\cdots\otimes v_s\\ &\quad+f_1\otimes\cdots\otimes f_r\otimes v_1\otimes\pi(g)v_2\otimes\cdots\otimes v_s+\cdots\\ &\quad+f_1\otimes\cdots\otimes f_r\otimes v_1\otimes\cdots\otimes\pi(g)v_s.\end{aligned}\tag{5.25}$$

We'll get a handle on these formulae by considering several examples.

Exercise 5.12. Let (Π, V) be a representation of some group G and (Π^*, V^*) its dual representation. Show that if the bases $\{e_i\}$ and $\{e^i\}$ are dual to each other, then so are $\{\Pi(g)e_i\}$ and $\{\Pi^*(g)e^i\}$ for any $g\in G$.

Exercise 5.13. Alternatively, we could have defined Π^r_s by thinking of tensors as functions on V and using the idea behind (5.9) to get

$$(\Pi^r_s(g)T)(v_1,\ldots,v_r,f_1,\ldots,f_s)\equiv T(\Pi_{g^{-1}}v_1,\ldots,\Pi_{g^{-1}}v_r,\Pi^*_{g^{-1}}f_1,\ldots\Pi^*_{g^{-1}}f_s).$$

Expand T in components and show that this is equivalent to the definition (5.24).

Example 5.13. *The dual representation*

Let's consider (5.24) with $r=1, s=0$, in which case the vector space at hand is just V^* and our representation is just the dual representation (5.23). What does this representation look like in terms of matrices? For arbitrary $f\in V^*,\ v\in V$ we have

$$\begin{aligned}(\Pi^*_g f)(v)&=[\Pi^*_g f]^T[v]\\ &=([\Pi^*_g][f])^T[v]\\ &=[f]^T[\Pi^*_g]^T[v]\end{aligned}$$

as well as

$$(\Pi_g^* f)(v) = f(\Pi_{g^{-1}} v)$$
$$= [f][\Pi_{g^{-1}}][v]$$

which implies

$$[\Pi_g^*]^T = [\Pi_{g^{-1}}] = [\Pi_g^{-1}] = [\Pi_g]^{-1}.$$

(Why are those last two equalities true?) We thus obtain

$$\boxed{[\Pi_g^*] = [\Pi_g]^{-1T}.} \tag{5.26}$$

In other words, the matrices representing G on the dual space are just the inverse transposes of the matrices representing G on the original vector space! This is just the active transformation version of (3.24b). Accordingly, if (Π, V) is the fundamental representation of $O(n)$ or $SO(n)$, then the matrices of the dual representation are identical to those of the fundamental, which is just another expression of the fact that dual vectors transform just like regular vectors under orthogonal transformations (passive *or* active).

If G is a matrix Lie group, then (Π, V) induces a representation (π, V) of $\mathfrak{g}$, and hence (Π^*, V^*) should induce a representation (π^*, V^*) of $\mathfrak{g}$ as well. What does this representation look like? Well, by the definition of π^*, we have (for any $X \in \mathfrak{g}$, $f \in V^*$, $v \in V$)

$$(\pi^*(X)f)(v) = \frac{d}{dt}(\Pi^*(e^{tX})f)(v)|_{t=0}$$
$$= \frac{d}{dt}f(\Pi(e^{-tX})v)|_{t=0}$$
$$= f(-\pi(X)v).$$

You will show below that in terms of matrices this means

$$\boxed{[\pi^*(X)] = -[\pi(X)]^T.} \tag{5.27}$$

Note again that if π is the fundamental representation of $O(n)$ or $SO(n)$ then $[\pi(X)] = X$ is antisymmetric and so the dual representation is identical to the original representation.

Exercise 5.14. Prove (5.27). This can be done a couple different ways, either by calculating the infinitesimal form of (5.26) (by letting $g = e^{tX}$ and differentiating at $t = 0$) or by a computation analogous to the derivation of (5.26).

Example 5.14. $(\Pi^1_1, \mathcal{L}(V))$ *The linear operator representation*

Now consider (5.24) with $(r, s) = (1, 1)$. Our vector space is then just $\mathcal{T}^1_1 = \mathcal{L}(V)$, the space of linear operators on V! Thus, any representation of a group G on a vector space V leads naturally to a representation of G on the space of *operators* on V. What does this representation look like? We know from (5.24) that G acts on $\mathcal{T}^1_1$ by

$$\Pi^1_1(g)(f \otimes v) = (\Pi^*_g f) \otimes (\Pi_g v), \quad v \in V,\ f \in V^*,\ g \in G. \tag{5.28}$$

This isn't very enlightening, though. To interpret this, consider $f \otimes v$ as a linear operator T on V, so that

$$T(w) \equiv f(w)\, v, \quad w \in V.$$

Then (careful working your way through these equalities!)

$$\begin{aligned}
(\Pi^1_1(g)T)(w) &= ((\Pi^*_g f)(w))\, \Pi_g v \\
&= f(\Pi_{g^{-1}} w)\, \Pi_g v && \text{by definition of } \Pi^*_g \\
&= \Pi_g(f(\Pi_{g^{-1}} w) v) && \text{since } \Pi_g \text{ linear} \\
&= \Pi_g(T(\Pi_g^{-1}(w))) && \text{by definition of } T \\
&= (\Pi_g T \Pi_g^{-1})(w)
\end{aligned}$$

so we have

$$\boxed{\Pi^1_1(g)T = \Pi_g T \Pi_g^{-1}.} \tag{5.29}$$

It's easy to check that this computation also holds for an arbitrary $T \in \mathcal{L}(V)$, since any such T can be written as a linear combination of terms of the form $f \otimes v$. Thus, (5.29) tells us that the tensor product representation of G on $V^* \otimes V = \mathcal{L}(V)$ is just the original representation acting on operators by similarity transformations! This should not be too surprising, and you perhaps could have guessed that this is how the action of G on V would extend to $\mathcal{L}(V)$. Representing (5.29) by matrices yields

$$[\Pi^1_1(g)T] = [\Pi_g][T][\Pi_g]^{-1} \tag{5.30}$$

which is just the active version of (3.27).

If G is a matrix Lie group, we can also consider the induced Lie algebra rep $(\pi^1_1, \mathcal{L}(V))$, which according to (5.21) acts by

$$(\pi^1_1(X))(f \otimes v) = (\pi^*(X) f) \otimes v + f \otimes \pi(X) v.$$

Since we saw in Chap. 4 that the adjoint representation of G induces the adjoint representation of $\mathfrak{g}$, we expect that π_1^1 should act by the commutator. This is in fact the case, and you will show below that

$$\boxed{\pi_1^1(X)T = [\pi(X), T].} \tag{5.31}$$

This, of course, reduces to a commutator of matrices when a basis is chosen. □

Exercise 5.15. Prove (5.31). As in Exercise 5.14, this can be done by either computing the infinitesimal form of (5.29), or performing a calculation similar to the one above (5.29), starting with the Lie algebra representation associated with (5.28).

It should come as no surprise that the examples above reproduced the formulae from Sect. 3.2, and in fact it's easy to show that the general tensor product representation (5.24) is just the active version of our tensor transformation law (3.17). Using the fact that

$$(\Pi_g^*)_i{}^j = (\Pi_{g^{-1}})_i{}^j, \tag{5.32}$$

which you will prove below, we have for an arbitrary (r, s) tensor T,

$$\begin{aligned} \Pi_s^r(g)T &= \Pi_s^r(g)(T_{i_1\dots i_r}{}^{j_1\dots j_s} e^{i_1} \otimes \dots \otimes e^{i_r} \otimes e_{j_1} \otimes \dots \otimes e_{j_s}) \\ &= T_{i_1\dots i_r}{}^{j_1\dots j_s} \Pi_g^* e^{i_1} \otimes \dots \otimes \Pi_g^* e^{i_r} \otimes \Pi_g e_{j_1} \otimes \dots \otimes \Pi_g e_{j_s} \\ &= (\Pi_g^{-1})_{k_1}{}^{i_1} \cdots (\Pi_g^{-1})_{k_r}{}^{i_r} (\Pi_g)_{j_1}{}^{l_1} \cdots (\Pi_g)_{j_s}{}^{l_s} T_{i_1\dots i_r}{}^{j_1\dots j_s} e^{k_1} \otimes \cdots \\ &\quad \otimes e^{k_r} \otimes e_{l_1} \otimes \dots \otimes e_{l_s} \end{aligned} \tag{5.33}$$

so that, relabeling dummy indices, we have

$$(\Pi_s^r(g)T)_{i_1\dots i_r}{}^{j_1\dots j_s} = (\Pi_g^{-1})_{i_1}{}^{k_1} \cdots (\Pi_g^{-1})_{i_r}{}^{k_r} (\Pi_g)_{l_1}{}^{j_1} \cdots (\Pi_g)_{l_s}{}^{j_s} T_{k_1\dots k_r}{}^{l_1\dots l_s} \tag{5.34}$$

which is just the active version of (3.17), with $\Pi(g)$ replacing A and $\Pi(g)^{-1}$ replacing A^{-1}.

Exercise 5.16. Verify (5.32).

5.5 Symmetric and Antisymmetric Tensor Product Representations

With the tensor product representation now in place, we can now consider symmetric and antisymmetric tensor product representations. These are important for a few reasons. The symmetric tensor product, when applied to the fundamental representation of $SU(2)$, actually yields all the spin s representations of $SU(2)$.

Meanwhile, the antisymmetric tensor product, when applied to the fundamental representations of $O(3)$ and $O(3,1)$, yields the pseudovector and pseudoscalar representations of these groups.

Before proceeding to these (and other) examples, we must convince ourselves that the spaces of symmetric and antisymmetric tensors on a representation space are in fact representations in their own right. Note that for tensors of type $(r,0)$ or $(0,r)$ the tensor product representation (5.24) is symmetric, in the sense that all the factors in the tensor product are treated equally. A moment's thought then shows that if we have a completely symmetric tensor $T \in S^r(V)$, then $\Pi^0_r(g)T$ is also in $S^r(V)$. The same is true for the completely antisymmetric tensors $\Lambda^r V$, and for the spaces $S^r(V^*)$ and $\Lambda^r V^*$. Thus these subspaces of $\mathcal{T}^0_r(V)$ and $\mathcal{T}^r_0(V)$ indeed furnish representations of G in their own right.

Example 5.15. $S^l(\mathbb{C}^2)$

Consider $S^l(\mathbb{C}^2)$, the completely symmetric $(0,l)$ tensors on $\mathbb{C}^2$. By way of example, when $l=3$ a basis for this space is given by

$$\begin{aligned}
v_0 &\equiv e_1 \otimes e_1 \otimes e_1 \\
v_1 &\equiv e_2 \otimes e_1 \otimes e_1 + e_1 \otimes e_2 \otimes e_1 + e_1 \otimes e_1 \otimes e_2 \\
v_2 &\equiv e_2 \otimes e_2 \otimes e_1 + e_1 \otimes e_2 \otimes e_2 + e_2 \otimes e_1 \otimes e_2 \\
v_3 &\equiv e_2 \otimes e_2 \otimes e_2
\end{aligned} \tag{5.35}$$

and in general we have

$$\begin{aligned}
v_0 &\equiv e_1 \otimes e_1 \otimes \cdots \otimes e_1 \\
v_1 &\equiv e_2 \otimes e_1 \otimes \cdots \otimes e_1 + \text{permutations} \\
v_2 &\equiv e_2 \otimes e_2 \otimes e_1 \otimes \cdots \otimes e_1 + \text{permutations} \\
&\;\;\vdots \\
v_{l-1} &\equiv e_2 \otimes e_2 \otimes \cdots \otimes e_1 + \text{permutations} \\
v_l &\equiv e_2 \otimes e_2 \otimes \cdots \otimes e_2 .
\end{aligned} \tag{5.36}$$

Now consider the fundamental representation $(\Pi, \mathbb{C}^2)$ of $SU(2)$. The tensor product representation $(\Pi^0_l, \mathcal{T}^0_l(\mathbb{C}^2))$ restricts to $S^l(\mathbb{C}^2) \subset \mathcal{T}^0_l(\mathbb{C}^2)$, yielding a representation of $SU(2)$ which we'll denote as $(S^l\Pi, S^l(\mathbb{C}^2))$. Taking (5.36) as a basis for this space, one can easily compute the corresponding $\mathfrak{su}(2)$ representation, denoted $(S^l\pi, S^l(\mathbb{C}^2))$. It's then easy to show that

$$[(S^l\pi)(S_z)] = -i\begin{pmatrix} l/2 & & & \\ & l/2-1 & & \\ & & \dots & \\ & & & -l/2 \end{pmatrix} \tag{5.37}$$

which is the same as (5.11) (up to a sign; this can be eliminated by reversing the order of the basis vectors). This suggests that $S^l(\mathbb{C}^2)$ is the same representation as $P_l(\mathbb{C}^2)$, and is thus also the same as the spin $l/2$ representation of $\mathfrak{su}(2)$. We will soon see that this is indeed the case. Note that we have a correspondence here between symmetric $(0, l)$ tensors on a vector space and degree l polynomials on a vector space, just as we did in Example 3.24.

Exercise 5.17. Verify (5.37).

Example 5.16. *Antisymmetric 2nd rank tensors and the adjoint representation of $O(n)$*

Consider the fundamental representation $(\Pi, \mathbb{R}^n)$ of $O(n)$. We can restrict the tensor product representation $(\Pi_2^0, \mathbb{R}^n \otimes \mathbb{R}^n)$ to the subspace $\Lambda^2\mathbb{R}^n$ to get the antisymmetric tensor product representation of $O(n)$ on $\Lambda^2\mathbb{R}^n$, which we'll denote as $\Lambda^2\Pi$. Let $X = X^{ij} e_i \otimes e_j \in \Lambda^2\mathbb{R}^n$ with the standard basis. Then by (5.24), we have (for $R \in O(n)$)

$$\Lambda^2\Pi(R)(X) = X^{ij}\, Re_i \otimes Re_j = \sum_{k,l} X^{ij} R_{ki} R_{lj}\, e_k \otimes e_l$$

which in terms of matrices reads

$$[\Lambda^2\Pi(R)X] = R[X]R^T = R[X]R^{-1}. \tag{5.38}$$

So far we have just produced the active transformation law for a $(0, 2)$ tensor, and we have not yet made use of the fact that X is antisymmetric. Taking the antisymmetry of X into account, however, means that $[X]$ is an antisymmetric matrix, and (5.38) tells us that it transforms under $O(n)$ by similarity transformations. This, however, is an exact description of the adjoint representation of $O(n)$! So we conclude that the adjoint representation of $O(n)$ (and hence of $SO(n)$ and $\mathfrak{so}(n)$) are the same as the tensor product representation $(\Lambda^2\Pi, \Lambda^2\mathbb{R}^n)$. □

Example 5.17. *Antisymmetric tensor representations of $O(3)$*

Consider the antisymmetric tensor representations $(\Lambda^k\Pi, \Lambda^k\mathbb{R}^3)$, $k = 1, 2, 3$ of $O(3)$, obtained by restricting $(\Pi_k^0, \mathcal{T}_k^0(\mathbb{R}^3))$ to $\Lambda^k\mathbb{R}^3 \subset \mathcal{T}_k^0(\mathbb{R}^3)$. For convenience we'll define $\Lambda^0\mathbb{R}^3$ to be the trivial representation on $\mathbb{R}$ (also known as the ***scalar*** representation). Now, we already know that $\Lambda^1\mathbb{R}^3 = \mathbb{R}^3$ is the fundamental representation, and from the previous example we know that $\Lambda^2\mathbb{R}^3$ is the adjoint representation, also known as the pseudovector representation. What about $\Lambda^3\mathbb{R}^3$? We know that this vector space is one-dimensional (why?), so is it just the trivial

Table 5.3 The antisymmetric tensor representations $\Lambda^k\mathbb{R}^3$ of $SO(3)$ and $O(3)$

V	$O(3)$	$SO(3)$	$\{I,-I\} \simeq \mathbb{Z}_2$
$\Lambda^0\mathbb{R}^3$	Scalar	Scalar	Trivial
$\Lambda^1\mathbb{R}^3$	Vector	Vector	Alt
$\Lambda^2\mathbb{R}^3$	Pseudovector	Vector	Trivial
$\Lambda^3\mathbb{R}^3$	Pseudoscalar	Scalar	Alt

representation? Not quite. Taking the Levi–Civita tensor $e_1 \wedge e_2 \wedge e_3$ as our basis vector, we have:

$$\begin{aligned}(\Lambda^3\Pi(R))(e_1 \wedge e_2 \wedge e_3) &= (Re_1) \wedge (Re_2) \wedge (Re_3)\\ &= |R|\, e_1 \wedge e_2 \wedge e_3 \qquad \text{by (3.90)}\\ &= \pm e_1 \wedge e_2 \wedge e_3.\end{aligned}$$

Thus $e_1 \wedge e_2 \wedge e_3$ is invariant under *rotations* but is still not a scalar, since it changes sign under inversion. An object that transforms this way is known as a ***pseudoscalar***, and $\Lambda^3\mathbb{R}^3$ is thus known as the *pseudoscalar representation* of $O(3)$. Note that if we restrict to $SO(3)$, then $|R| = 1$ and $\Lambda^3\mathbb{R}^3$ *is* then just the scalar (trivial) representation, just as $\Lambda^2\mathbb{R}^3$ is just the vector representation of $SO(3)$. Also, as we mentioned before, if we restrict our representations to $\mathbb{Z}_2 \simeq \{I, -I\} \subset O(3)$ then they reduce to either the trivial or the alternating representation. We summarize all this in the Table 5.3.

Example 5.18. *Antisymmetric tensor representations of* $O(3,1)$

Here we repeat the analysis from the previous example but in the case of $O(3,1)$. As above, $\Lambda^0\mathbb{R}^4 \equiv \mathbb{R}$ is the trivial (scalar) representation, $\Lambda^1\mathbb{R}^4 = \mathbb{R}^4$ is the vector representation, and $\Lambda^2\mathbb{R}^4$ is just known as the 2nd rank antisymmetric tensor representation (or sometimes just *tensor* representation). How about $\Lambda^3\mathbb{R}^4$? To get a handle on that, we'll compute matrix representations for the corresponding Lie algebra representation, using the following basis $\mathcal{B}$ for $\Lambda^3\mathbb{R}^4$:

$$\begin{aligned}f_1 &\equiv e_2 \wedge e_3 \wedge e_4\\ f_2 &\equiv -e_1 \wedge e_3 \wedge e_4\\ f_3 &\equiv e_1 \wedge e_2 \wedge e_4\\ f_4 &\equiv e_1 \wedge e_2 \wedge e_3.\end{aligned}$$

You will check below that in this basis, the operators $(\Lambda^3\pi)(\tilde{L}_i)$ and $(\Lambda^3\pi)(K_i)$ are given by

$$\begin{aligned}[(\Lambda^3\pi)(\tilde{L}_i)]_{\mathcal{B}} &= \tilde{L}_i\\ [(\Lambda^3\pi)(K_i)]_{\mathcal{B}} &= K_i\end{aligned} \tag{5.39}$$

so $(\Lambda^3\pi, \Lambda^3\mathbb{R}^4)$ is the same as the fundamental (vector) representation of $\mathfrak{so}(3,1)$! When we consider the group representation $\Lambda^3\Pi$, though, there is a slight difference between $\Lambda^3\mathbb{R}^4$ and the vector representation; in the vector representation, the parity operator takes the usual form

$$\Pi(P) = P = \begin{pmatrix} -1 & 0 & 0 & 0 \\ 0 & -1 & 0 & 0 \\ 0 & 0 & -1 & 0 \\ 0 & 0 & 0 & 1 \end{pmatrix}, \tag{5.40}$$

whereas on $\Lambda^3\mathbb{R}^4$, we have (as you will check below)

$$[\Lambda^3\Pi(P)]_{\mathcal{B}} = \begin{pmatrix} 1 & 0 & 0 & 0 \\ 0 & 1 & 0 & 0 \\ 0 & 0 & 1 & 0 \\ 0 & 0 & 0 & -1 \end{pmatrix} \tag{5.41}$$

which is equal to $-P$. Thus the elements of $\Lambda^3\mathbb{R}^4$ transform like four-vectors under infinitesimal Lorentz transformations (and, as we'll show, under *proper* Lorentz transformations), but they transform with the wrong sign under parity. In analogy to the three-dimensional Euclidean case, $\Lambda^3\mathbb{R}^4$ is thus known as the ***pseudovector*** representation of $O(3,1)$. As in the Euclidean case, if one restricts to $SO(3,1)_o$, then parity is excluded and then $\Lambda^3\mathbb{R}^4$ and $\mathbb{R}^4$ are identical, but only as representations of the *proper* Lorentz group $SO(3,1)_o$.

The next representation, $\Lambda^4\mathbb{R}^4$, is one-dimensional, but (as in the previous example) is not quite the trivial representation. As above, we compute the action of $\Lambda^4\Pi(A)$, $A \in O(3,1)$, on the Levi–Civita tensor $e_1 \wedge e_2 \wedge e_3 \wedge e_4$:

$$\begin{aligned} (\Lambda^4\Pi(A))(e_1 \wedge e_2 \wedge e_3 \wedge e_4) &= (Ae_1) \wedge (Ae_2) \wedge (Ae_3) \wedge (Ae_4) \\ &= |A|\, e_1 \wedge e_2 \wedge e_3 \wedge e_4 \\ &= \pm e_1 \wedge e_2 \wedge e_3 \wedge e_4. \end{aligned}$$

Thus $e_1 \wedge e_2 \wedge e_3 \wedge e_4$ is invariant under proper Lorentz transformations but changes sign under improper Lorentz transformations. As in the Euclidean case, such an object is known as a ***pseudoscalar***, and so $(\Lambda^4\Pi, \Lambda^4\mathbb{R}^4)$ is known as the *pseudoscalar representation* of $O(3,1)$.

As in the previous example, we can use $\{I, P\} \simeq \mathbb{Z}_2$ to distinguish between $\Lambda^3\mathbb{R}^4$ and $\mathbb{R}^4$, as well as between $\Lambda^4\mathbb{R}^4$ and the trivial representation. The operators $\Pi(P)$ and $\Lambda^3\Pi(P)$ can be distinguished in a basis-independent way by noting that $\Pi(P)$ is diagonalizable with eigenvalues $\{-1,-1,-1,1\}$, whereas $\Lambda^3\Pi(P)$ has eigenvalues $\{1,1,1,-1\}$. We again summarize in a Table 5.4, where in the last column we write the eigenvalues of $\Lambda^k\Pi(P)$ in those cases where $\Lambda^k\Pi(P) \neq \pm I$:

Table 5.4 The antisymmetric tensor representations $\Lambda^k\mathbb{R}^4$ of $SO(3,1)$ and $O(3,1)$

V	$O(3,1)$	$SO(3,1)_o$	$\{I,P\}\simeq\mathbb{Z}_2$
$\Lambda^0\mathbb{R}^4$	Scalar	Scalar	Trivial
$\Lambda^1\mathbb{R}^4$	Vector	Vector	$\{-1,-1,-1,1\}$
$\Lambda^2\mathbb{R}^4$	Tensor	Tensor	$\{1,1,1,-1,-1,-1\}$
$\Lambda^3\mathbb{R}^4$	Pseudovector	Vector	$\{1,1,1,-1\}$
$\Lambda^4\mathbb{R}^4$	Pseudoscalar	Scalar	Alt

Exercise 5.18. By explicit computation, verify (5.39) and (5.41). Also, show that $\Lambda^2\Pi(P)$ has eigenvalues $\{-1,-1,-1,1,1,1\}$. You may need to choose a basis for $\Lambda^2\mathbb{R}^4$ to do this.

5.6 Equivalence of Representations

In the previous section we noted that the vector and dual vector representations of $O(n)$ (and hence $SO(n)$ and $\mathfrak{so}(n)$) were "the same," as were the adjoint representation of $O(n)$ and the antisymmetric tensor product representation $(\Lambda^2\Pi, \Lambda^2\mathbb{R}^n)$. We had also noted a few equivalences in Sect. 4.1, where we pointed out that the adjoint representation of $SO(3)$ was, in a certain matrix representation, identical to the vector representation of $SO(3)$, and where we claimed that the adjoint representation of $\mathfrak{so}(3,1)$ is equivalent to the antisymmetric 2nd rank tensor representation. However, we never made precise what we meant when we said that two representations were "the same" or "equivalent"; in the cases where we attempted to prove such a claim, we usually just showed that the matrices of two representations were identical when particular bases were chosen. Defining equivalence in such a way is adequate but somewhat undesirable, as it requires a choice of basis; we'd like an alternative definition that is more intrinsic and conceptual and that doesn't require a choice of coordinates. The desired definition goes as follows: Suppose we have two representations (Π_1, V_1) and (Π_2, V_2) of a group G. A linear map $\phi : V_1 \to V_2$ which satisfies

$$\boxed{\Pi_2(g)(\phi(v)) = \phi(\Pi_1(g)v) \quad \forall\, v \in V_1,\ g \in G} \tag{5.42}$$

is said to be an ***intertwining map*** or ***intertwiner***. This just means that the action of G via the representations commutes with the action of ϕ. If in addition ϕ is a vector space isomorphism, then (Π_1, V_1) and (Π_2, V_2) are said to be ***equivalent***. Occasionally we'll denote equivalence by $(\Pi_1, V_1) \simeq (\Pi_2, V_2)$. Equivalence of Lie algebra representations and their corresponding intertwining maps are defined similarly, by the equation

$$\pi_2(X)(\phi(v)) = \phi(\pi_1(X)v) \quad \forall\, v \in V_1,\ X \in \mathfrak{g}. \tag{5.43}$$

Another way to write (5.42) is as an equality between maps that go from V_1 to V_2:

$$\Pi_2(g) \circ \phi = \phi \circ \Pi_1(g) \quad \forall\, g \in G \tag{5.44}$$

and likewise for Lie algebra representations. When ϕ is an isomorphism, this can be interpreted as saying that $\Pi_1(g)$ and $\Pi_2(g)$ are the "same" map, once we use the intertwiner ϕ to identify V_1 and V_2. Another way to interpret this is to choose bases for V_1 and V_2 and then write (5.44) as

$$[\Pi_2(g)][\phi] = [\phi][\Pi_1(g)] \quad \forall\, g \in G$$

or

$$[\Pi_2(g)] = [\phi][\Pi_1(g)][\phi]^{-1} \quad \forall\, g \in G$$

which says that the matrices $[\Pi_2(g)]$ and $[\Pi_1(g)]$ are related by a similarity transformation. You will use this below to show that our definition of equivalence of representations is equivalent to the statement that there exists bases for V_1 and V_2 such that $[\Pi_1(g)] = [\Pi_2(g)]\ \forall\, g \in G$ (or the analogous statement for Lie algebras).

Exercise 5.19. Show that two representations (Π_1, V_1) and (Π_2, V_2) are equivalent if and only if there exist bases $\mathcal{B}_1 \subset V_1$ and $\mathcal{B}_2 \subset V_2$ such that

$$[\Pi_1(g)]_{\mathcal{B}_1} = [\Pi_2(g)]_{\mathcal{B}_2} \quad \forall\, g \in G. \tag{5.45}$$

Exercise 5.20. Let (Π_i, V_i), $i = 1, 2$ be two equivalent representations of a group G, and let $H \subset G$ be a subgroup. Prove that restricting $\Pi_i : G \to GL(V_i)$ to maps $\Pi_i : H \to GL(V_i)$ yield representations of H, and that these representations of H are also equivalent. Thus, for example, equivalent representations of $O(n)$ yield equivalent representations of $SO(n)$, as one would expect.

Before we get to some examples, there are some immediate questions that arise. For instance, do equivalent representations of a matrix Lie group G give rise to equivalent representations of $\mathfrak{g}$, and conversely, do equivalent representations of $\mathfrak{g}$ come from equivalent representations of G? As to the first question, we would expect heuristically that since $\mathfrak{g}$ consists of "infinitesimal" group elements, equivalent group representations should yield equivalent Lie algebra representations. This is in fact the case:

Proposition 5.2. *Let G be a matrix Lie group, and let (Π_i, V_i) $i = 1, 2$ be two equivalent representations of G with intertwining map ϕ. Then ϕ is also an intertwiner for the induced Lie algebra representations (π_i, V_i), and so the induced Lie algebra representations are equivalent as well.*

Proof. We proceed by direct calculation. Since ϕ is an intertwiner between (Π_1, V_1) and (Π_2, V_2) we have

$$\Pi_2(e^{tX})(\phi(v)) = \phi(\Pi_1(e^{tX})v) \quad \forall\, v \in V_1,\ X \in \mathfrak{g},\ t \in \mathbb{R}.$$

Taking the time derivative of the above equation, evaluating at $t = 0$, and using the definition of the induced Lie algebra representations π_i, as well as the fact that ϕ commutes with derivatives (cf. Exercise 4.37), we obtain

$$\pi_2(X)(\phi(v)) = \phi(\pi_1(X)v) \quad \forall\, v \in V_1,\ X \in \mathfrak{g}$$

and so π_1 and π_2 are equivalent. (You should explicitly confirm this as an exercise if more detail is needed.) □

The second question, of whether or not equivalent representations of $\mathfrak{g}$ come from equivalent representations of G, is a bit trickier. After all, we noted in the introduction to this chapter (and will see very concretely in the case of $\mathfrak{so}(3)$) that not every representation of $\mathfrak{g}$ necessarily comes from a representation of G. However, *if* we know that two equivalent Lie algebra representations $(\pi_i, V_i),\ i = 1,2$ actually do come from two group representations (Π_i, V_i), *and* we know the group is connected, then it is true that Π_1 and Π_2 are equivalent.

Proposition 5.3. *Let G be a connected Lie group and let $(\Pi_i, V_i),\ i = 1,2$ be two representations of G, with associated Lie algebra representations (π_i, V_i). If the Lie algebra representations (π_i, V_i) are equivalent, then so are the group representations (Π_i, V_i) from which they came.*

Proof. The argument relies on the following fact, which we will not prove (see Hall [11] for details): if G is a *connected* matrix Lie group, then any $g \in G$ can be written as a product of exponentials. That is, for any $g \in G$ there exist $X_i \in \mathfrak{g},\ t_i \in \mathbb{R},\ i = 1, \ldots, n$ such that

$$g = e^{t_1 X_1} e^{t_2 X_2} \cdots e^{t_n X_n}. \tag{5.46}$$

(In fact, for all the connected matrix Lie groups we've met besides $SL(2, \mathbb{C})$, every group element can be written as a *single* exponential. For $SL(2, \mathbb{C})$, the polar decomposition theorem [see Problem 4-6] guarantees that any group element can be written as a product of *two* exponentials.) With this fact in hand we can show that Π_1 and Π_2 are equivalent. Let ϕ be an intertwining map between (π_1, V_1) and (π_2, V_2). Then for any $g \in G$ and $v \in V_1$ we have

$$\begin{aligned}
\Pi_2(g)(\phi(v)) &= \Pi_2(e^{t_1 X_1} e^{t_2 X_2} \cdots e^{t_n X_n})(\phi(v)) \\
&= \Pi_2(e^{t_1 X_1})\Pi_2(e^{t_2 X_2}) \cdots \Pi_2(e^{t_n X_n})(\phi(v)) \quad \text{since } \Pi_2 \text{ is a homomorphism} \\
&= (e^{t_1 \pi_2(X_1)} e^{t_2 \pi_2(X_2)} \cdots e^{t_n \pi_2(X_n)})(\phi(v)) \quad \text{by definition of } \pi_2 \\
&= \phi(e^{t_1 \pi_1(X_1)} e^{t_2 \pi_1(X_2)} \cdots e^{t_n \pi_1(X_n)} v) \quad \text{by Exercise 5.21 below} \\
&= \phi(\Pi_1(e^{t_1 X_1})\Pi_1(e^{t_2 X_2}) \cdots \Pi_1(e^{t_n X_n}) v) \quad \text{by definition of } \pi_1
\end{aligned}$$

$$= \phi(\Pi_1(e^{t_1X_1}e^{t_2X_2}\cdots e^{t_nX_n})v) \qquad \text{since } \Pi_1 \text{ is a homomorphism}$$
$$= \phi(\Pi_1(g)v)$$

and so Π_1 and Π_2 are equivalent. □

Proposition 5.3 is useful in that it allows us to prove equivalence of group representations by examining the associated Lie algebra representations, which are (by virtue of linearity) often easier to work with.

Exercise 5.21. Let π_1 and π_2 be two equivalent representations of a Lie algebra $\mathfrak{g}$ with intertwining map ϕ. Prove by expanding the exponential in a power series that

$$e^{t\pi_2(X)} \circ \phi = \phi \circ e^{t\pi_1(X)} \quad \forall\, X \in \mathfrak{g}.$$

Exercise 5.22. We claimed above that when a matrix Lie group G is connected, then any group element can be written as a product of exponentials as in (5.46). To see why the hypothesis of connectedness is important, consider the disconnected matrix Lie group $O(3)$ and find an element of $O(3)$ that *cannot* be written as a product of exponentials.

Now it's time for some examples.

Example 5.19. *Equivalence of* $\mathfrak{so}(3)$ *and* $\mathbb{R}^3$ *as* $SO(3)$ *representations*

We know that the adjoint representation and the fundamental representation of $\mathfrak{so}(3)$ are equivalent, by Exercise 5.19 and the fact that in the $\mathcal{B} = \{L_i\}_{i=1,2,3}$ basis we have $[\mathrm{ad}_{L_i}] = L_i$. What, then, is the intertwining map between $\mathbb{R}^3$ and $\mathfrak{so}(3)$? Simply the map

$$\begin{aligned} \phi : \quad & \mathfrak{so}(3) \to \mathbb{R}^3 \\ & \begin{pmatrix} 0 & -z & y \\ z & 0 & -x \\ -y & x & 0 \end{pmatrix} \mapsto (x, y, z) \end{aligned}$$

which is just the map $X \mapsto [X]_{\mathcal{B}}$. To verify that this is an intertwiner we'll actually work on the Lie algebra level, and then use Proposition 5.3 to conclude that the representations are equivalent on the group level. To verify that ϕ satisfies (5.43), one need to only prove that the equation holds for an arbitrary *basis* element of $\mathfrak{so}(3)$; since the π_i are linear maps, we can expand any $X \in \mathfrak{so}(3)$ in terms of our basis and the calculation will reduce to verifying the equality just for the basis elements. (This is the advantage of working with Lie algebras; facts about representations [such as equivalence] are usually much easier to establish directly for Lie algebras than for the corresponding groups, since we can use linearity.) We thus calculate, for any $Y \in \mathfrak{so}(3)$,

$$(\phi \circ \mathrm{ad}_{L_i})(Y) = \phi(\mathrm{ad}_{L_i} Y)$$

$$= [\mathrm{ad}_{L_i} Y]$$
$$= [\mathrm{ad}_{L_i}][Y]$$
$$= L_i[Y]$$

while

$$(L_i \circ \phi)(Y) = L_i[Y].$$

Hence $L_i \circ \phi = \phi \circ \mathrm{ad}_{L_i}$, and so ϕ is an intertwining map and the fundamental and adjoint representations of $\mathfrak{so}(3)$ are equivalent.

By Proposition 5.3, we can then conclude that the adjoint and fundamental representations of $SO(3)$ are equivalent, since $SO(3)$ is connected. What about the adjoint and fundamental representations of $O(3)$? You will recall that $O(3)$ is *not* connected (and in fact has two separate connected components), so we cannot conclude that its adjoint and fundamental representations are equivalent. In fact, as we mentioned before, we know that these representations are *not* equivalent, since if $\phi : \mathfrak{so}(3) \to \mathbb{R}^3$ were an intertwining map, we would have for any $Y \in \mathfrak{so}(3)$,

$$\phi(\mathrm{Ad}_{-I} Y) = \phi(Y) \qquad \text{since } \mathrm{Ad}_{-I} \text{ is the identity}$$

as well as

$$\phi(\mathrm{Ad}_{-I} Y) = (-I)\phi(Y) \qquad \text{since } \phi \text{ is an intertwiner}$$
$$= -\phi(Y),$$

a contradiction. Thus ϕ cannot exist, and the fundamental and adjoint representations of $O(3)$ are inequivalent.

We summarize all this in Table 5.5 below, which is in part a subset of Table 5.3. Here, we emphasize that the equivalences in the first two columns imply each other, as a result of Propositions 5.2 and 5.3. Furthermore, the inequivalence in the case of $O(3)$ shows the necessity of the connectedness hypothesis in Proposition 5.3, and also that the inverse to Proposition 5.2 is not true.

Exercise 5.23. Use Example 5.17 and Exercise 5.19 to deduce that $\mathbb{R}$ and $\Lambda^3\mathbb{R}^3$ are equivalent as $SO(3)$ representations (in fact, they are both the trivial representation). Can you find an intertwiner? Do the same for the $SO(3,1)_o$ representations $\mathbb{R}$ and $\Lambda^4\mathbb{R}^4$, using the results of Example 5.18. Using Proposition 5.3, also conclude that $\mathbb{R}^4$ and $\Lambda^3\mathbb{R}^4$ are equivalent as $SO(3,1)_o$ representations.

Table 5.5 Summary of the fundamental and adjoint representations and their equivalences for $\mathfrak{so}(3)$, $SO(3)$, and $O(3)$

	$\mathfrak{so}(3)$	$SO(3)$	$O(3)$
$\mathbb{R}^3$ (fundamental)	Vector	Vector	Vector
	$\cong$	$\cong$	$\ncong$
$\mathfrak{so}(3)$ (adjoint)	Vector	Vector	Pseudovector

Exercise 5.24. Reread Sect. 3.10 in light of the last few sections. What would we now call the map J that we introduced in that section?

Example 5.20. *Vector spaces with metrics and their duals*

Let V be a vector space equipped with a metric g (recall that a metric is any symmetric, non-degenerate bilinear form), and let (Π, V) be a representation of G whereby G acts by isometries, i.e. $\Pi(g) \in \mathrm{Isom}(V)\ \forall\, g \in G$. Examples of this include the fundamental representation of $O(n)$ on $\mathbb{R}^n$ equipped with the Euclidean metric, or the fundamental representation of $O(n-1, 1)$ on $\mathbb{R}^n$ with the Minkoswki metric, but *not* $U(n)$ or $SU(n)$ acting on $\mathbb{C}^n$ with the standard Hermitian inner product (why not?). If the assumptions above are satisfied, then (Π, V) is equivalent to the dual representation (Π^*, V^*) and the intertwiner is nothing but our old friend

$$\begin{aligned} L : V &\to V^* \\ v &\mapsto g(v, \cdot). \end{aligned}$$

You will verify in Exercise 5.25 that L is indeed an interwiner, proving the asserted equivalence. Thus, in particular we again reproduce (this time in a basis-independent way) the familiar fact that dual vectors on n-dimensional Euclidean space transform just like ordinary vectors under orthogonal transformations. Additionally, we see that dual vectors on n-dimensional *Minkowski* space transform just like ordinary vectors under Lorentz transformations! See Exercise 5.26 for the matrix manifestation of this.

The reason we've excluded complex vector spaces with Hermitian inner products from this example is that in such circumstances, the map L is not linear (why not?) and thus can't be an intertwiner. In fact, the fundamental representation of $SU(n)$ on $\mathbb{C}^n$ for $n \geq 3$ is *not* equivalent to its dual,[10] and the dual representation in such circumstances can sometimes be interpreted as an *antiparticle* if the original representation represents a particle. For instance, a quark can be thought of as a vector in the fundamental representation $\mathbb{C}^3$ of $SU(3)$ (this is sometimes denoted as $\mathbf{3}$ in the physics literature), and then the dual representation $\mathbb{C}^{3*}$ corresponds to the antiquarks (this representation is often denoted as $\bar{\mathbf{3}}$). For $SU(2)$, however, the fundamental (spinor) representation *is* equivalent to its dual, though that doesn't follow from the discussion above. This equivalence is the subject of the next example.

Exercise 5.25. Let V be a vector space equipped with a metric g and let (Π, V) be a representation of a group G by isometries. Consider $L : V \to V^*$ as defined above. Prove that $L \circ \Pi_g = \Pi_g^* \circ L$ for all $g \in G$. Since $(L \circ \Pi_g)(v) \in V^*$ for any $v \in V$, you must show that

$$(L \circ \Pi_g)(v) = (\Pi_g^* \circ L)(v)$$

as *dual* vectors on V, which means showing that they have the same action on an arbitrary second vector w.

[10] We won't prove this here; see Hall [11] for details.

Exercise 5.26. Let $(\Pi, \mathbb{R}^4)$ be the fundamental representation of $O(3,1)$ on $\mathbb{R}^4$, and let $\mathcal{B} = \{e_i\}_{i=1\text{–}4}$ be the standard basis for $\mathbb{R}^4$ and $\mathcal{B}^* = \{e^i\}_{i=1\text{–}4}$ the corresponding dual basis. Find another basis $\mathcal{B}^{*\prime}$ for $\mathbb{R}^{4*}$ such that

$$[\Pi^*(g)]_{\mathcal{B}^{*\prime}} = [\Pi(g)]_{\mathcal{B}}.$$

The map L might help you here. Does this generalize to the fundamental representation of $O(n-1,1)$ on n-dimensional Minkowski space?

Example 5.21. *The fundamental (spinor) representation of $SU(2)$ and its dual*

Consider the fundamental representation $(\Pi, \mathbb{C}^2)$ of $SU(2)$ and its dual representation $(\Pi^*, \mathbb{C}^{2*})$. We'll show that these representations are equivalent, by first showing that the induced Lie algebra representations are equivalent and then invoking Proposition 5.3. We'll show equivalence of the Lie algebra representations by exhibiting a basis for $\mathbb{C}^{2*}$ that yields the same matrix representations for $(\pi^*, \mathbb{C}^{2*})$ as for $(\pi, \mathbb{C}^2)$. For an intrinsic (coordinate-independent) proof, see Problem 5-3.

As should be familiar by now, the fundamental representation of $\mathfrak{su}(2)$ is just the identity:

$$\pi(S_x) = S_x = \frac{1}{2}\begin{pmatrix} 0 & -i \\ -i & 0 \end{pmatrix}$$

$$\pi(S_y) = S_y = \frac{1}{2}\begin{pmatrix} 0 & -1 \\ 1 & 0 \end{pmatrix}$$

$$\pi(S_z) = S_z = \frac{1}{2}\begin{pmatrix} -i & 0 \\ 0 & i \end{pmatrix}.$$

This, combined with (5.27), tells us that in the standard dual basis $\mathcal{B}^*$,

$$[\pi^*(S_x)] = \frac{1}{2}\begin{pmatrix} 0 & i \\ i & 0 \end{pmatrix}$$

$$[\pi^*(S_y)] = \frac{1}{2}\begin{pmatrix} 0 & -1 \\ 1 & 0 \end{pmatrix}$$

$$[\pi^*(S_z)] = \frac{1}{2}\begin{pmatrix} i & 0 \\ 0 & -i \end{pmatrix}.$$

Now define a new basis $\mathcal{B}^{*\prime} = \{e^{1'}, e^{2'}\}$ for $\mathbb{C}^{2*}$ by

$$e^{1'} \equiv -e^2$$
$$e^{2'} \equiv e^1.$$

You can check that the corresponding change of basis matrix A is

$$A = \begin{pmatrix} 0 & -1 \\ 1 & 0 \end{pmatrix}. \tag{5.47}$$

You can also check that in this new basis, the operators $\pi^*(S_i)$ are given by

$$\begin{aligned} [\pi^*(S_x)]_{\mathcal{B}^{*\prime}} &= A[\pi^*(S_x)]_{\mathcal{B}^*}A^{-1} = S_x \\ [\pi^*(S_y)]_{\mathcal{B}^{*\prime}} &= A[\pi^*(S_y)]_{\mathcal{B}^*}A^{-1} = S_y \\ [\pi^*(S_z)]_{\mathcal{B}^{*\prime}} &= A[\pi^*(S_z)]_{\mathcal{B}^*}A^{-1} = S_z \end{aligned} \tag{5.48}$$

and thus $(\pi^*, \mathbb{C}^{2*})$ is equivalent to $(\pi, \mathbb{C}^2)$. The connectedness of $SU(2)$ and Proposition 5.3 then imply that $(\Pi^*, \mathbb{C}^{2*})$ is equivalent to $(\Pi, \mathbb{C}^2)$, as claimed.

Exercise 5.27. Verify (5.47) and (5.48).

Exercise 5.28. Extend the above argument to show that the fundamental representation of $SL(2, \mathbb{C})$ is equivalent to its dual.

Example 5.22. *Antisymmetric 2nd rank tensors and the adjoint representation of orthogonal and Lorentz groups*

We claimed in Sect. 4.1 that for the Lorentz group, the adjoint representation is "the same" as the antisymmetric 2nd rank tensor representation (the latter being the representation to which the electromagnetic field tensor $F^{\mu\nu}$ belongs). We also argued in Example 5.16 that the same is true for the orthogonal group $O(n)$. Now it is time to precisely and rigorously prove these claims. Consider $\mathbb{R}^n$ equipped with a metric g, where g is either the Euclidean metric or the Minkowski metric. The isometry group $G = \text{Isom}(V)$ is then either $O(n)$ or $O(n-1, 1)$, and $\mathfrak{g}$ is then either $\mathfrak{so}(n)$ or $\mathfrak{so}(n-1, 1)$. To prove equivalence, we need an intertwiner $\phi : \Lambda^2\mathbb{R}^n \to \mathfrak{g}$. Here we'll define ϕ abstractly and use coordinate-free language to prove that it's an intertwiner. The proof is a little long, requires some patience, and may be skipped on a first reading, but it is a good exercise for getting acquainted with the machinery of this chapter; for an illuminating coordinate proof, see Problem 5-2.

To define ϕ abstractly we interpret $\mathfrak{g}$ not as a space of matrices but rather as linear operators on $\mathbb{R}^n$. Since $\Lambda^2\mathbb{R}^n$ is just the set of antisymmetric $(0, 2)$ tensors on $\mathbb{R}^n$, we can then define ϕ by just using the map L to "lower an index" on an element $T \in \Lambda^2\mathbb{R}^n$, converting the $(0, 2)$ tensor into a $(1, 1)$ tensor, i.e. a linear operator. More precisely, we define the linear operator $\phi(T)$ as

$$\phi(T)(v, f) \equiv T(L(v), f) \quad v \in \mathbb{R}^n, \; f \in \mathbb{R}^{n*}.$$

We must check, though, that $\phi(T) \in \mathfrak{g}$. This will be facilitated by characterizing $X \in \mathfrak{g}$ by (4.55), which in this context we write as

$$g(Xv, w) + g(v, Xw) = 0. \tag{5.49}$$

Thus, to show that the range of ϕ really is $\mathfrak{g}$, we just need to show that $\phi(T)$ as defined above satisfies (5.49):

$$\begin{aligned} & g(\phi(T)v, w) + g(v, \phi(T)w) \\ &= \phi(T)(v, L(w)) + \phi(T)(w, L(v)) \text{ by definition of } L \\ &= T(L(v), L(w)) + T(L(w), L(v)) \text{ by definition of } \phi(T) \\ &= 0 \qquad\qquad \text{by antisymmetry of } T. \end{aligned}$$

Thus $\phi(T)$ really is in $\mathfrak{g}$. To show that ϕ is an intertwiner, we need to show that $\phi \circ \Lambda^2\Pi(R) = \mathrm{Ad}(R) \circ \phi$ for all $R \in G$. To do this, we'll employ the alternate definition of the tensor product representation given in Exercise 5.13, which in this case says

$$(\Lambda^2\Pi(R)T)(f, h) = T(\Pi^*(R^{-1})f, \Pi^*(R^{-1})h), \quad R \in G, \ f, h \in \mathbb{R}^{n*}. \tag{5.50}$$

We then have, for all $v \in \mathbb{R}^n$, $f \in \mathbb{R}^{n*}$ (careful with all the parentheses!),

$$\begin{aligned} ((\phi \circ \Lambda^2\Pi(R))(T))(v, f) &= (\Lambda^2\Pi(R)(T))(L(v), f) \\ &= T(\Pi^*(R^{-1})L(v), \Pi^*(R^{-1})f) \\ &= T(L(R^{-1}v), \Pi^*(R^{-1})f), \end{aligned}$$

where in the last equality we used the fact that L is an intertwiner between $\mathbb{R}^n$ and $\mathbb{R}^{n*}$. Again employing the alternate definition of the tensor product representation, but this time for Π^1_1, we also have

$$\begin{aligned} ((\mathrm{Ad}(R) \circ \phi)(T))(v, f) &= \phi(T)(R^{-1}v, \Pi^*(R^{-1})f) \\ &= T(L(R^{-1}v), \Pi^*(R^{-1})f) \end{aligned}$$

and we can thus conclude that $\phi \circ (\Lambda^2\Pi(R)) = \mathrm{Ad}(R) \circ \phi$, as desired.

So what does all this tell us? The conclusion that the tensor product representation of G on antisymmetric 2nd rank tensors coincides with the adjoint representation of G on $\mathfrak{g}$ is not at all surprising in the Euclidean case, because there $\mathfrak{g}$ is just $\mathfrak{so}(n)$, the set of all antisymmetric matrices! In the Lorentzian case, however, we might be a little surprised, since the matrices in $\mathfrak{so}(n-1, 1)$ are not all antisymmetric. These matrices, however, represent linear operators, and if we use L to convert them into $(0, 2)$ tensors (via ϕ^{-1}), then they *are* antisymmetric! In other words,

The Lie algebra of a Lie group of metric-preserving operators can always be viewed as antisymmetric tensors,

though we may have to raise or lower an index (via L) to make this manifest.

This is actually quite easy to show in coordinates: using the standard basis for $\mathbb{R}^n$ and the corresponding components of $X = X_i{}^j e^i \otimes e_j \in \mathfrak{g}$ (still viewed as a linear operator on $\mathbb{R}^n$), we can plug two basis vectors into (5.49) and obtain

$$\begin{aligned} 0 &= g(Xe_i, e_j) + g(e_i, Xe_j) \\ &= X_i{}^k g(e_k, e_j) + X_j{}^k g(e_i, e_k) \\ &= X_i{}^k g_{kj} + X_j{}^k g_{ik} \\ &= X_{ij} + X_{ji} \end{aligned}$$

and so the (2,0) tensor corresponding to $X \in \mathfrak{g}$ is antisymmetric! □

Our last example takes the form of an exercise, in which you will show that some of the representations of $SU(2)$ on $P_l(\mathbb{C}^2)$, as described in Example 5.7, are equivalent to more basic $SU(2)$ representations.

Exercise 5.29. Prove (by exhibiting an intertwining map) that $(\pi_1, P_1(\mathbb{C}^2))$ from Example 5.7 is equivalent to the fundamental representation of $\mathfrak{su}(2)$. Conclude (since $SU(2)$ is connected) that $(\Pi_1, P_1(\mathbb{C}^2))$ is equivalent to the fundamental representation of $SU(2)$. Do the same for $(\pi_2, P_2(\mathbb{C}))$ and the adjoint representation of $\mathfrak{su}(2)$ (you will need to consider a new basis that consists of *complex* linear combinations of the elements of $\mathcal{B}_2$).

5.7 Direct Sums and Irreducibility

One of our goals in this chapter is to organize the various representations we've met into a coherent scheme, and to see how they are all related. Defining a notion of equivalence was the first step, so that we would know when two representations are "the same." With that in place, we would now like to determine all the (inequivalent) representations of a given group or Lie algebra. In general this is a difficult problem, but for most of the matrix Lie groups we've met so far and their associated Lie algebras, there is a very nice way to do this: for each group or algebra, there exists a denumerable set of inequivalent representations (known as the "irreducible" representations) out of which all other representations can be built. Once these irreducible representations are known, any other representation can be broken down into a kind of "sum" of its irreducible components. In this section we'll present the notions of irreducibility and sum of vector spaces, and in subsequent sections we'll enumerate all the irreducible representations of $SU(2)$, $SO(3)$, $SO(3,1)$, $SL(2,\mathbb{C})$, and their associated Lie algebras.

To motivate the discussion, consider the vector space $M_n(\mathbb{R})$ of all $n \times n$ real matrices. There is a representation Π of $O(n)$ on this vector space given by similarity transformations:

$$\Pi(R)A \equiv RAR^{-1} \quad R \in O(n),\ A \in M_n(\mathbb{R}).$$

If we consider $M_n(\mathbb{R})$ to be the matrices corresponding to elements of $\mathcal{L}(\mathbb{R}^n)$, then this is just the matrix version of the linear operator representation Π_1^1 described above, with $V = \mathbb{R}^n$. Alternatively, it can be viewed as the matrix version of the representation Π_2^0 on $\mathbb{R}^n \otimes \mathbb{R}^n$, with identical matrix transformation law

$$[\Pi_2^0(R)(T)] = R[T]R^T = R[T]R^{-1}, \quad T \in \mathbb{R}^n \otimes \mathbb{R}^n.$$

Now, it turns out that there are some special properties that $A \in M_n(\mathbb{R})$ could have that would be preserved by $\Pi(R)$. For instance, if A is symmetric or antisymmetric, then RAR^{-1} is also (you can check this directly, or see it as a corollary of the discussion at the beginning of Sect. 5.5). Furthermore, if A has zero trace, then so does RAR^{-1}. In fact, we can decompose A into a symmetric piece and an antisymmetric piece, and then further decompose the symmetric piece into a traceless piece and a piece proportional to the identity, as follows:

$$\begin{aligned} A &= \frac{1}{2}(A + A^T) + \frac{1}{2}(A - A^T) \\ &= \frac{1}{n}(\mathrm{Tr}A)\, I + \frac{1}{2}\left(A + A^T - \frac{2}{n}(\mathrm{Tr}A)\, I\right) + \frac{1}{2}\left(A - A^T\right). \end{aligned} \tag{5.51}$$

(You should check explicitly that the first term in (5.51) is proportional to the identity, the second is symmetric and traceless, and the third is antisymmetric.) Furthermore, this decomposition is unique, as you will show below. If we recall the definitions

$$\begin{aligned} S_n(\mathbb{R}) &= \{M \in M_n(\mathbb{R}) \mid M = M^T\} \\ A_n(\mathbb{R}) &= \{M \in M_n(\mathbb{R}) \mid M = -M^T\}, \end{aligned}$$

and add the new definitions

$$\begin{aligned} S_n'(\mathbb{R}) &\equiv \{M \in M_n(\mathbb{R}) \mid M = M^T,\ \mathrm{Tr}\, M = 0\} \\ \mathbb{R}I &\equiv \{M \in M_n(\mathbb{R}) \mid M = cI,\ c \in \mathbb{R}\}, \end{aligned}$$

then this means that any $A \in M_n(\mathbb{R})$ can be written uniquely as a sum of elements of S_n', A_n, and $\mathbb{R}I$, all of which are subspaces of $M_n(\mathbb{R})$. This type of situation turns up frequently, so we formalize it with the following definition: If V is a vector space with subspaces $W_1, W_2, \ldots, W_k$ such that any $v \in V$ can be written *uniquely* as $v = w_1 + w_2 + \cdots + w_k$, where $w_i \in W_i$, then we say that V is the ***direct sum*** of the W_i and we write

$$V = W_1 \oplus W_2 \oplus \cdots \oplus W_k \qquad \text{or} \qquad V = \bigoplus_{i=1}^{k} W_i.$$

The decomposition of a vector v is sometimes written as $v = (w_1, \ldots, w_k)$. In the situation above we have, as you can check,

$$M_n(\mathbb{R}) = S_n(\mathbb{R}) \oplus A_n(\mathbb{R}) = \mathbb{R}I \oplus S_n'(\mathbb{R}) \oplus A_n(\mathbb{R}). \tag{5.52}$$

Exercise 5.30. Show that if $V = \bigoplus_{i=1}^{k} W_i$, then $W_i \cap W_j = \{0\}\ \forall i \neq j$, i.e. the intersection of two different W_i is just the zero vector. Verify this explicitly in the case of the two decompositions in (5.52).

Exercise 5.31. Show that $V = \bigoplus_{i=1}^{k} W_i$ is equivalent to the statement that the set

$$\mathcal{B} = \mathcal{B}_1 \cup \cdots \cup \mathcal{B}_k,$$

where each $\mathcal{B}_i$ is an arbitary basis for W_i, is a basis for V.

Exercise 5.32. Show that $M_n(\mathbb{C}) = H_n(\mathbb{C}) \oplus \mathfrak{u}(n)$.

Our discussion above shows that all the subspaces that appear in (5.52) are ***invariant***, in the sense that the action of $\Pi(R)$ on any element of one of the subspaces produces another element of the same subspace (i.e., if A is symmetric then $\Pi(R)A$ is also, etc.). In fact, we can define an ***invariant subspace*** of a group representation (Π, V) as a subspace $W \subset V$ such that $\Pi(g)w \in W$ for all $g \in G$, $w \in W$. Invariant subspaces of Lie algebra representations are defined analogously. Notice that the entire vector space V, as well as the zero vector $\{0\}$, are always (trivially) invariant subspaces. An invariant subspace $W \subset V$ that is neither equal to V nor to $\{0\}$ is said to be a ***nontrivial*** invariant subspace. Notice also that the invariance of W under the $\Pi(g)$ means we can restrict each $\Pi(g)$ to W (that is, interpret each $\Pi(g)$ as an operator $\Pi(g)|_W \in GL(W)$) and so obtain a representation $(\Pi|_W, W)$ of G on W. Given a representation V, the symmetric and antisymmetric subspaces $S^r(V)$ and $\Lambda^r(V)$ of the tensor product representation $\mathcal{T}_r^0(V)$ are nice examples of nontrivial invariant subspaces (recall, of course, that $\Lambda^r(V)$ is only nontrivial when $r \leq \dim V$).

Nontrivial invariant subspaces are very important in representation theory, as they allow us to block diagonalize the matrices corresponding to our operators. If we have a representation (Π, V) of a group G where V decomposes into a direct sum of invariant subspaces W_i,

$$V = W_1 \oplus W_2 \oplus \cdots \oplus W_k,$$

and if $\mathcal{B}_i$ are bases for W_i and we take the union of the $\mathcal{B}_i$ as a basis for V, then you should check that the matrix representation of the operator $\Pi(g)$ in this basis will look like

$$[\Pi(g)]_{\mathcal{B}} = \begin{pmatrix} [\Pi(g)]_{\mathcal{B}_1} & & & \\ & [\Pi_2(g)]_{\mathcal{B}_2} & & \\ & & \cdots & \\ & & & [\Pi_k(g)]_{\mathcal{B}_k} \end{pmatrix}, \tag{5.53}$$

where each $[\Pi(g)]_{\mathcal{B}_i}$ is the matrix of $\Pi(g)$ restricted to the subspace W_i.

Thus, given a finite-dimensional representation (Π, V) of a group or Lie algebra, we can try to get a handle on it by decomposing V into a direct sum of (two or more) invariant subspaces. Each of these invariant subspaces forms a representation in its own right, and we could then try to further decompose these representations, iterating until we get a decomposition $V = W_1 \oplus \cdots \oplus W_k$ in which each of the W_i has no nontrivial invariant subspaces (if they did, we might be able to decompose further). These elementary representations play an important role in the theory, as we'll see, and so we give them a name: we say that a representation W that has no nontrivial invariant subspaces is an ***irreducible*** representation (or ***irrep***, for short). Furthermore, if a representation (Π, V) admits a decomposition $V = W_1 \oplus \cdots \oplus W_k$ where each of the W_i is irreducible, then we say that (Π, V) is ***completely reducible***.[11] If the decomposition consists of two or more summands, then we say that V is ***decomposable***.[12] Thus we have the following funny-sounding sentence: if (Π, V) is completely reducible, then it is either decomposable or irreducible!

You may be wondering at this point how a finite-dimensional representations could not be completely reducible; after all, it is either irreducible or it contains a nontrivial invariant subspace W; can't we then decompose V into W and some subspace W' complementary to W? We can, but the potential problem is that W' may not be invariant; that is, the group or Lie algebra action might take vectors in W' to vectors that don't lie in W'. For an example of this, see Problem 5-8. However, there do exist many groups and Lie algebras for whom every finite-dimensional representation *is* completely reducible. Such groups and Lie algebras are said to be ***semi-simple***.[13] (Most of the matrix Lie groups we've met and their associated Lie algebras are semi-simple, but some of the abstract Lie algebras we've seen [like the Heisenberg algebra], as well as the matrix Lie group $U(n)$, are not). Thus, an arbitrary finite-dimensional representation of a semi-simple group or Lie algebra can always be written as a direct sum of irreducible representations. (This is what we did when we wrote $M_n(\mathbb{R})$ as $M_n(\mathbb{R}) = \mathbb{R}I \oplus S_n'(\mathbb{R}) \oplus A_n(\mathbb{R})$, though we can't yet prove that the summands are irreducible.) If we know all the irreducible representations of a given semi-simple group or Lie algebra, we then have a complete classification of *all* the finite-dimensional representations of that group or algebra, since any

[11] Note that an irreducible representation is, trivially, completely reducible, since $V = V$ is a decomposition into irreducibles. Thus "irreducible" and "completely reducible" are not mutually exclusive categories, even if they may sound like it!

[12] This terminology is not standard but will prove useful.

[13] Semi-simplicity can be defined in a number of equivalent ways, all of which are important. For more, see Hall [11] or Varadarajan [20].

representation decomposes into a finite sum of irreps. This makes the determination of irreps an important task, which we'll complete in this chapter for our favorite Lie algebras.

Before moving on to this task, however, there is more to say about decomposable and irreducible representations. We'll begin with examples of decomposable representations, which can arise in a number of ways. One of the most common is by taking tensor products. Say we have a semi-simple group or Lie algebra and two irreps V_1 and V_2. The tensor product representation $V_1 \otimes V_2$ is usually *not* irreducible, but since our group or Lie algebra is semi-simple we can decompose $V_1 \otimes V_2$ as

$$V_1 \otimes V_2 = W_1 \oplus \cdots \oplus W_k,$$

where the W_i are irreducible. In fact, the last decomposition in (5.52) is just the matrix version of the $O(n)$-invariant decomposition

$$\mathcal{T}^0_2(\mathbb{R}^n) = \mathbb{R}^n \otimes \mathbb{R}^n = \mathbb{R}g \oplus S^{2'}(\mathbb{R}^n) \oplus \Lambda^2(\mathbb{R}^n), \tag{5.54}$$

where

$$g = \sum_i e_i \otimes e_i \tag{5.55a}$$

$$S^{2'}(\mathbb{R}^n) = \{T \in S^2(\mathbb{R}^n) \mid \delta_{ij} T^{ij} = 0\} \quad \text{where } \delta_{ij} \text{ is Kronecker delta.} \tag{5.55b}$$

Another more general instance of the decomposition of a tensor product representation into irreducibles, of great interest to physicists, is given by the next example.

Example 5.23. *Decomposition of the tensor product of $SU(2)$ representations, or "addition of angular momentum"*

Consider the spin j representation $\mathbb{C}^{2j+1} \equiv V_j$ and spin j' representation $\mathbb{C}^{2j'+1} \equiv V_{j'}$. Taking their tensor product yields a decomposable representation, which (as mentioned in Example 3.21) decomposes as

$$V_j \otimes V_{j'} = V_{j+j'} \oplus V_{j+j'-1} \oplus \cdots \oplus V_{|j-j'|+1} \oplus V_{|j-j'|}. \tag{5.56}$$

It's not hard to see why this might be true. Intuitively, adding two angular momentum vectors of length j and j' can only yield vectors with lengths between $j + j'$ and $|j - j'|$. We can also sketch a more formal argument as follows. Using the notation of Example 3.21, the highest m value in $V_j \otimes V_{j'}$ must be $j + j'$ with corresponding eigenvector $|j\rangle \otimes |j'\rangle$, and so $V_{j+j'}$ must be a summand in the decomposition. The next highest m value is $j + j' - 1$, but now there are *two* possible eigenvectors, $|j\rangle \otimes |j' - 1\rangle$ and $|j - 1\rangle \otimes |j'\rangle$. These span a two-dimensional space, one-dimension of which must belong to $V_{j+j'}$ and the other to $V_{j+j'-1}$, so the latter must also be a summand. Considering the next highest m value $j + j' - 2$ yields a three-dimensional subspace spanned by

$\{|j-2\rangle \otimes |j'\rangle\,,\ |j-1\rangle \otimes |j'-1\rangle\,,\ |j\rangle \otimes |j'-2\rangle\}$, implying that $V_{j+j'-2}$ must also be a summand. One can continue in this way until $m = |j - j'|$, at which point the m eigenspaces stop growing in dimensions. At this point, however, a simple dimension count (see Exercise 5.33 below) confirms that all vectors are accounted for and hence that (5.56) is true.

More detailed proofs of (5.56) can be found in Appendix B of Sakurai [17] and Appendix D of Hall [11]. The further process of explicitly decomposing the various m eigenspaces into vectors belonging to the different summands in (5.56) is known in the physics literature as the "addition of angular momentum," and is tantamount to computing Clebsch–Gordan coefficients. See Sakurai [17] for details. □

Exercise 5.33. Check that the dimensions on both sides of (5.56) are equal.

Another source of decomposable representations is to take an irrep of a group G and then consider it as a representation of a subgroup $H \subset G$. This is the context of our next example.

Example 5.24. *Decomposition of* $\Lambda^2\mathbb{R}^4$ *as an* $O(3)$ *representation*

Consider the 2nd rank antisymmetric tensor representation of $O(3,1)$ on $\Lambda^2\mathbb{R}^4$, but restrict the representation to $O(3)$ where we view $O(3) \subset O(3,1)$ in the obvious way. This representation of $O(3)$ is reducible, since clearly $\Lambda^2\mathbb{R}^3 \subset \Lambda^2\mathbb{R}^4$ is an $O(3)$-invariant subspace spanned by

$$\begin{aligned} f_1 &\equiv e_2 \wedge e_3 \\ f_2 &\equiv e_3 \wedge e_1 \\ f_3 &\equiv e_1 \wedge e_2. \end{aligned} \tag{5.57}$$

There is also a complementary invariant subspace, spanned by

$$\begin{aligned} f_4 &\equiv e_1 \wedge e_4 \\ f_5 &\equiv e_2 \wedge e_4 \\ f_6 &\equiv e_3 \wedge e_4. \end{aligned} \tag{5.58}$$

This second subspace is clearly equivalent to the vector representation of $O(3)$ since $O(3)$ leaves e_4 unaffected (the unconvinced reader can quickly check that the map $\phi : e_i \wedge e_4 \mapsto e_i$ is an intertwiner). Thus, as an $O(3)$ representation, $\Lambda^2\mathbb{R}^4$ decomposes into the vector and pseudovector representations, i.e.

$$\Lambda^2\mathbb{R}^4 \simeq \mathbb{R}^3 \oplus \Lambda^2\mathbb{R}^3. \tag{5.59}$$

We can interpret this physically in the case of the electromagnetic field tensor, which lives in $\Lambda^2\mathbb{R}^4$; in that case, (5.59) says that **under (proper and improper) rotations, some components of the field tensor transform amongst themselves**

as a vector, and others as a pseudovector. The components that transform as a vector comprise the electric field, and those that transform like a pseudovector comprise the magnetic field. To see this explicitly, one can think about which basis vectors from (5.58) and (5.57) go with which components of the matrix in (3.49).

One could, of course, further restrict this representation to $SO(3) \subset O(3,1)$; one would get the same decomposition (5.59), except that for $SO(3)$ the representations $\mathbb{R}^3$ and $\Lambda^2\mathbb{R}^3$ are equivalent, and so under (proper) rotations the field tensor transforms as a pair of vectors, the electric field "vector" and the magnetic field "vector." Of course, *how these objects transform depends on what transformation group you're considering*. The electric field and magnetic field both transform as vectors under rotations, but under improper rotations the electric field transforms as a vector and the magnetic field as a pseudovector. Furthermore, under Lorentz transformations the electric and magnetic fields cannot be meaningfully distinguished, as they transform together as the components of an antisymmetric second rank tensor! Physically, this corresponds to the fact that boosts can turn electric fields in one reference frame into magnetic fields in another, and vice-versa. □

Yet another source of decomposable representations are function spaces, as in the next example.

Example 5.25. *Decomposition of $L^2(S^2)$ into irreducibles*

A nice example of a direct sum decomposition of a representation into its irreducible components is furnished by $L^2(S^2)$. As we noted in Example 5.9, the spectral theorems of functional analysis tell us that the eigenfunctions Y^l_m of the spherical laplacian Δ_{S^2} form an orthogonal basis for the Hilbert space $L^2(S^2)$. We already know, though, that the Y^l_m of fixed l form a basis for $\tilde{\mathcal{H}}_l$, the spherical harmonics of degree l. We'll show in the next section that each of the $\tilde{\mathcal{H}}_l$ is an irreducible $SO(3)$ representation, so we can decompose $L^2(S^2)$ into irreducible representations as

$$L^2(S^2) = \bigoplus_l \tilde{\mathcal{H}}_l = \tilde{\mathcal{H}}_1 \oplus \tilde{\mathcal{H}}_2 \oplus \tilde{\mathcal{H}}_3 \oplus \cdots.$$

□

Before concluding this section we should point out that the notion of direct sum is useful not only in decomposing a given vector space into mutually exclusive subspaces, but also in "adding" vector spaces together. That is, given two vector spaces V and W, we can define their ***direct sum*** $V \oplus W$ to be the set $V \times W$ with vector addition and scalar multiplication defined by

$$\begin{aligned}(v_1, w_1) + (v_1, w_2) &= (v_1 + v_2, w_1 + w_2) \\ c(v, w) &= (cv, cw).\end{aligned}$$

It's straightforward to check that with vector addition and scalar multiplication so defined, $V \oplus W$ is a bona fide vector space. Also, V can be considered a subset of $V \oplus W$, as just the set of all vectors of the form $(v, 0)$, and likewise for W. With this identification it's clear that any element $(v, w) \in V \oplus W$ can be written uniquely as $v + w$ with $v \in V$ and $w \in W$, so this notion of a direct sum is consistent with our earlier definition: if we take the direct sum of V and W, the resulting vector space really can be decomposed into the subspaces V and W.

This notion[14] of direct sum may also be extended to representations; that is, given two representations $(\Pi_i, V_i),\ i = 1, 2$, we may construct the ***direct sum representation*** $(\Pi_1 \oplus \Pi_2, V_1 \oplus V_2)$ defined by

$$((\Pi_1 \oplus \Pi_2)(g))(v_1, v_2) \equiv (\Pi_1(g)v_1, \Pi_2(g)v_2) \quad \forall\, (v_1, v_2) \in V_1 \oplus V_2,\ g \in G.$$

These constructions may seem trivial, but they have immediate physical application, as we'll now see.

Example 5.26. *The Dirac Spinor*

Consider the left- and right-handed spinor representations of $SL(2,\mathbb{C})$, $(\Pi, \mathbb{C}^2)$ and $(\bar{\Pi}, \mathbb{C}^2)$. We define the ***Dirac spinor representation*** of $SL(2,\mathbb{C})$ to be the direct sum representation $(\Pi \oplus \bar{\Pi}, \mathbb{C}^2 \oplus \mathbb{C}^2)$, which is then given by

$$((\Pi \oplus \bar{\Pi})(A))(v, w) \equiv (Av, A^{\dagger-1}w) \quad \forall\, (v, w) \in \mathbb{C}^2 \oplus \mathbb{C}^2,\ A \in SL(2,\mathbb{C}).$$

Making the obvious identification of $\mathbb{C}^2 \oplus \mathbb{C}^2$ with $\mathbb{C}^4$, we can write $(\Pi \oplus \bar{\Pi})(A)$ in block matrix form as

$$(\Pi \oplus \bar{\Pi})(A) = \begin{pmatrix} A & 0 \\ 0 & A^{\dagger-1} \end{pmatrix} \tag{5.60}$$

as in (5.53).

This representation is, by construction, decomposable. Why deal with a decomposable representation rather than its irreducible components? There are a few different ways to answer this in the case of the Dirac spinor, but one rough answer has to do with parity. We will show in Sect. 5.12 that it is impossible to define a consistent action of the parity operator on either the left-handed or right-handed spinors individually, and that what the parity operator naturally wants to do is *interchange* the two representations. Thus, to have a representation of $SL(2,\mathbb{C})$ spinors on which parity naturally acts, we must combine both the left- and right-handed spinors into a Dirac spinor, and in the most natural cases parity is represented by

$$(\Pi \oplus \bar{\Pi})(\text{Parity}) = \pm \begin{pmatrix} 0 & I \\ I & 0 \end{pmatrix} \tag{5.61}$$

[14]In some texts our first notion of direct sum, in which we decompose a vector space into mutually exclusive subspaces, is called an ***internal direct sum***, and our second notion of direct sum, in which we take distinct vector spaces and add them together, is known as an ***external direct sum***.

which obviously just interchanges the left- and right-handed spinors. From this we see that the Dirac spinor is reducible under $SL(2, \mathbb{C})$, but not under larger groups which include parity.

In addition to the Weyl and Dirac spinors you may have heard of the ***Majorana*** spinor, which is a real version of the Dirac representation. One way to obtain the Majorana representation is to perform a similarity transformation on (5.60) which produces a purely real matrix. This is certainly not possible for general complex matrices, but can done for matrices in the Dirac spinor representation. The first step is to use the similarity transformation from Example 5.21 and Exercise 5.28 on the second block to turn $\begin{pmatrix} A & 0 \\ 0 & A^{\dagger -1} \end{pmatrix}$ into $\begin{pmatrix} A & 0 \\ 0 & \bar{A} \end{pmatrix}$. Any matrix of this form can then be transformed into a purely real matrix, as you will show in Problem 5-6. With a purely real matrix in hand we are then free to restrict our vector components to be real, yielding a representation of $SL(2, \mathbb{C})$ on $\mathbb{R}^4$. This is the Majorana representation. You will compute the induced Lie algebra representation explicitly in Problem 5-6(c).

5.8 More on Irreducibility

In the last section we introduced the notion of an irreducible representation, but we didn't prove that any of the representations we've met are irreducible. In this section we'll remedy that and also learn a bit more about irreducibility along the way. In proving the irreducibility of a given representation, the following proposition about the irreps of matrix Lie groups and their Lie algebras is often useful. The proposition just says that for a *connected* matrix Lie group, the irreps of the group are the same as the irreps of the Lie algebra. This may seem unsurprising and perhaps even trivial, but the conclusion does not hold when the group is disconnected. We'll have more to say about this later. The proof of this proposition is also a nice exercise in using some of the machinery we've developed so far.

Proposition 5.4. *A representation* (Π, V) *of a connected matrix Lie group* G *is irreducible if and only if the induced Lie algebra representation* (π, V) *of* $\mathfrak{g}$ *is irreducible as well.*

Proof. First, assume that (Π, V) is an irrep of G. Then consider the induced Lie algebra representation (π, V), and suppose that this representation has an invariant subspace W. We'll show that W must be an invariant subspace of (Π, V) as well, which by the irreducibility of (Π, V) will imply that W is either V or $\{0\}$, which will then show that (π, V) is irreducible. For all $X \in \mathfrak{g}$, $w \in W$, we have

$$\begin{aligned} &\pi(X)w \in W \\ &\Rightarrow e^{\pi(X)}w \in W \\ &\Rightarrow \Pi(e^X)w \in W. \end{aligned} \tag{5.62}$$

Now, using the fact that any element of G can be written as a product of exponentials (cf. Proposition 5.3), we then have

$$\begin{aligned}\Pi(g)w &= \Pi(e^{t_1X_1}e^{t_2X_2}\cdots e^{t_nX_n})w \\ &= \Pi(e^{t_1X_1})\Pi(e^{t_2X_2})\cdots\Pi(e^{t_nX_n})w\end{aligned}$$

which must be in W by repeated application of (5.62). Thus W is also an invariant subspace of (Π, V), but we assumed (Π, V) was irreducible and so W must be equal to V or $\{0\}$, which then proves that (π, V) is irreducible as well.

Conversely, assume that (π, V) is irreducible, and let W be an invariant subspace of (Π, V). Then for all $t \in \mathbb{R}$, $X \in \mathfrak{g}$, $w \in W$ we know that $\Pi(e^{tX})w \in W$, and hence

$$\begin{aligned}\pi(X)w &= \frac{d}{dt}\Pi(e^{tX})w\,|_{t=0} \\ &= \lim_{t\to 0}\frac{\Pi(e^{tX})w - w}{t}\end{aligned}$$

must be in W as well, so W is invariant under π and hence must be equal to V or $\{0\}$. Thus (Π, V) is irreducible. □

Note that the assumption of connectedness was crucial in the above proof, as otherwise we could not have used Proposition 5.3. Furthermore, we'll soon meet irreducible representations of disconnected groups (like $O(3,1)$) that yield Lie algebra representations which *do* have nontrivial invariant subspaces, and are thus not irreducible. The above proposition is still very useful, however, as we'll see in this next example.

Example 5.27. *The $SU(2)$ representation on $P_l(\mathbb{C}^2)$, revisited*

In this example we'll prove that the $\mathfrak{su}(2)$ representations $(\pi_l, P_l(\mathbb{C}^2))$ of Example 5.7 are all irreducible. Proposition 5.4 will then tell us that the $SU(2)$ representations $(\Pi_l, P_l(\mathbb{C}^2))$ are all irreducible as well. Later on, we'll see that these representations are in fact *all* the finite-dimensional irreducible representations of $SU(2)$!

Recall that the $\pi_l(S_i)$ are given by

$$\begin{aligned}\pi_l(S_1) &= \frac{i}{2}\left(z_2\frac{\partial}{\partial z_1} + z_1\frac{\partial}{\partial z_2}\right) \\ \pi_l(S_2) &= \frac{1}{2}\left(z_2\frac{\partial}{\partial z_1} - z_1\frac{\partial}{\partial z_2}\right) \\ \pi_l(S_3) &= \frac{i}{2}\left(z_1\frac{\partial}{\partial z_1} - z_2\frac{\partial}{\partial z_2}\right).\end{aligned}$$

Define now the "raising" and "lowering" operators

$$Y_l \equiv \pi_l(S_2) + i\pi_l(S_1) = -z_1 \frac{\partial}{\partial z_2}$$

$$X_l \equiv -\pi_l(S_2) + i\pi_l(S_1) = -z_2 \frac{\partial}{\partial z_1}.$$

It's easy to see that Y_l trades a factor of z_2 for a factor of z_1, thereby raising the $\pi_l(S_3)$ eigenvalue of a single term by i. Likewise, X_l trades a z_1 for a z_2 and lowers the eigenvalue by i.[15] Consider a nonzero invariant subspace $W \subset P_l(\mathbb{C}^2)$. If we can show that $W = P_l(\mathbb{C}^2)$, then we know that $P_l(\mathbb{C}^2)$ is irreducible. Being nonzero, W contains at least one element of the form

$$w = a_l z_1^l + a_{l-1} z_1^{l-1} z_2 + a_{l-2} z_1^{l-2} z_2^2 + \cdots + a_0 z_2^l,$$

where at least one of the a_k is not zero. Let k_0 be the biggest value of k such that a_k is nonzero, so that $a_{k_0} z_1^{k_0} z_2^{l-k_0}$ is the term in w with the highest power of z_1. Then applying $(X_l)^{k_0}$ to w lowers the z_1 degree by k_0, killing all the terms except $a_{k_0} z_1^{k_0} z_2^{l-k_0}$. In fact, one can compute easily that

$$X_l^{k_0}(a_{k_0} z_1^{k_0} z_2^{l-k_0}) = (-1)^{k_0} k_0!\, a_{k_0} z_2^l.$$

This is proportional to z_2^l, and since W is invariant W must then contain z_2^l. But then we can successively apply the raising operator Y_l to get monomials of the form $z_1^k z_2^{l-k}$ for $0 \le k \le l$, and so these must be in W as well. These, however, form a basis for $P_l(\mathbb{C}^2)$, hence W must equal $P_l(\mathbb{C}^2)$, and so $P_l(\mathbb{C}^2)$ is irreducible. □

Before moving on to further examples, we need to state and prove Schur's lemma, one of the most basic and crucial facts about irreducible representations. Roughly speaking, the upshot of it is that if a linear operator on the representation space of an irrep commutes with the group action (i.e., is an intertwiner), then it must be proportional to the identity. The precise statement is as follows:

Proposition 5.5 (Schur's Lemma). *Let (Π_i, V_i), $i = 1, 2$ be two irreducible representations of a group or Lie algebra, and let $\phi : V_1 \to V_2$ be an intertwiner. Then either $\phi = 0$ or ϕ is a vector space isomorphism. Furthermore, if $(\Pi_1, V_1) = (\Pi_2, V_2)$ and V_1 is a finite-dimensional* complex *vector space, then ϕ is a multiple of the identity, i.e. $\phi = cI$ for some $c \in \mathbb{C}$.*

[15] You may object to the use of i in our definition of these operators; after all, $\mathfrak{su}(2)$ is a real Lie algebra, and so the expression $S_2 + iS_1$ has no meaning as an element of $\mathfrak{su}(2)$, and so one can't say that, for instance, $Y_l = \pi(S_2 + iS_1)$. Thus X_l and Y_l are not in the image of $\mathfrak{su}(2)$ under π_l. This is a valid objection, and to deal with it one must introduce the notion of the *complexification* of a Lie algebra. A discussion of this here would lead us too far astray from our main goals of applications in physics, however, so we relegate this material to the appendix, which you can consult at leisure.

Proof. We'll prove this for group representations Π_i ; the Lie algebra case follows immediately with the obvious notational changes. Let K be the kernel or null space of ϕ. Then $K \subset V_1$ is an invariant subspace of Π_1, since for any $v \in K,\ g \in G$,

$$\begin{aligned}\phi(\Pi_1(g)v) &= \Pi_2(g)(\phi(v)) \\ &= \Pi_2(g)(0) \\ &= 0 \\ &\Rightarrow \Pi_1(g)v \in K.\end{aligned}$$

However, since (Π_1, V_1) is irreducible, the only invariant subspaces are 0 and V_1, so K must be one of those. If $K = V_1$, then $\phi = 0$, so henceforth we assume that $K = 0$, which means that ϕ is one-to-one, and so $\phi(V_1) \subset V_2$ is isomorphic to V_1. Furthermore, $\phi(V_1)$ is an invariant subspace of (Π_2, V_2), since for any $\phi(v) \in \phi(V_1),\ g \in G$ we have

$$(\Pi_2(g))(\phi(v)) = \phi(\Pi_1(g)v) \in \phi(V_1).$$

But (Π_2, V_2) is also irreducible, so $\phi(V_1)$ must equal 0 or V_2. We already assumed that $\phi(V_1) \neq 0$, though, so we conclude that $\phi(V_1) = V_2$ and hence ϕ is an isomorphism.

Now assume that $(\Pi_1, V_1) = (\Pi_2, V_2) \equiv (\Pi, V)$ and that V is a *finite-dimensional complex* vector space of dimension n (notice that we didn't assume finite-dimensionality at the outset of the proof). Then ϕ is a linear operator and the eigenvalue equation

$$\det(\phi - \lambda I) = 0$$

is an nth degree complex polynomial in λ. By the fundamental theorem of algebra,[16] this polynomial has at least one root $c \in \mathbb{C}$, hence $\phi - cI$ has determinant 0 and is thus noninvertible. This means that $\phi - cI$ has a nontrivial kernel K. K is invariant, though; for all $v \in K$,

$$\begin{aligned}(\phi - cI)(\Pi(g)v) &= \phi(\Pi(g)v) - c\Pi(g)v \\ &= \Pi(g)(\phi(v) - cv) \\ &= 0 \\ &\Rightarrow \Pi(g)v \in K\end{aligned}$$

and since we know $K \neq 0$, we then conclude by irreducibility of V that $K = V$. This means that $(\phi - cI)(v) = 0\ \forall\ v \in V$, which means $\phi = cI$, as desired. □

[16] See Herstein [12], for instance.

Exercise 5.34. Prove the following corollary of Schur's lemma: *If* (Π_i, V_i), $i = 1, 2$ *are two complex irreducible representations of a group or Lie algebra, and* $\phi, \psi : V_1 \to V_2$ *are two intertwiners with* $\phi \neq 0$*, then* $\psi = c\phi$ *for some* $c \in \mathbb{C}$.

Before applying Schur's lemma, we should note that it embodies the connection between symmetry and degeneracy that is often mentioned in quantum mechanics texts. If (Π, V) is an irrep of some symmetry group G and $H \in GL(V)$ is an intertwiner (i.e., commutes with $\Pi(g)$ for all $g \in G$), then Schur's lemma says that $H = cI$. This means that all vectors in V have the same H-eigenvalue, i.e. they are *degenerate*. Thus if (Π, V) is the angular momentum j representation of $G = SU(2)$ and H is a quantum-mechanical Hamiltonian, then this means that all the spin j states will have the same energy, so there will be a $\dim V = (2j+1)$-fold degeneracy.

It should be noted, however, that symmetry does not *always* imply degeneracy. For instance, the double delta function potential well problem (see, e.g., Gasiorowicz [7]) has a parity-symmetric Hamiltonian (i.e., $[H, P] = 0$) but only two energy eigenfunctions, with differing energies. The fact that parity symmetry does not imply degeneracy can be seen as a consequence of the following proposition, which is our first application of Schur's lemma:

Proposition 5.6. *An irreducible finite-dimensional complex representation of an* abelian *group or Lie algebra is one-dimensional.*

Proof. Again we prove only the group case. Since G is abelian, each $\Pi(g)$ commutes with $\Pi(h)$ for all $h \in G$, hence each $\Pi(g) : V \to V$ is an intertwiner! By Schur's lemma, this implies that every $\Pi(g)$ is proportional to the identity (with possibly varying proportionality constants), and so *every* subspace of V is an invariant one. Thus the only way V could have no nontrivial invariant subspaces is to have no nontrivial subspaces at all, which means it must be one-dimensional. □

Exercise 5.35. Show that the fundamental representation of $SO(2)$ on $\mathbb{R}^2$ is irreducible. Prove this by contradiction, showing that if the fundamental representation were reducible then the $SO(2)$ generator

$$X = \begin{pmatrix} 0 & -1 \\ 1 & 0 \end{pmatrix}$$

would be diagonalizable over the real numbers, which you should show it is not. This shows that one really needs the hypothesis of a *complex* vector space in the above proposition.

Example 5.28. *The irreducible representations of* $\mathbb{Z}_2$

Proposition 5.6 allows us to easily enumerate all the irreducible representations of $\mathbb{Z}_2$. Since $\mathbb{Z}_2$ is abelian any irreducible representation (Π_{irr}, V) must be one-dimensional (i.e., $V = \mathbb{R}$ or $\mathbb{C}$), and Π_{irr} must also satisfy

$$(\Pi_{\text{irr}}(-1))^2 = \Pi_{\text{irr}}((-1)^2) = \Pi_{\text{irr}}(1) = 1,$$

which means that $\Pi_{\text{irr}}(-1) = \pm 1$, i.e. Π_{irr} is either the alternating representation or the trivial representation! Furthermore, $\mathbb{Z}_2$ is semi-simple (as you will show in Problem 5-9), and so *any* representation (Π, V) of $\mathbb{Z}_2$ is completely reducible and thus decomposes into one-dimensional irreducible subspaces, on which $\Pi(-1)$ equals either 1 or -1. Thus $\Pi(-1)$ is diagonalizable, with eigenvalues ± 1 (cf. Example 5.18 for examples of this). Let's say the $\mathbb{Z}_2$ in question is $\mathbb{Z}_2 \simeq \{I, P\} \subset O(3)$, and that we're working in a quantum-mechanical context with a Hilbert space $\mathcal{H}$. Then there exists a basis for $\mathcal{H}$ consisting of eigenvectors of $\Pi(P)$; for a given eigenvector ψ, its eigenvalue of ± 1 is known as its ***parity***. If the eigenvalue is $+1$, then ψ is said to have ***even*** parity, and if the eigenvalue is -1, then ψ is said to have ***odd*** parity. If $[H, \Pi(P)] = 0$, then the energy eigenfunctions can be taken to be parity eigenvectors, but as mentioned above this does not necessarily imply degeneracy of the energy eigenvalues. □

5.9 The Irreducible Representations of $\mathfrak{su}(2)$, $SU(2)$, and $SO(3)$

In this section we'll construct (up to equivalence) *all* the finite-dimensional irreducible complex representations of $\mathfrak{su}(2)$. Besides being of intrinsic interest, our results will also allow us to classify all the irreducible representations of $SO(3)$, $SU(2)$, and even the apparently unrelated representations of $\mathfrak{so}(3,1)$, $SO(3,1)$, and $SL(2,\mathbb{C})$. The construction we'll give is more or less the same as that found in the physics literature under the heading "theory of angular momentum," except that we're using different language and notation. Our strategy will be to use the commutation relations to deduce the possible structures of $\mathfrak{su}(2)$ irreps, and then show that we've already constructed representations which exhaust these possibilities, thus yielding a complete classification.

Let (π, V) be a finite-dimensional complex irreducible representation of $\mathfrak{su}(2)$. It will be convenient to use the following shorthand, familiar from the physics literature:

$$\begin{aligned} J_z &\equiv i\pi(S_z) \\ J_+ &\equiv i\pi(S_x) - \pi(S_y) \\ J_- &\equiv i\pi(S_x) + \pi(S_y). \end{aligned} \tag{5.63}$$

These "raising" and "lowering" operators obey the following commutation relations, as you can check:

$$\begin{aligned} [J_z, J_\pm] &= \pm J_\pm \\ [J_+, J_-] &= 2J_z. \end{aligned}$$

Now, as discussed in our proof of Schur's lemma, the fact that V is complex means that every operator on V has at least one eigenvector. In particular, this means that J_z has an eigenvector v with eigenvalue b. The above commutation relations then imply that

$$J_z(J_\pm v) = [J_z, J_\pm]\, v \;+\; J_\pm(J_z\, v) = (b \pm 1) J_\pm\, v$$

so that if $J_\pm v$ is not zero (which it might be!), then it is another eigenvector of J_z with eigenvalue $b \pm 1$. Now, we can repeatedly apply J_+ to v to yield more and more eigenvectors of J_z, but since V is finite-dimensional and eigenvectors with different eigenvalues are linearly independent (see Exercise 5.36 below), this process must end somewhere, say at N applications of J_+. Let v_0 be this vector with the highest eigenvalue (also known as the ***highest weight vector*** of the representation), so that we have

$$\begin{aligned} v_0 &= (J_+)^N v \\ J_+ v_0 &= 0. \end{aligned}$$

Then v_0 has a J_z eigenvalue of $b + N \equiv j$ (note that so far we haven't proved anything about b or j, but we will soon see that they must be integral or half-integral). Starting with v_0, then, we can repeatedly "lower" with J_- to get eigenvectors with lower eigenvalues. In fact, we can define

$$v_k \equiv (J_-)^k v_0$$

which has J_z eigenvalue $j - k$. This chain must also end, though, so there must exist an integer l such that

$$v_l = (J_-)^l v_0 \neq 0 \;\text{ but }\; v_{l+1} = (J_-)^{l+1} v_0 = 0.$$

How can we find l ? For this we'll need the following formula, which you will prove in Exercise 5.37 below:

$$J_+(v_k) = [2jk - k(k-1)]\, v_{k-1}. \tag{5.64}$$

Applying this to $v_{l+1} = 0$ gives

$$0 = J_+(v_{l+1}) = [2j(l+1) - (l+1)\, l\,]\, \mathrm{v}_l$$

and since $v_l \neq 0$ we conclude that

$$[2j(l+1) - (l+1)l] = 0 \iff j = l/2.$$

Thus j is a nonnegative integer or half integer! (In fact, j is just the "spin" of the representation.) We further conclude that V contains $2j+1$ vectors $\{v_k \mid 0 \le k \le 2j\}$, all of which are eigenvectors of J_z with eigenvalue $j-k$. Furthermore, since

$$\pi(S_x) = -\frac{i}{2}(J_+ + J_-)$$
$$\pi(S_y) = \frac{1}{2}(J_- - J_+),$$

the action of S_x and S_y take a given v_k into a linear combination of other v_k, so the span of v_k is a nonzero invariant subspace of V. We assumed V was irreducible, however, so we must have $V = \text{Span}\,\{v_k\}$, and since the v_k are linearly independent, they form a basis for V!

To summarize, any finite-dimensional irreducible complex representation of $\mathfrak{su}(2)$ has dimension $2j+1$, $2j \in \mathbb{N}$, and a basis $\{v_k\}_{k=0-2j}$ which satisfies

$$\begin{aligned} J_+(v_0) &= 0. \\ J_-(v_k) &= v_{k+1} \qquad k < 2j \\ J_z(v_k) &= (j-k)v_k \\ J_-(v_{2j}) &= 0 \\ J_+(v_k) &= [2jk - k(k-1)]v_{k-1} \quad k \neq 0. \end{aligned} \tag{5.65}$$

What's more, we can actually use the above equations to *define* representations (π_j, V_j), where V_j is a $2j+1$ dimensional vector space with basis $\{v_k\}_{k=0-2j}$ and the action of the operators $\pi_j(S_i)$ is defined by (5.65). It's straightforward to check that this defines a representation of $\mathfrak{su}(2)$ (see exercise below), and one can prove irreducibility in the same way that we did in Example 5.27. Furthermore, any irrep (π, V) of $\mathfrak{su}(2)$ must be equivalent to (π_j, V_j) for some j, since we can find a basis w_k for V satisfying (5.65) for some j and then define an intertwiner by

$$\begin{aligned} \phi : V &\to V_j \\ w_k &\mapsto v_k \end{aligned}$$

and extending linearly. We have thus proved the following:

Proposition 5.7. *The $\mathfrak{su}(2)$ representations (π_j, V_j), $2j \in \mathbb{N}$ defined above are all irreducible, and any other finite-dimensional complex irreducible representation of $\mathfrak{su}(2)$ is equivalent to (π_j, V_j) for some j, $2j \in \mathbb{N}$.*

In other words, the (π_j, V_j) are, up to equivalence, *all* the finite-dimensional complex irreducible representations of $\mathfrak{su}(2)$. They are also all the finite-dimensional complex irreducible representations of $\mathfrak{so}(3)$, since $\mathfrak{su}(2) \simeq \mathfrak{so}(3)$.

If you look back over our arguments you'll see that we deduced (5.65) from just the $\mathfrak{su}(2)$ commutation relations, the finite-dimensionality of V, and the existence of a highest weight vector v_0 satisfying $J_+(v_0) = 0$ and $J_z(v_0) = jv_0$. Thus, if we have an arbitrary (i.e., not necessarily irreducible) finite-dimensional $\mathfrak{su}(2)$ representation (π, V) and can find a highest weight vector v_0 for some j, we can lower with J_- to generate a basis $\{v_k\}_{k=0-2j}$ satisfying (5.65) and conclude that V has an invariant subspace equivalent to (π_j, V_j). We can then repeat this until V is completely decomposed into irreps. If we know that (π, V) is irreducible from the start, then we don't even have to find the vector v_0, we just use the fact that (π, V) must be equivalent to (π_j, V_j) for some j and note that j is given by $j = \frac{1}{2}(\dim V - 1)$. These observations make it easy to identify which (π_j, V_j) occur in any given $\mathfrak{su}(2)$ representation.

Note that if we have a finite-dimensional complex irreducible $\mathfrak{su}(2)$ representation that is also *unitary*, then we could work in an orthonormal basis. In that case, it turns out that the v_k defined above are *not* orthonormal, and are thus not ideal basis vectors to work with. They are orthogonal, but are not normalized to have unit length. In quantum-mechanical contexts the $\mathfrak{su}(2)$ representations usually are unitary, and so in that setting one works with the orthonormal basis vectors $|m\rangle$, $-j \le m \le j$. The vector $|m\rangle$ is proportional to our v_{j-m}, but is normalized. See Sakurai [17] for details on the normalization procedure.

Exercise 5.36. Let $S = \{v_i\}_{i=1-k}$ be a set of eigenvectors of some linear operator T on a vector space V. Show that if each of the v_i has *distinct* eigenvalues, then S is a linearly independent set. (Hint: One way to do this is by induction on k. Another is to argue by assuming that S is linearly dependent and reaching a contradiction. In this case you may assume without loss of generality that the v_i, $1 \le i \le k-1$ are linearly independent, so that v_k is the vector that spoils the assumed linear independence.)

Exercise 5.37. Prove (5.64). Proceed by induction, i.e. first prove the formula for $k = 1$, then assume it is true for k and show that it must be true for $k + 1$.

Exercise 5.38. Show that (5.65) *defines* a representation of $\mathfrak{su}(2)$. This consists of showing that the operators J_+, J_-, J_z satisfy the appropriate commutation relations. Then show that this representation is irreducible, using an argument similar to the one from Example 5.27.

Example 5.29. *$P_l(\mathbb{C}^2)$, revisited again*

In Example 5.7 we met the representations $(\pi_l, P_l(\mathbb{C}^2))$ of $\mathfrak{su}(2)$ on the space of degree l polynomials in two complex variables. In Example 5.27 we saw that these representations are all irreducible, and so by setting $\dim P_l(\mathbb{C}^2) = l + 1$ equal to $2j + 1$ we deduce that

$$(\pi_l, P_l(\mathbb{C}^2)) \simeq (\pi_{l/2}, V_{l/2}) \tag{5.66}$$

and so the $(\pi_l, P_l(\mathbb{C}^2))$, $l \in \mathbb{N}$ also yield all the complex finite-dimensional irreps of $\mathfrak{su}(2)$. What's more, this allows us to enumerate all the finite-dimensional complex irreps of the associated group $SU(2)$. Any irrep (Π, V) of $SU(2)$ yields an irrep (π, V) of $\mathfrak{su}(2)$, by Proposition 5.4. This irrep must be equivalent to $(\pi_l, P_l(\mathbb{C}^2))$ for some $l \in \mathbb{N}$, however, and so by Proposition 5.3 (Π, V) is equivalent to $(\Pi_l, P_l(\mathbb{C}^2))$. Thus, **the representations** $(\Pi_l, P_l(\mathbb{C}^2))$, $l \in \mathbb{N}$ **are (up to equivalence) all the finite-dimensional complex irreducible representations of** $SU(2)$**!**

It's instructive to construct the equivalence (5.66) explicitly. Recall that the raising and lowering operators (which we called Y_l and X_l in Example 5.27) are given by

$$J_- = -z_1 \frac{\partial}{\partial z_2}$$

$$J_+ = -z_2 \frac{\partial}{\partial z_1}$$

and that

$$J_z = i\pi_l(S_z) = \frac{1}{2}\left(z_2 \frac{\partial}{\partial z_2} - z_1 \frac{\partial}{\partial z_1} \right).$$

It's easy to check that $v_0 \equiv z_2^l$ is a highest weight vector with $j = l/2$, and so the basis that satisfies (5.65) is given by

$$v_k \equiv (J_-)^k (z_2^l) = (-1)^k \frac{l!}{(l-k)!} z_1^k z_2^{l-k} \quad 0 \le k \le l. \tag{5.67}$$

Exercise 5.39. Show by direct calculation that $(\pi_{2j}, P_{2j}(\mathbb{C}^2))$ satisfies (5.65) with basis vectors given by (5.67).

Example 5.30. $S^{2j}(\mathbb{C}^2)$ *as irreps of* $\mathfrak{su}(2)$

We suggested in Example 5.15 that the tensor product representation of $\mathfrak{su}(2)$ on $S^{2j}(\mathbb{C}^2)$, the totally symmetric $(0, 2j)$ tensors on $\mathbb{C}^2$, is equivalent to $(\pi_{2j}, P_{2j}(\mathbb{C}^2)) \simeq (\pi_j, V_j)$, $2j \in \mathbb{N}$. With the classification of $\mathfrak{su}(2)$ irreps in place, we can now prove this fact. It's easy to verify that

$$v_0 \equiv \underbrace{e_1 \otimes \cdots \otimes e_1}_{2j \text{ times}} \in S^{2j}(\mathbb{C}^2)$$

is a highest weight vector with eigenvalue j, so there is an irreducible invariant subspace of $S^{2j}(\mathbb{C}^2)$ equivalent to V_j. Since $\dim V_j = \dim S^{2j}(\mathbb{C}^2) = 2j + 1$ (as you can check), we conclude that V_j is all of $S^{2j}(\mathbb{C}^2)$, and hence that

$$(S^{2j}\pi, S^{2j}(\mathbb{C}^2)) \simeq (\pi_j, V_j) \simeq (\pi_{2j}, P_{2j}(\mathbb{C}^2))$$

which also implies $(S^{2j}\Pi, S^{2j}(\mathbb{C}^2)) \simeq (\Pi_{2j}, P_{2j}(\mathbb{C}^2))$. Thus, we see that:

> **Every irreducible representation of $\mathfrak{su}(2)$ and *SU(2)* can be obtained by taking a symmetric tensor product of the fundamental representation.**

Thus, using nothing more than the fundamental representation (which corresponds to $j = 1/2$, as expected) and the tensor product, we can generate all the representations of $\mathfrak{su}(2)$ and $SU(2)$. We'll see in the next section that the same is true for $\mathfrak{so}(3,1)$ and $SO(3,1)_o$. □

Our results about the finite-dimensional irreps of $\mathfrak{su}(2)$ and $SU(2)$ are not only interesting in their own right; they also allow us to determine all the irreps of $SO(3)$! To see this, first consider the degree l harmonic polynomial representation $(\Pi_l, \mathcal{H}_l(\mathbb{R}^3))$ of $SO(3)$. This induces a representation $(\pi_l, \mathcal{H}_l(\mathbb{R}^3))$ of $\mathfrak{so}(3) \simeq \mathfrak{su}(2)$. It's easy to check (see exercise below) that if we define, in analogy to the $\mathfrak{su}(2)$ case,

$$J_z \equiv i\pi_l(L_z) = i\left(y\frac{\partial}{\partial x} - x\frac{\partial}{\partial y}\right)$$

$$J_+ \equiv i\pi(L_x) - \pi(L_y) = i\left(z\frac{\partial}{\partial y} - y\frac{\partial}{\partial z}\right) - \left(x\frac{\partial}{\partial z} - z\frac{\partial}{\partial x}\right)$$

$$J_- \equiv i\pi(L_x) + \pi(L_y) = i\left(z\frac{\partial}{\partial y} - y\frac{\partial}{\partial z}\right) + \left(x\frac{\partial}{\partial z} - z\frac{\partial}{\partial x}\right)$$

then the vector $f_0 \equiv (x+iy)^l$ is a highest weight vector with eigenvalue l, and so $\mathcal{H}_l(\mathbb{R}^3)$ has an invariant subspace equivalent to (π_l, V_l). We will argue in Problem 5-10 that $\dim \mathcal{H}_l(\mathbb{R}^3) = 2l+1$, so we conclude that $(\pi_l, \mathcal{H}_l(\mathbb{R}^3)) \simeq (\pi_l, V_l)$, and is hence an irrep of $\mathfrak{so}(3)$. Proposition 5.4 then implies that $(\Pi_l, \mathcal{H}_l(\mathbb{R}^3))$ is an irrep of $SO(3)$!

Are these *all* the irreps of $SO(3)$? To find out, let (Π, V) be an arbitrary finite-dimensional complex irrep of $SO(3)$. Then the induced $\mathfrak{so}(3) \simeq \mathfrak{su}(2)$ representation (π, V) must be equivalent to (π_j, V_j) for some integral or half-integral j. We just saw that any integral j value is possible, by taking $(\Pi, V) = (\Pi_j, \mathcal{H}_j(\mathbb{R}^3))$. What about half-integral values of j? In this case, we have (careful not to confuse the number π with the representation π!)

$$\begin{aligned} e^{2\pi\cdot\pi(L_z)}v_0 &= e^{-i2\pi J_z}v_0 \\ &= e^{-i2\pi j}v_0 && \text{since } v_0 \text{ has eigenvalue } j \\ &= -v_0 && \text{since } j \text{ half-integral.} \end{aligned}$$

However, we also have

$$\begin{aligned} e^{2\pi\cdot\pi(L_z)}v_0 &= (\Pi(e^{2\pi\cdot L_z}))v_0 \\ &= (\Pi(I))v_0 && \text{by (4.72)} \\ &= v_0, \end{aligned}$$

a contradiction. Thus, j cannot be half-integral, so (π, V) must be equivalent to (π_j, V_j) for j integral. But this implies that $(\pi, V) \simeq (\pi_j, \mathcal{H}_j(\mathbb{R}^3))$, and so by Proposition 5.3, (Π, V) is equivalent to $(\Pi_j, \mathcal{H}_j(\mathbb{R}^3))$! Thus:

The representations $(\Pi_j, \mathcal{H}_j(\mathbb{R}^3))$, $j \in \mathbb{N}$ are (up to equivalence) all the finite-dimensional complex irreducible representations of $SO(3)$.

An important lesson to take away from this is that for a matrix Lie group G with Lie algebra $\mathfrak{g}$, **not all representations of $\mathfrak{g}$ necessarily come from representations of G.** If $G = SU(2)$, then there *is* a one-to-one correspondence between Lie algebra representations and group representations, but in the case of $SO(3)$ there are Lie algebra representations (corresponding to half-integral values of j) that *don't* come from $SO(3)$ representations. We won't say much more about this here, except to note that this is connected to the fact that there are non-isomorphic matrix Lie groups that have isomorphic Lie algebras, as is the case with $SU(2)$ and $SO(3)$. For a more complete discussion, see Hall [11].

Exercise 5.40. Verify that $J_+ f_0 = 0$ and $J_z f_0 = l f_0$.

5.10 Real Representations and Complexifications

So far we have classified all the *complex* finite-dimensional irreps of $\mathfrak{su}(2)$, $SU(2)$, and $SO(3)$, but we haven't said anything about *real* representations, despite the fact that many of the most basic representations of these groups and Lie algebras (like the fundamental of $SO(3)$ and all its various tensor products) are real. Fortunately, there is a way to turn every real representation into a complex representation, so that we can then apply our classification of complex irreps. Given any real vector space V, we can define the ***complexification*** of V as $V_{\mathbb{C}} \equiv \mathbb{C} \otimes V$, where $\mathbb{C}$ and V are both thought of as real vector spaces ($\mathbb{C}$ being a two-dimensional real vector space with basis $\{1, i\}$), so that if $\{e_i\}$ is a basis for V then $\{1 \otimes e_i, i \otimes e_i\}$ is a (real) basis for $V_{\mathbb{C}}$. Note that $V_{\mathbb{C}}$ also carries the structure of a *complex* vector space, with multiplication by i defined by

$$i(z \otimes v) = (iz) \otimes v, \quad z \in \mathbb{C}, \ v \in V.$$

A complex basis for $V_{\mathbb{C}}$ is then given by $\{1 \otimes e_i\}$, and the complex dimension of $V_{\mathbb{C}}$ is equal to the real dimension of V.

We can then define the ***complexification*** of a real representation (Π, V) to be the (complex) representation $(\Pi_{\mathbb{C}}, V_{\mathbb{C}})$ defined by

$$(\Pi_{\mathbb{C}}(g))(z \otimes v) \equiv z \otimes \Pi(g)v. \tag{5.68}$$

We can then get a handle on the real representation (Π, V) by applying our classification scheme to its complexification $(\Pi_{\mathbb{C}}, V_{\mathbb{C}})$. You should be aware, however, that the irreducibility of (Π, V) does not guarantee the irreducibility of $(\Pi_{\mathbb{C}}, V_{\mathbb{C}})$ (see Exercise 5.42 and Example 5.39), though in many cases $(\Pi_{\mathbb{C}}, V_{\mathbb{C}})$ will end up being irreducible.

Example 5.31. *The complexifications of* $\mathbb{R}^n$ *and* $M_n(\mathbb{R})$

As a warm-up to considering complexifications of representations, we consider the complexification of a simple vector space. Consider the complexification $\mathbb{R}^n_{\mathbb{C}}$ of $\mathbb{R}^n$. We can define the (obvious) complex-linear map

$$\begin{aligned} \phi : \mathbb{R}^n_{\mathbb{C}} &\to \mathbb{C}^n \\ 1 \otimes (a^j e_j) + i \otimes (b^j e_j) &\mapsto (a^j + ib^j)e_j, \quad a^j, b^j \in \mathbb{R} \end{aligned}$$

which is easily seen to be a vector space isomorphism, so we can identify $\mathbb{R}^n_{\mathbb{C}}$ with $\mathbb{C}^n$. One can also extend this argument in the obvious way to show that

$$(M_n(\mathbb{R}))_{\mathbb{C}} = M_n(\mathbb{C}).$$

Example 5.32. *The fundamental representation of* $\mathfrak{so}(3)$

Now consider the complexification $(\pi_{\mathbb{C}}, \mathbb{R}^3_{\mathbb{C}})$ of the fundamental representation of $\mathfrak{so}(3)$. As explained above, $\mathbb{R}^3_{\mathbb{C}}$ can be identified with $\mathbb{C}^3$, and a moment's consideration of (5.68) will show that the complexification of the fundamental representation of $\mathfrak{so}(3)$ is just given by the usual $\mathfrak{so}(3)$ matrices acting on $\mathbb{C}^3$ rather than $\mathbb{R}^3$. You should check that J_z and J_+ are given by

$$J_z = \begin{pmatrix} 0 & -i & 0 \\ i & 0 & 0 \\ 0 & 0 & 0 \end{pmatrix}, \qquad J_+ = \begin{pmatrix} 0 & 0 & -1 \\ 0 & 0 & -i \\ 1 & i & 0 \end{pmatrix} \tag{5.69}$$

and that the vector

$$v_0 \equiv e_1 + ie_2 = \begin{pmatrix} 1 \\ i \\ 0 \end{pmatrix}$$

is a highest weight vector with $j = 1$. This, along with the fact that $\dim \mathbb{C}^3 = 3$, allows us to conclude that $(\pi_{\mathbb{C}}, \mathbb{R}^3_{\mathbb{C}}) \simeq (\pi_1, V_1)$. This is why regular three-dimensional vectors are said to be "spin-one." Notice that our highest weight vector $v_0 = e_1 + ie_2$ is just the analog of the function $f_0 = x + iy \in \mathcal{H}_1(\mathbb{R}^3)$, which is an equivalent representation, and that in terms of $SO(3)$ reps we have

$$(\Pi_{\mathbb{C}}, \mathbb{R}^3_{\mathbb{C}}) \simeq (\Pi_1, \mathcal{H}_1(\mathbb{R}^3)).$$

Recall also that the adjoint representation of $\mathfrak{su}(2) \simeq \mathfrak{so}(3)$ is equivalent to the fundamental representation of $\mathfrak{so}(3)$. This combined with the above results implies

$$(\mathrm{ad}_{\mathbb{C}}, \mathfrak{so}(3)_{\mathbb{C}}) \simeq (\mathrm{ad}_{\mathbb{C}}, \mathfrak{su}(2)_{\mathbb{C}}) \simeq (\pi_1, V_1)$$

so the adjoint representation of $\mathfrak{su}(2)$ is "spin-one" as well.

Example 5.33. *Symmetric traceless tensors*

Consider the space $S^{2'}(\mathbb{R}^3)$ of symmetric traceless 2nd rank tensors defined in (5.55b). This is an $SO(3)$ invariant subspace of $\mathbb{R}^3 \otimes \mathbb{R}^3$, and so furnishes a representation $(\Pi_2^0, S^{2'}(\mathbb{R}^3))$ of $SO(3)$. What representation is this? To find out, consider the complexification of the associated $\mathfrak{so}(3)$ representation, $(\pi^0_{2'\mathbb{C}}, S^{2'}(\mathbb{C}^3))$, where $S^{2'}(\mathbb{C}^3)$ is also defined by (5.55b), just with $\mathbb{C}^3$ replacing $\mathbb{R}^3$. Then it's straightforward to verify that

$$v_0 \equiv (e_1 + ie_2) \otimes (e_1 + ie_2) \tag{5.70}$$

is a highest weight vector with $j = 2$, and this, along with fact that $\dim S^{2'}(\mathbb{C}^3) = 5$ (check!), implies that

$$(\pi^0_{2'\mathbb{C}}, S^{2'}(\mathbb{C}^3)) \simeq (\pi_2, V_2).$$

This is why symmetric traceless 2nd rank tensors on $\mathbb{R}^3$ are sometimes said to be "spin-two."

Recall that $S^{2'}(\mathbb{R}^3)$ was defined in (5.55b) as part of the decomposition

$$\mathbb{R}^3 \otimes \mathbb{R}^3 = \mathbb{R}g \oplus \Lambda^2(\mathbb{R}^3) \oplus S^{2'}(\mathbb{R}^3)$$

which has matrix counterpart

$$M_3(\mathbb{R}) = \mathbb{R}I \oplus A_3(\mathbb{R}) \oplus S_3'(\mathbb{R}).$$

Complexifying and using the fact (cf. Example 5.22) that $\Lambda^2(\mathbb{R}^3)$ is equivalent to the adjoint representation and is hence "spin-one," we obtain the following decomposition:

$$V_1 \otimes V_1 \simeq \mathbb{C}^3 \otimes \mathbb{C}^3 \simeq M_3(\mathbb{C}) \simeq V_0 \oplus V_1 \oplus V_2.$$

This is an instance of (5.56) and should be familiar from angular momentum addition in quantum mechanics.

Exercise 5.41. Using the standard basis for $\mathbb{C}^3$, write down the matrix $[v_0]$ for $v_0 = (e_1 + ie_2) \otimes (e_1 + ie_2)$. Then use the appropriate matrix transformation law to show that $[v_0]$ is an eigenvector of $i\pi^0_{2'\mathbb{C}}(L_z)$ with eigenvalue 2. Equation (5.69) may come in handy here.

Exercise 5.42. Consider the fundamental representation of $SO(2)$ on $\mathbb{R}^2$, which we know is irreducible by Exercise 5.35. The complexification of this representation is just given by the same $SO(2)$ matrices acting on $\mathbb{C}^2$ rather than $\mathbb{R}^2$. Show that this representation is reducible, by diagonalizing the $SO(2)$ generator

$$X = \begin{pmatrix} 0 & -1 \\ 1 & 0 \end{pmatrix}.$$

Note that this diagonalization can now be done because both complex eigenvalues and complex basis transformations are allowed, in contrast to the real case.

5.11 The Irreducible Representations of $\mathfrak{sl}(2,\mathbb{C})_{\mathbb{R}}$, $SL(2,\mathbb{C})$, and $SO(3,1)_o$

In this section we'll use the techniques and results of the previous section to classify all the finite-dimensional complex irreps of $\mathfrak{sl}(2,\mathbb{C})_{\mathbb{R}} \simeq \mathfrak{so}(3,1)$, and then use this to find the irreps of the associated groups $SL(2,\mathbb{C})$ and $SO(3,1)_o$.

Let (π, V) be a finite-dimensional complex irreducible representation of $\mathfrak{sl}(2,\mathbb{C})_{\mathbb{R}}$. Define the operators

$$\boxed{\begin{aligned} M_i &\equiv \frac{1}{2}(\pi(S_i) - i\pi(\tilde{K}_i)) \quad i = 1,2,3 \\ N_i &\equiv \frac{1}{2}(\pi(S_i) + i\pi(\tilde{K}_i)) \quad i = 1,2,3, \end{aligned}} \tag{5.71}$$

where $\{S_i, \tilde{K}_i\}_{i=1,2,3}$ is our usual basis for $\mathfrak{sl}(2,\mathbb{C})_{\mathbb{R}}$. One can check that the Ms and Ns commute between each other, as well as satisfy the $\mathfrak{su}(2)$ commutation relations internally, i.e.

$$\begin{aligned} [M_i, N_j] &= 0 \\ [M_i, M_j] &= \sum_{k=1}^{3} \epsilon_{ijk} M_k \\ [N_i, N_j] &= \sum_{k=1}^{3} \epsilon_{ijk} N_k. \end{aligned} \tag{5.72}$$

We have thus taken the *complex* span of the set $\{\pi(\tilde{K}_i), \pi(S_i)\}$ [notice the factors of i in (5.71)] and found a new basis for this Lie algebra of operators that makes it look like two *commuting* copies of $\mathfrak{su}(2)$. We can thus define the usual raising and lowering operators

$$\begin{aligned} N_{\pm} &\equiv iN_1 \mp N_2 \\ M_{\pm} &\equiv iM_1 \mp M_2 \end{aligned}$$

which then have the usual commutation relations between themselves and iM_z, iN_z:

$$
\begin{aligned}
[iM_z, M_\pm] &= \pm M_\pm \\
[iN_z, N_\pm] &= \pm N_\pm \\
[M_+, M_-] &= 2iM_z \\
[N_+, N_-] &= 2iN_z.
\end{aligned}
$$

With this machinery set up we can now use the strategy from the last section. First, pick a vector $v \in V$ that is an eigenvector of both iM_z and iN_z (that such a vector exists is guaranteed by Problem 5-11). Then by applying M_+ and N_+ we can raise the iM_z and iN_z eigenvalues until the raising operators give us the zero vector; let $v_{0,0}$ denote the vector with the highest eigenvalues (which we'll again refer to as a "highest weight vector"), and let (j_1, j_2) denote those eigenvalues under iM_z and iN_z respectively, so that

$$
\begin{aligned}
M_+(v_{0,0}) &= 0 \\
N_+(v_{0,0}) &= 0 \\
iM_z(v_{0,0}) &= j_1\, v_{0,0} \\
iN_z(v_{0,0}) &= j_2\, v_{0,0}\,.
\end{aligned}
\tag{5.73}
$$

We can then lower the eigenvalues with N_- and M_- to get vectors

$$v_{k_1,k_2} \equiv (M_-)^{k_1}(N_-)^{k_2} v_{0,0}$$

which are eigenvectors of iM_z and iN_z with eigenvalues $j_1 - k_1$ and $j_2 - k_2$ respectively. By finite-dimensionality of V this chain of vectors must eventually end, though, so there exists nonnegative integers l_1, l_2 such that $v_{l_1,l_2} \neq 0$ but

$$M_-(v_{l_1,l_2}) = N_-(v_{l_1,l_2}) = 0.$$

Calculations identical to those from the $\mathfrak{su}(2)$ case show that $l_i = 2j_i$, and that the action of the operators M_i, N_i is given by

$$
\begin{aligned}
M_+(v_{0,0}) &= 0 \\
N_+(v_{0,0}) &= 0 \\
M_-(v_{k_1,k_2}) &= v_{k_1+1,k_2} \qquad k_1 < 2j_1 \\
N_-(v_{k_1,k_2}) &= v_{k_1,k_2+1} \qquad k_2 < 2j_2 \\
iM_z(v_{k_1,k_2}) &= (j_1 - k_1)v_{k_1,k_2} \\
iN_z(v_{k_1,k_2}) &= (j_2 - k_2)v_{k_1,k_2}
\end{aligned}
\tag{5.74}
$$

$$\begin{aligned}
M_-(v_{2j_1,k_2}) &= 0 \;\; \forall\, k_2 \\
N_-(v_{k_1,2j_2}) &= 0 \;\; \forall\, k_1 \\
M_+(v_{k_1,k_2}) &= [2j_1k_1 - k_1(k_1-1)]v_{k_1-1,k_2} \quad k_1 \neq 0 \\
N_+(v_{k_1,k_2}) &= [2j_2k_2 - k_2(k_2-1)]v_{k_1,k_2-1} \quad k_2 \neq 0
\end{aligned}$$

(this is just two copies of (5.65), one each for the M_i and the N_i). As in the $\mathfrak{su}(2)$ case, we note that the $\{v_{k_1,k_2}\}$ are linearly independent and span an invariant subspace of V, hence must span all of V since we assumed V was irreducible. Thus, we conclude that any complex finite-dimensional irrep of $\mathfrak{sl}(2,\mathbb{C})_\mathbb{R}$ is of the form $(\pi_{(j_1,j_2)}, V_{(j_1,j_2)})$ where $V_{(j_1,j_2)}$ has a basis

$$\mathcal{B} = \{v_{k_1,k_2} \,|\, 0 \le k_1 \le 2j_1,\; 0 \le k_2 \le 2j_2\}$$

and the operators $\pi_{(j_1,j_2)}(S_i)$, $\pi_{(j_1,j_2)}(\tilde{K}_i)$ satisfy (5.74), with $2j_1, 2j_2 \in \mathbb{N}$. This tells us that

$$\dim V_{(j_1,j_2)} = (2j_1+1)(2j_2+1).$$

Let's abbreviate the representations $(\pi_{(j_1,j_2)}, V_{(j_1,j_2)})$ as simply $(\mathbf{j_1},\mathbf{j_2})$, as is done in the physics literature. As in the $\mathfrak{su}(2)$ case, we can show that (5.74) actually *defines* a representation of $\mathfrak{sl}(2,\mathbb{C})_\mathbb{R}$, and using the same arguments that we did in the $\mathfrak{su}(2)$ case we conclude that

Proposition 5.8. *The representations* $(\mathbf{j_1},\mathbf{j_2})$, $2j_1, 2j_2 \in \mathbb{N}$ *are, up to equivalence, all the complex finite-dimensional irreducible representations of* $\mathfrak{sl}(2,\mathbb{C})_\mathbb{R}$.

As in the $\mathfrak{su}(2)$ case, we deduced (5.74) from just the $\mathfrak{sl}(2,\mathbb{C})_\mathbb{R}$ commutation relations, the finite-dimensionality of V, and the existence of a highest weight vector $v_{0,0}$ satisfying (5.73). Thus if we're given a finite-dimensional $\mathfrak{sl}(2,\mathbb{C})_\mathbb{R}$ representation and can find a highest weight vector $v_{0,0}$ for some (j_1, j_2), we can conclude that the representation space contains an invariant subspace equivalent to $(\mathbf{j_1},\mathbf{j_2})$.

Example 5.34. $(\pi, \mathbb{C}^2)$ *The fundamental (left-handed spinor) representation*

Consider the left-handed spinor representation $(\pi, \mathbb{C}^2)$ of $\mathfrak{sl}(2,\mathbb{C})_\mathbb{R}$, which is also just the fundamental of $\mathfrak{sl}(2,\mathbb{C})_\mathbb{R}$, i.e.

$$\begin{aligned}
\pi(S_1) &= S_1 = \frac{1}{2}\begin{pmatrix} 0 & -i \\ -i & 0 \end{pmatrix}, \\
\pi(S_2) &= S_2 = \frac{1}{2}\begin{pmatrix} 0 & -1 \\ 1 & 0 \end{pmatrix}, \\
\pi(S_3) &= S_3 = \frac{1}{2}\begin{pmatrix} -i & 0 \\ 0 & i \end{pmatrix}
\end{aligned}$$

$$\pi(\tilde{K}_1) = \tilde{K}_1 = \frac{1}{2}\begin{pmatrix} 0 & 1 \\ 1 & 0 \end{pmatrix},$$

$$\pi(\tilde{K}_2) = \tilde{K}_2 = \frac{1}{2}\begin{pmatrix} 0 & -i \\ i & 0 \end{pmatrix},$$

$$\pi(\tilde{K}_3) = \tilde{K}_3 = \frac{1}{2}\begin{pmatrix} 1 & 0 \\ 0 & -1 \end{pmatrix}.$$

In this case we have (check!)

$$M_i = S_i, \qquad N_i = 0$$

and hence

$$M_+ = \begin{pmatrix} 0 & 1 \\ 0 & 0 \end{pmatrix}, \quad M_- = \begin{pmatrix} 0 & 0 \\ 1 & 0 \end{pmatrix}, \quad N_+ = N_- = 0.$$

With this in hand you can easily check that $(1, 0) \in \mathbb{C}^2$ is a highest weight vector with $j_1 = 1/2,\ j_2 = 0$. Thus the fundamental representation of $\mathfrak{sl}(2, \mathbb{C})_{\mathbb{R}}$ is just $(\frac{\mathbf{1}}{\mathbf{2}}, \mathbf{0})$.

Example 5.35. $(S^{2j}\pi, S^{2j}(\mathbb{C}^2))$ *Symmetric tensor products of left-handed spinors*

As with $\mathfrak{su}(2)$, we can build other irreps by taking symmetric tensor products. Consider the $2j$th symmetric tensor product representation $(S^{2j}\pi, S^{2j}(\mathbb{C}^2))$, $2j \in \mathbb{N}$ and the vector

$$v_{0,0} \equiv \underbrace{e_1 \otimes \cdots \otimes e_1}_{2j \text{ times}} \in S^{2j}(\mathbb{C}^2).$$

Using (5.25) it's straightforward to check, as you did in Example 5.30, that this is a highest weight vector with eigenvalue $(j, 0)$, and so we conclude that $S^{2j}(\mathbb{C}^2)$ contains an invariant subspace equivalent to $(\mathbf{j}, \mathbf{0})$. Noting that

$$\dim S^{2j}(\mathbb{C}^2) = \dim(\mathbf{j}, \mathbf{0}) = 2j + 1$$

we conclude that $(S^{2j}\pi, S^{2j}(\mathbb{C}^2)) \simeq (\mathbf{j}, \mathbf{0})$.

Example 5.36. $(\bar{\pi}, \mathbb{C}^2)$ *The right-handed spinor representation*

Consider the right-handed spinor representation $(\bar{\Pi}, \mathbb{C}^2)$ from Example 5.6. A quick calculation (do it!) reveals that the induced Lie algebra representation $(\bar{\pi}, \mathbb{C}^2)$ is given by

$$\bar{\pi}(X) = -X^\dagger, \qquad X \in \mathfrak{sl}(2, \mathbb{C})_{\mathbb{R}}.$$

In particular, then, we have $\bar{\pi}(S_i) = S_i$ since the S_i are anti-Hermitian, as well as

$$\bar{\pi}(\tilde{K}_i) = \bar{\pi}(iS_i) = -i\bar{\pi}(S_i) = -iS_i.$$

We then have

$$M_i = 0, \qquad N_i = S_i$$

and hence

$$N_+ = \begin{pmatrix} 0 & 1 \\ 0 & 0 \end{pmatrix}, \qquad N_- = \begin{pmatrix} 0 & 0 \\ 1 & 0 \end{pmatrix}, \qquad M_+ = M_- = 0.$$

You can again check that $(1,0) \in \mathbb{C}^2$ is a highest weight vector, but this time with $j_1 = 0,\ j_2 = 1/2$, and so the right-handed spinor representation of $\mathfrak{sl}(2,\mathbb{C})_{\mathbb{R}}$ is just $(\mathbf{0}, \frac{\mathbf{1}}{\mathbf{2}})$.

Example 5.37. $(\bar{\pi}_{2j}, S^{2j}(\mathbb{C}^2))$ *Symmetric tensor products of right-handed spinors*

As before, we can build other irreps by taking symmetric tensor products. Again, consider the $2j$th symmetric tensor product representation $(S^{2j}\bar{\pi}, S^{2j}(\mathbb{C}^2))$, $2j \in \mathbb{N}$ and the vector

$$v_{0,0} \equiv \underbrace{e_1 \otimes \cdots \otimes e_1}_{2j \text{ times}} \in S^{2j}(\mathbb{C}^2).$$

Again, it's straightforward to check that this is a highest weight vector with eigenvalues $(0, j)$, and we can conclude as before that $(S^{2j}\bar{\pi}, S^{2j}(\mathbb{C}^2))$ is equivalent to $(\mathbf{0}, \mathbf{j})$. □

So far we have used symmetric tensor products of the left-handed and right-handed spinor representations to build the $(\mathbf{j}, \mathbf{0})$ and $(\mathbf{0}, \mathbf{j})$ irreps. From here, getting the general irrep $(\mathbf{j_1}, \mathbf{j_2})$ is easy; we just take the tensor product of $(\mathbf{j_1}, \mathbf{0})$ and $(\mathbf{0}, \mathbf{j_2})$! To see this, let $v_{0,0} \in (\mathbf{j}, \mathbf{0})$ and $\bar{v}_{0,0} \in (\mathbf{0}, \mathbf{k})$ be highest weight vectors. Then it's straightforward to check that

$$v_{0,0} \otimes \bar{v}_{0,0} \in (\mathbf{j}, \mathbf{0}) \otimes (\mathbf{0}, \mathbf{k})$$

is a highest weight vector with $j_1 = j,\ j_2 = k$, and so we conclude that $(\mathbf{j}, \mathbf{0}) \otimes (\mathbf{0}, \mathbf{k})$ contains an invariant subspace equivalent to $(\mathbf{j}, \mathbf{k})$. However, since

$$\dim[(\mathbf{j}, \mathbf{0}) \otimes (\mathbf{0}, \mathbf{k})] = \dim(\mathbf{j}, \mathbf{k}) = (2j+1)(2k+1)$$

we conclude that these representations are equivalent, and so in general (switching notation a little),

$$\boxed{(\mathbf{j_1}, \mathbf{j_2}) \simeq (\mathbf{j_1}, \mathbf{0}) \otimes (\mathbf{0}, \mathbf{j_2}).}$$

Thus, **all the irreps of $\mathfrak{sl}(2,\mathbb{C})_{\mathbb{R}}$ can be built out of the left-handed spinor (fundamental) representation, the right-handed spinor representation, and various tensor products of the two.**

As before, this classification of the complex finite-dimensional irreps of $\mathfrak{sl}(2,\mathbb{C})_{\mathbb{R}}$ also yields, with minimal effort, the classification of the complex finite-dimensional irreps of $SL(2,\mathbb{C})$. Any complex finite-dimensional irrep of $SL(2,\mathbb{C})$ yields a complex finite-dimensional irrep of $\mathfrak{sl}(2,\mathbb{C})_{\mathbb{R}}$, which must be equivalent to $(\mathbf{j_1},\mathbf{j_2})$ for some j_1, j_2. Since

$$(\mathbf{j_1},\mathbf{j_2}) \simeq (S^{2j_1}\pi \otimes S^{2j_2}\bar{\pi}\,,\ S^{2j_1}(\mathbb{C}^2) \otimes S^{2j_2}(\mathbb{C}^2))$$

we then conclude that our original $SL(2,\mathbb{C})$ irrep is equivalent to $(S^{2j_1}\Pi \otimes S^{2j_2}\bar{\Pi}\,,\ S^{2j_1}(\mathbb{C}^2) \otimes S^{2j_2}(\mathbb{C}^2))$ for some j_1, j_2. Thus, the representations

$$(S^{2j_1}\Pi \otimes S^{2j_2}\bar{\Pi}\,,\ S^{2j_1}(\mathbb{C}^2) \otimes S^{2j_2}(\mathbb{C}^2)),\quad 2j_1, 2j_2 \in \mathbb{N}$$

are (up to equivalence) all the complex finite-dimensional irreducible representations of $SL(2,\mathbb{C})$.

How about representations of $SO(3,1)_o$? We saw that in the case of $SO(3)$, not all representations of the associated Lie algebra actually arise from representations of the group, and the same is true here. Say we have a complex finite-dimensional irrep (Π, V) of $SO(3,1)_o$, and consider its induced Lie algebra representation (π, V), which must be equivalent to $(\mathbf{j_1},\mathbf{j_2})$ for some j_1, j_2. Noting that

$$iM_z + iN_z = i\pi(\tilde{L}_z)$$

we have (again, be sure to distinguish π the number from π the representation!)

$$e^{i2\pi\cdot(iM_z+iN_z)}\, v_{0,0} = e^{i2\pi(j_1+j_2)}\, v_{0,0}$$

as well as

$$\begin{aligned} e^{i2\pi\cdot(iM_z+iN_z)}\, v_{0,0} &= e^{-2\pi\cdot\pi(\tilde{L}_z)}\, v_{0,0} \\ &= \Pi(e^{-2\pi\tilde{L}_z})\, v_{0,0} \\ &= \Pi(I)\, v_{0,0} \\ &= v_{0,0} \end{aligned}$$

so we conclude that

$$e^{i2\pi(j_1+j_2)} = 1 \iff j_1 + j_2 \in \mathbb{N}, \tag{5.75}$$

and thus only representations $(\mathbf{j_1},\mathbf{j_2})$ satisfying this condition can arise from $SO(3,1)_o$ representations. (It's also true that for any j_1, j_2 satisfying this condition, there exists an $SO(3,1)_o$ representation with induced Lie algebra representation $(\mathbf{j_1},\mathbf{j_2})$, though we won't prove that here.)

Example 5.38. $\mathbb{R}^4$ *The four-vector representation of* $SO(3,1)_o$

The fundamental representation $(\Pi, \mathbb{R}^4)$ is the most familiar $SO(3,1)_o$ representation, corresponding to four-dimensional vectors in Minkowski space. What $(\mathbf{j_1},\mathbf{j_2})$ does it correspond to? To find out, we first complexify the representation to $(\Pi_{\mathbb{C}}, \mathbb{C}^4)$ and then consider the induced $\mathfrak{sl}(2,\mathbb{C})_{\mathbb{R}}$ representation $(\pi_{\mathbb{C}}, \mathbb{C}^4)$. Straightforward calculations show that

$$iM_z = \frac{1}{2}\begin{pmatrix} 0 & -i & 0 & 0 \\ i & 0 & 0 & 0 \\ 0 & 0 & 0 & 1 \\ 0 & 0 & 1 & 0 \end{pmatrix}, \quad iN_z = \frac{1}{2}\begin{pmatrix} 0 & -i & 0 & 0 \\ i & 0 & 0 & 0 \\ 0 & 0 & 0 & -1 \\ 0 & 0 & -1 & 0 \end{pmatrix}$$

and this, along with expressions for M_+ and N_+ that you should derive, can be used to show that

$$v_{0,0} = (e_1 + ie_2) = \begin{pmatrix} 1 \\ i \\ 0 \\ 0 \end{pmatrix}$$

is a highest weight vector with $(j_1, j_2) = (1/2, 1/2)$. Noting that

$$\dim \mathbb{C}^4 = \dim \left(\mathbf{\frac{1}{2}}, \mathbf{\frac{1}{2}}\right) = 4$$

we conclude that

$$(\pi_{\mathbb{C}}, \mathbb{C}^4) \simeq \left(\mathbf{\frac{1}{2}}, \mathbf{\frac{1}{2}}\right).$$

Note that $j_1 + j_2 = 1/2 + 1/2 = 1 \in \mathbb{N}$, in accordance with (5.75). □

Before moving on to our next example, we need to discuss tensor products of $\mathfrak{sl}(2,\mathbb{C})_{\mathbb{R}}$ irreps. The nice thing here is that we can use what we know about the tensor product of $\mathfrak{su}(2)$ irreps to compute the decomposition of the tensor product of $\mathfrak{sl}(2,\mathbb{C})_{\mathbb{R}}$ irreps. In fact, the following is true:

Proposition 5.9. *The decomposition into irreps of the tensor product of two* $\mathfrak{sl}(2,\mathbb{C})_{\mathbb{R}}$ *irreps* $(\mathbf{j_1},\mathbf{j_2})$ *and* $(\mathbf{k_1},\mathbf{k_2})$ *is given by*

$$(\mathbf{j_1}, \mathbf{j_2}) \otimes (\mathbf{k_1}, \mathbf{k_2}) = \bigoplus (\mathbf{l_1}, \mathbf{l_2}) \;\; \textit{where} \; |j_1 - k_1| \le l_1 \le j_1 + k_1, \quad |j_2 - k_2| \le l_2 \le j_2 + k_2 \tag{5.76}$$

and each $(\mathbf{l_1}, \mathbf{l_2})$ *consistent with the above inequalities occurs exactly once in the direct sum decomposition.*

Notice the restrictions on l_1 and l_2, which correspond to the decomposition of tensor products of $\mathfrak{su}(2)$ representations. We relegate a proof of this formula to the appendix, but it should seem plausible. We've seen that one can roughly think of an $\mathfrak{sl}(2, \mathbb{C})_{\mathbb{R}}$ representations as a "product" of two $\mathfrak{su}(2)$ representations, and so the tensor product of two $\mathfrak{sl}(2, \mathbb{C})_{\mathbb{R}}$ representations (which can both be "factored" into $\mathfrak{su}(2)$ representations) should just be given by the various "products" of $\mathfrak{su}(2)$ representations that occur when taking the tensor product of the factors. We'll apply this formula and make this concrete in the next example.

Example 5.39. $\Lambda^2\mathbb{R}^4$ *The antisymmetric tensor representation of* $SO(3,1)_o$

This is an important example since the electromagnetic field tensor $F^{\mu\nu}$ lives in this representation. To classify this representation, we first note that $\Lambda^2\mathbb{R}^4$ occurs in the $O(3,1)$-invariant decomposition

$$\mathbb{R}^4 \otimes \mathbb{R}^4 = \mathbb{R}\,\eta^{-1} \oplus \Lambda^2(\mathbb{R}^4) \oplus S^{2'}(\mathbb{R}^4), \tag{5.77}$$

where $\eta^{-1} = \eta^{\mu\nu} e_\mu \otimes e_\nu$ is the inverse of the Minkowski metric and

$$S^{2'}(\mathbb{R}^4) = \{T \in S^2(\mathbb{R}^4) \,|\, \eta_{\mu\nu} T^{\mu\nu} = 0\} \tag{5.78}$$

is the set of symmetric "traceless" 2nd rank tensors, where the trace is effected by the Minkowski metric η. Note that this is just the $O(3,1)$ analog of the $O(n)$ decomposition in (5.54). You should check that each of the subspaces in (5.77) really is $O(3,1)$ invariant. Complexifying this yields

$$\mathbb{C}^4 \otimes \mathbb{C}^4 = \mathbb{C}\eta^{-1} \oplus \Lambda^2(\mathbb{C}^4) \oplus S^{2'}(\mathbb{C}^4).$$

Now, we can also decompose $\mathbb{C}^4 \otimes \mathbb{C}^4$ using Proposition 5.9, which yields

$$\left(\mathbf{\frac{1}{2}}, \mathbf{\frac{1}{2}}\right) \otimes \left(\mathbf{\frac{1}{2}}, \mathbf{\frac{1}{2}}\right) = (\mathbf{0}, \mathbf{0}) \oplus (\mathbf{1}, \mathbf{0}) \oplus (\mathbf{0}, \mathbf{1}) \oplus (\mathbf{1}, \mathbf{1}). \tag{5.79}$$

Now, clearly $\mathbb{C}\eta^{-1}$ corresponds to $(\mathbf{0}, \mathbf{0})$ since the former is a one-dimensional representation and $(\mathbf{0}, \mathbf{0})$ is the only one-dimensional irrep in the decomposition (5.79). What about $S^{2'}(\mathbb{C}^4)$? Well, it's straightforward to check using the results of the previous example that

$$v_{0,0} = (e_1 + i e_2) \otimes (e_1 + i e_2) \in S^{2'}(\mathbb{C}^4)$$

is a highest weight vector with $(j_1, j_2) = (1, 1)$ [it's also instructive to verify that $v_{0,0}$ actually satisfies the condition in (5.78)]. Checking dimensions then tells us that $S^{2'}(\mathbb{C}^4) \simeq \mathbf{(1, 1)}$, so we conclude that

$$\Lambda^2(\mathbb{C}^4) \simeq \mathbf{(1, 0)} \oplus \mathbf{(0, 1)}.$$

This representation is decomposable, but remember that this does *not* imply that $\Lambda^2(\mathbb{R}^4)$ is decomposable! In fact, $\Lambda^2(\mathbb{R}^4)$ is irreducible. This is an unavoidable subtlety of the relationship between complex representations and real representations.[17] For an interpretation of the representations $\mathbf{(1, 0)}$ and $\mathbf{(0, 1)}$ individually, see Problem 5-13.

5.12 Irreducibility and the Representations of $O(3, 1)$ and Its Double Covers

In this section we'll examine the constraints that parity and time-reversal place on representations of $O(3, 1)$ and its double covers. In particular, we will clarify our discussion of the Dirac spinor from Example 5.26 and explain why such $\mathfrak{so}(3, 1)$ decomposable representations seem to occur so naturally.

To start, consider the adjoint representation $(\text{Ad}, \mathfrak{so}(3, 1))$ of $O(3, 1)$. It is easily checked that the parity operator acts as

$$\text{Ad}_P(\tilde{L}_i) = \tilde{L}_i, \quad \text{Ad}_P(K_i) = -K_i.$$

Now say that we have a double-cover of $O(3, 1)$, call it H (these certainly exist and are non-unique; see Sternberg [19] and the comments at the end of this section). H will have multiple components, just as $O(3, 1)$ does, and the component containing the identity will be isomorphic to $SL(2, \mathbb{C})$.[18] Since H is a double-cover, there exists a two-to-one group homomorphism $\Phi : H \to O(3, 1)$, which induces the usual Lie algebra isomorphism $\phi : \mathfrak{sl}(2, \mathbb{C})_{\mathbb{R}} \to \mathfrak{so}(3, 1)$. Now let $\tilde{P} \in H$ cover $P \in O(3, 1)$, so that $\Phi(\tilde{P}) = P$. Then from the identity

$$\phi(\text{Ad}_h(X)) = \text{Ad}_{\Phi(h)}(\phi(X)) \quad \forall\, h \in H,\ X \in \mathfrak{sl}(2, \mathbb{C})_{\mathbb{R}} \tag{5.80}$$

which you will prove below, we have

$$\phi(\text{Ad}_{\tilde{P}}(S_i)) = \text{Ad}_P(\tilde{L}_i) = \tilde{L}_i$$

---

[17] For the whole story on this relationship, see Onischik [14].

[18] This should seem plausible, but proving it rigorously would require homotopy theory and would take us too far afield. See Frankel [6] for a nice discussion of this topic.

as well as

$$\phi(\mathrm{Ad}_{\tilde{P}}(\tilde{K}_i)) = -K_i.$$

Since ϕ is an isomorphism we conclude that

$$\begin{aligned}\mathrm{Ad}_{\tilde{P}}(S_i) &= \tilde{P} S_i \tilde{P}^{-1} = S_i \\ \mathrm{Ad}_{\tilde{P}}(\tilde{K}_i) &= \tilde{P} \tilde{K}_i \tilde{P}^{-1} = -\tilde{K}_i.\end{aligned}$$

If (Π, V) is a representation of H and (π, V) the induced $\mathfrak{sl}(2,\mathbb{C})_{\mathbb{R}}$ representation, then this implies that

$$\begin{aligned}\Pi_{\tilde{P}} M_i \Pi_{\tilde{P}}^{-1} &= N_i \\ \Pi_{\tilde{P}} N_i \Pi_{\tilde{P}}^{-1} &= M_i.\end{aligned} \tag{5.81}$$

Let's examine some consequences of this. Let $W \subset V$ be an irreducible subspace of (π, V) equivalent to $(\mathbf{j_1}, \mathbf{j_2})$, spanned by our usual basis of the form

$$\mathcal{B} = \{v_{k_1,k_2} \,|\, 0 \le k_1 \le 2j_1,\ 0 \le k_2 \le 2j_2\}.$$

We then have

$$\begin{aligned}iM_z \Pi_{\tilde{P}} v_{0,0} &= i\Pi_{\tilde{P}} N_z v_{0,0} = j_2 \Pi_{\tilde{P}} v_{0,0} \\ iN_z \Pi_{\tilde{P}} v_{0,0} &= i\Pi_{\tilde{P}} M_z v_{0,0} = j_1 \Pi_{\tilde{P}} v_{0,0} \\ M_+ \Pi_{\tilde{P}} v_{0,0} &= \Pi_{\tilde{P}} N_+ v_{0,0} = 0 \\ N_+ \Pi_{\tilde{P}} v_{0,0} &= \Pi_{\tilde{P}} M_+ v_{0,0} = 0\end{aligned}$$

and thus $\Pi_{\tilde{P}} v_{0,0}$ is a highest weight vector for $(\mathbf{j_2}, \mathbf{j_1})$! We have thus proven the following proposition:

Proposition 5.10. *Let H be a double-cover of $O(3,1)$ and (Π, V) a complex representation of H with induced $\mathfrak{sl}(2,\mathbb{C})_{\mathbb{R}}$ representation (π, V). If $W \subset V$ is an irreducible subspace of (π, V) equivalent to $(\mathbf{j_1}, \mathbf{j_2})$ and $j_1 \neq j_2$, then there exists another irreducible subspace W' of (π, V) equivalent to $(\mathbf{j_2}, \mathbf{j_1})$.*

It should be clear from the above that the operator $\Pi(\tilde{P})$ corresponding to parity takes us back and forth between W and W'. This means that even though W is an invariant subspace of the Lie algebra representation (π, V), W is *not* invariant under the Lie *group* representation (Π, V), since $\Pi(\tilde{P})$ takes vectors in W to vectors in W'! If W and W' make up all of V, i.e. if $V = W \oplus W'$, this means that V is irreducible under the H representation Π but not under the $\mathfrak{so}(3,1)$ representation π, and so we have an irreducible Lie group representation whose induced Lie algebra representation is *not* irreducible! We'll meet two examples of this type of representation below. Note that this does not contradict Proposition 5.4, as the group H doesn't satisfy the required hypothesis of connectedness.

Exercise 5.43. Let $\Phi : H \to G$ be a Lie group homomorphism with induced Lie algebra homomorpishm $\phi : \mathfrak{h} \to \mathfrak{g}$. Use the definition of the adjoint mapping and of ϕ to show that

$$\phi(\mathrm{Ad}_h(X)) = \mathrm{Ad}_{\Phi(h)}(\phi(X)) \quad \forall\, h \in H,\ X \in \mathfrak{h}.$$

Exercise 5.44. Verify Eq. (5.81). You will need the result of the previous exercise!

Example 5.40. *The Dirac Spinor Revisited*

As a particular application of Proposition 5.10, suppose our representation (Π, V) of H contains a subspace equivalent to the left-handed spinor $\left(\frac{1}{2}, \mathbf{0}\right)$; then, it must also contain a subspace equivalent to the right-handed spinor $\left(\mathbf{0}, \frac{1}{2}\right)$. This is why the Dirac spinor is $\left(\frac{1}{2}, \mathbf{0}\right) \oplus \left(\mathbf{0}, \frac{1}{2}\right)$. It is not irreducible as an $SL(2,\mathbb{C})$ representation, but it is irreducible as a representation of a group H which extends $SL(2,\mathbb{C})$ and covers $O(3,1)$. □

The Dirac spinor representation is a representation of H, not $O(3,1)$, but many of the other $\mathfrak{sl}(2,\mathbb{C})_{\mathbb{R}} \simeq \mathfrak{so}(3,1)$ representations of interest do come from $O(3,1)$ representations (for instance, the fundamental representation of $\mathfrak{so}(3,1)$ and its various tensor products). Since any $O(3,1)$ representation $\Pi : O(3,1) \to GL(V)$ yields an H representation $\Pi \circ \Phi : H \to GL(V)$, Proposition 5.10 and the comments following it hold for $O(3,1)$ representations as well as representations of H. In the $O(3,1)$ case, though, we can do even better:

Proposition 5.11. *Let (Π, V) be a finite-dimensional complex irreducible representation of $O(3,1)$. Then the induced $\mathfrak{so}(3,1)$ representation (π, V) is equivalent to one of the following:*

$$(\mathbf{j},\mathbf{j}),\ 2j \in \mathbb{N} \quad \text{or} \quad (\mathbf{j_1},\mathbf{j_2}) \oplus (\mathbf{j_2},\mathbf{j_1}),\ 2j_1, 2j_2 \in \mathbb{N},\ j_1 + j_2 \in \mathbb{N},\ j_1 \neq j_2. \tag{5.82}$$

Proof. Since $\mathfrak{so}(3,1)$ is semi-simple, (π, V) is completely reducible, i.e. equivalent to a direct sum of irreducible representations. Let $W \subset V$ be one such representation, equivalent to $(\mathbf{j_1},\mathbf{j_2})$, with highest weight vector $v_{0,0}$. Then $\Pi(P)v_{0,0} \equiv v_P$ is a highest weight vector with eigenvalues (j_2, j_1), and the same arguments show that $\Pi(T)v_{0,0} \equiv v_T$ is also a highest weight vector with eigenvalues (j_2, j_1) (it is not necessarily equal to v_P, though). The same arguments also show that $\Pi(PT)v_{0,0} \equiv v_{PT}$ is a highest weight vector with eigenvalues (j_1, j_2) [recall that T is the time-reversal operator defined in (4.35)]. Now consider the vectors

$$w_{0,0} \equiv v_{0,0} + v_{PT}$$
$$u_{0,0} \equiv v_p + v_T.$$

$w_{0,0}$ is clearly a highest weight vector with eigenvalues (j_1, j_2), and $u_{0,0}$ is clearly a highest weight vector with eigenvalues (j_2, j_1). Using the fact that P and T commute, you can easily check that

$$\Pi(P)w_{0,0} = \Pi(T)w_{0,0} = u_{0,0}$$
$$\Pi(P)u_{0,0} = \Pi(T)u_{0,0} = w_{0,0}.$$

We can then define basis vectors

$$w_{k_1,k_2} \equiv (M_-)^{k_1}(N_-)^{k_2}w_{0,0}, \quad 0 \le k_1 \le 2j_1,\ 0 \le k_2 \le 2j_2$$
$$u_{k_1,k_2} \equiv (M_-)^{k_1}(N_-)^{k_2}u_{0,0}, \quad 0 \le k_1 \le 2j_2,\ 0 \le k_2 \le 2j_1$$

which span $\mathfrak{so}(3,1)$ irreducible subspaces $W \simeq (\mathbf{j_1}, \mathbf{j_2})$ and $U \simeq (\mathbf{j_2}, \mathbf{j_1})$. By Proposition 5.4 and the connectedness of $SO(3,1)_o$, W and U are also irreducible under $SO(3,1)_o$, and from (5.75) we know that $j_1 + j_2 \in \mathbb{N}$. Furthermore, by the definition of the w_{k_1,k_2} and u_{k_1,k_2}, as well as (5.81), we have

$$\Pi(P)w_{k_1,k_2} = \Pi(T)w_{k_1,k_2} = u_{k_2,k_1}$$
$$\Pi(P)u_{k_1,k_2} = \Pi(T)u_{k_1,k_2} = w_{k_2,k_1}$$

and so $\mathrm{Span}\{w_{k_1,k_2}, u_{l_1,l_2}\}$ is invariant under $O(3,1)$. We assumed V was irreducible, though, so we conclude that $V = \mathrm{Span}\{w_{k_1,k_2}, u_{l_1,l_2}\}$. Does that mean we can conclude that $V \simeq (\mathbf{j_1}, \mathbf{j_2}) \oplus (\mathbf{j_2}, \mathbf{j_1})$? Not quite, because we never established that $w_{0,0}$ and $u_{0,0}$ were linearly independent! In fact, they might be linearly *de*pendent, in which case they would be proportional, which would imply that each w_{k_1,k_2} is proportional to u_{k_1,k_2} (why?), and also that $j_1 = j_2 \equiv j$. In this case, we obtain

$$V = \mathrm{Span}\{w_{k_1,k_2}, u_{l_1,l_2}\} = \mathrm{Span}\{w_{k_1,k_2}\} \simeq (\mathbf{j}, \mathbf{j})$$

which is one of the alternatives mentioned in the proposition. If $w_{0,0}$ and $u_{0,0}$ are linearly *in*dependent, however, then so is the set $\{w_{k_1,k_2}, u_{l_1,l_2}\}$ and so

$$V = \mathrm{Span}\{w_{k_1,k_2}, u_{l_1,l_2}\} = W \oplus U \simeq (\mathbf{j_1}, \mathbf{j_2}) \oplus (\mathbf{j_2}, \mathbf{j_1})$$

which is the other alternative. All that remains is to show that $j_1 \neq j_2$, which you will do in Exercise 5.45 below. This concludes the proof. □

Exercise 5.45. Assume that $w_{0,0}$ and $u_{0,0}$ are linearly independent and that $j_1 = j_2$. Use this to construct a nontrivial $O(3,1)$-invariant subspace of V, contradicting the irreducibility of V. Thus if V is irreducible and $w_{0,0}$ and $u_{0,0}$ are linearly independent, then $j_1 \neq j_2$ as desired.

Example 5.41. $O(3, 1)$ *Representations Revisited*

In this example we just point out that all the $O(3, 1)$ representations we've met have the form (5.82). Below is a table of some of these representations, along with their complexifications and the corresponding $\mathfrak{so}(3, 1)$ representations.

Name	V	$V_{\mathbb{C}}$	$\mathfrak{so}(3, 1)$ rep
Scalar (trivial)	$\mathbb{R}$	$\mathbb{C}$	$(\mathbf{0}, \mathbf{0})$
Vector (fundamental)	$\mathbb{R}^4$	$\mathbb{C}^4$	$(\frac{\mathbf{1}}{\mathbf{2}}, \frac{\mathbf{1}}{\mathbf{2}})$
Antisymmetric tensor (adjoint)	$\Lambda^2\mathbb{R}^4$	$\Lambda^2\mathbb{C}^4$	$(\mathbf{1}, \mathbf{0}) \oplus (\mathbf{0}, \mathbf{1})$
Pseudovector	$\Lambda^3\mathbb{R}^4$	$\Lambda^3\mathbb{C}^4$	$(\frac{\mathbf{1}}{\mathbf{2}}, \frac{\mathbf{1}}{\mathbf{2}})$
Pesudoscalar	$\Lambda^4\mathbb{R}^4$	$\Lambda^4\mathbb{C}^4$	$(\mathbf{0}, \mathbf{0})$
Symmetric traceless tensor	$S^{2'}(\mathbb{R}^4)$	$S^{2'}(\mathbb{C}^4)$	$(\mathbf{1}, \mathbf{1})$

Note that the pseudovector and pseudoscalar representations yield the same $\mathfrak{so}(3, 1)$ representations as the vector and scalar, respectively, as discussed in Example 5.18. Thus we have a pair of examples in which two equivalent Lie algebra representations come from two *non*-equivalent matrix Lie group representations! Again, this doesn't contradict Proposition 5.3 since $O(3, 1)$ is not connected. Note also that the only representation in the above table that decomposes into more than one $\mathfrak{so}(3, 1)$ irrep is the antisymmetric tensor representation; see Problem 5-13 for the action of the parity operator on this representation, and how it takes one back and forth between the $\mathfrak{so}(3, 1)$-irreducible subspaces $(\mathbf{1}, \mathbf{0})$ and $(\mathbf{0}, \mathbf{1})$. □

Before concluding this chapter we should talk a little bit about this mysterious group H which is supposed to be a double-cover of $O(3, 1)$. It can be shown[19] that there are exactly eight non-isomorphic double covers of $O(3, 1)$ (in contrast to the case of $SO(3)$ and $SO(3, 1)_o$ which have the unique connected double covers $SU(2)$ and $SL(2, \mathbb{C})$). Most of these double covers are somewhat obscure and don't really crop up in the physics literature, but two of them are quite natural and well studied: these are the ***Pin*** groups Pin$(3, 1)$ and Pin$(1, 3)$ which appear in the study of ***Clifford Algebras***. Clifford algebras are rich and beautiful objects, and lead naturally to double covers of all the orthogonal and Lorentz groups. In the four-dimensional Lorentzian case in particular, one encounters the Dirac gamma matrices and the Dirac spinor, as well as the Pin groups which act naturally on the Dirac spinor. For details on the construction of the Pin groups and their properties, see Göckeler and Schücker [9]. We'll have a little more to say about Dirac gamma matrices in Sect. 6.3.

[19] See Sternberg [19].

Chapter 5 Problems

Note: Problems marked with an "*" tend to be longer, and/or more difficult, and/or more geared towards the completion of proofs and the tying up of loose ends. Though these problems are still worthwhile, they can be skipped on a first reading.

5-1. Generalize the results of Exercise 5.5 and Example 5.8 by redoing the calculation for arbitrary G, Π, V, and $C(V)$. That is, let (Π, V) be a finite-dimensional representation of G, and let $\tilde{\Pi}$ be the induced representation on some function space $C(V)$, which will further induce a Lie algebra representation $\tilde{\pi}$ on $C(V)$. Choose a basis $\mathcal{B} = \{e_i\}_{i=1\ldots n}$ for V and take the corresponding vector components v^i as coordinates on V. Show that with this basis and coordinates and for any $X \in \mathfrak{g}$, $\tilde{\pi}$ takes the form of the differential operator

$$\tilde{\pi}(X) = -\sum_{i,j} [\pi(X)]_{ij}\, v^j \frac{\partial}{\partial v^i}.$$

By specializing appropriately, reproduce (5.12) and (5.13).

5-2. (*) In this problem we'll develop a coordinate-based proof of our claim from Example 5.22.

(a) Let $X = X^{ij} e_i \otimes e_j \in \Lambda^2\mathbb{R}^n$, and define

$$\tilde{X} \equiv X^{ij} L(e_i) \otimes e_j = X^{ij} g_{ik} e^k \otimes e_j \in \mathcal{L}(\mathbb{R}^n)$$
$$\phi(X) \equiv [\tilde{X}] \in M_n(\mathbb{R}).$$

Find an expression for $\phi(X)$ in terms of $[X]$ and use it to show that $\phi(X) \in \mathfrak{g}$.

(b) Prove that $\mathrm{Ad}(R) \circ \phi = \phi \circ \Lambda^2\Pi(R)$ by evaluating both sides on an arbitrary $X \in \Lambda^2\mathbb{R}^n$ and showing that the components of the matrices are equal. You'll need the expansion of X given above, as well as the coordinate form (5.33) of $\Lambda^2\Pi$. For simplicity of matrix computation, you may wish to abandon the Einstein Summation Convention here and write the components of $R \in G$ as R_{ij}, even though you're interpreting R as a linear operator.

5-3. In this problem we'll develop a coordinate-free proof that the fundamental representation of $SL(2,\mathbb{C})$ is equivalent to its dual. This will also imply that the fundamental representation of $SU(2)$ is equivalent to its dual as well.

(a) Consider the epsilon tensor in $\mathcal{T}^2_0(\mathbb{C}^2)$,

$$\epsilon \equiv e^1 \wedge e^2.$$

Using the definition (3.72) of the determinant, show that ϵ is $SL(2,\mathbb{C})$ invariant, i.e. that

$$\epsilon(Av, Aw) = \epsilon(v, w) \quad \forall\, v, w \in \mathbb{C}^2,\ A \in SL(2,\mathbb{C}).$$

(b) Define a map

$$L_\epsilon : \mathbb{C}^2 \to \mathbb{C}^{2*}$$
$$v \mapsto \epsilon(v, \cdot)\,.$$

Using your result from a), show that L_ϵ is an intertwiner, and also show that L_ϵ is one-to-one. Conclude that $\mathbb{C}^2 \simeq \mathbb{C}^{2*}$ as $SL(2,\mathbb{C})$ irreps, and hence as $SU(2)$ irreps as well.

(c) Given your result from a), how would you now interpret the matrix A from (5.47)? Compute $[L_\epsilon]$ in the standard basis and dual basis and show that $[L_\epsilon] = A$.

(d) Assuming the very plausible but slightly annoying to prove fact that $(S^{2l}(\mathbb{C}^2))^* \simeq S^{2l}(\mathbb{C}^{2*})$, prove that *every* $SU(2)$ representation is equivalent to its dual. One might think that this is self-evident since taking the dual of an irrep doesn't change its dimension and for a given dimension there is only one $SU(2)$ irrep (up to equivalence), but this assumes that the dual of an irrep is itself an irrep. This is true, but needs to be proven, and is the subject of the next problem.

5-4. Show that if (Π, V) is a finite-dimensional irreducible group or Lie algebra representation, then so is (Π^*, V^*). Note that this together with Problem 2-5 also implies the converse, namely that if (Π^*, V^*) is an irrep then so is (Π, V). Use this to prove that every $\mathfrak{su}(2)$ irrep is equivalent to its dual.

5-5. If G is a matrix Lie group and (Π, V) its fundamental representation, one can sometimes generate new representations by considering the ***conjugate*** representation

$$\bar{\Pi} : G \to GL(V)$$
$$A \mapsto \bar{A},$$

where $\bar{A}$ denotes the matrix whose entries are just the complex conjugates of the entries of A.

(a) Verify that $\bar{\Pi}$ is a homomorphism, hence a bona fide representation. Show that for $G = SU(n)$, the conjugate representation is equivalent to the dual representation. What does this mean in the case of $SU(2)$?

(b) Let $G = SL(2,\mathbb{C})$. Show that the conjugate to $\left(\frac{1}{2}, \mathbf{0}\right)$ is $\left(\mathbf{0}, \frac{1}{2}\right)$. There are a few ways to do this. This justifies the notation $\bar{\Pi}$ for the representation homomorphism of $\left(\mathbf{0}, \frac{1}{2}\right)$.

5-6. (*) In this problem we'll show that any complex matrix of the block diagonal form $\begin{pmatrix} A & 0 \\ 0 & \bar{A} \end{pmatrix}$ is related to a purely *real* matrix by a similarity transformation. This then implies the existence of the Majorana spinor, as mentioned in Example 5.26.

(a) Consider $\mathbb{C}^n$ along with its standard basis $\mathcal{B} = \{e_i\}_{i=1\dots n}$ and linear operators $M_n(\mathbb{C})$. Let us decompose $v \in \mathbb{C}^n$ as well as $D \in M_n(\mathbb{C})$ into their real and imaginary parts, i.e. $v = x + iy$ and $D = E + iF$ where $x, y \in \mathbb{R}^n$ and $E, F \in M_n(\mathbb{R})$. If we now consider V as a $2n$-dimensional real vector space $V_{\mathbb{R}}$ with basis $\mathcal{B}_{\mathbb{R}} \equiv \{e_1, e_2, \dots, e_n, ie_1, ie_2, \dots, ie_n\}$, show that D is given (in $n \times n$ block diagonal form) by

$$[D]_{\mathcal{B}_{\mathbb{R}}} = \begin{pmatrix} E & -F \\ F & E \end{pmatrix}. \tag{5.83}$$

(b) Rewrite our basis vectors as $\mathcal{B}_{\mathbb{R}} = \{f_i = e_i,\ f_{n+i} = ie_i\}_{i=1\dots n}$. Now take a deep breath and *complexify* this space, to get a new complex vector space $(V_{\mathbb{R}})_{\mathbb{C}}$ with complex dimension $2n$ (twice that of our original space V). Now change bases in $(V_{\mathbb{R}})_{\mathbb{C}}$ from $\mathcal{B}_{\mathbb{R}} = \{f_i\}_{i=1\text{–}2n}$ to $\mathcal{B}_{\pm} \equiv \{f_i + if_{n+i}, f_i - if_{n+i}\}_{i=1\text{–}n}$. By making the appropriate (complex) similarity transformation, show that in this basis

$$[D]_{\mathcal{B}_{\pm}} = \begin{pmatrix} E + iF & 0 \\ 0 & E - iF \end{pmatrix}. \tag{5.84}$$

This implies that any complex matrix of the form (5.84) is related to a real matrix by a similarity transformation.

(c) If you are curious about the explicit form of the Majorana representation, compute the Dirac representation of $\mathfrak{sl}(2,\mathbb{C})_{\mathbb{R}}$ from that of $SL(2,\mathbb{C})$. Then use the similarity transformation from Example 5.21 to obtain matrices of the form (5.84) for all your $\mathfrak{sl}(2,\mathbb{C})_{\mathbb{R}}$ generators. Rewriting these as real matrices as per (5.83) gives the Majorana representation of $\mathfrak{sl}(2,\mathbb{C})_{\mathbb{R}}$.

5-7. Let (π, V) be a representation of a Lie algebra $\mathfrak{g}$ and assume that $\pi : \mathfrak{g} \to \mathfrak{gl}(V)$ is one-to-one (such Lie algebra representations are said to be ***faithful***). Show that there is an invariant subspace of $(\pi^1_1, V \otimes V^*)$ that is equivalent to the adjoint representation $(\mathrm{ad}, \mathfrak{g})$.

5-8. In this problem we'll meet a representation that is *not* completely reducible.

(a) Consider the representation $(\Pi, \mathbb{R}^2)$ of $\mathbb{R}$ given by

$$\Pi : \mathbb{R} \to GL(2,\mathbb{R})$$

$$a \mapsto \Pi(a) = \begin{pmatrix} 1 & a \\ 0 & 1 \end{pmatrix}.$$

Verify that $(\Pi, \mathbb{R}^2)$ is a representation.

(b) If Π was completely reducible then we'd be able to decompose $\mathbb{R}^2$ as $\mathbb{R}^2 = V \oplus W$ where V and W are one-dimensional. Show that such a decomposition is impossible.

5-9. In this problem we'll show that $\mathbb{Z}_2$ is semi-simple, i.e. that every finite-dimensional representation of $\mathbb{Z}_2$ is completely reducible. Our strategy will be to construct a $\mathbb{Z}_2$-invariant inner product on our vector space, and then show that for any invariant subspace W, its orthogonal complement $W^\perp$ is also invariant. We can then iterate this procedure to obtain a complete decomposition.

(a) Let (Π, V) be a representation of $\mathbb{Z}_2$. We first construct a $\mathbb{Z}_2$-invariant inner product. To do this, we start with an *arbitrary* inner product $(\cdot|\cdot)_0$ on V (which could be defined, for instance, as one for which some arbitrary set of basis vectors is orthonormal). We then define a "group averaged" inner product as

$$(v|w) \equiv \sum_{h \in \mathbb{Z}_2} (\Pi_h v | \Pi_h w)_0 .$$

Show that $(\cdot\,|\,\cdot)$ is $\mathbb{Z}_2$-invariant, i.e. that

$$(\Pi_g v | \Pi_g w) = (v|w) \quad \forall\, v, w \in V,\; g \in \mathbb{Z}_2.$$

(b) Assume now that there exists a nontrivial invariant subspace $W \subset V$ (if no such W existed, then V would be irreducible, hence completely reducible and we would be done). Define its ***orthogonal complement***

$$W^\perp \equiv \{v \in V \mid (v|w) = 0 \;\forall\, w \in W\}.$$

Argue that there exists an orthonormal basis

$$\mathcal{B} = \{w_1, \ldots, w_k, v_{k+1}, \ldots, v_n\} \quad \text{where } w_i \in W,\; i = 1, \ldots, k,$$

and conclude that $V = W \oplus W^\perp$.

(c) Show that $W^\perp$ is an invariant subspace, so that $V = W \oplus W^\perp$ is a decomposition into invariant subspaces. We still don't know that W is irreducible, though. Argue that we can nonetheless iterate our above argument until we obtain a decomposition into irreducibles, thus proving that (Π, V) is completely reducible.

5-10. (*) In this problem we'll sketch a proof that the space of harmonic polynomials $\mathcal{H}_l(\mathbb{R}^3)$ has dimension $2l + 1$.

(a) Recall our notation $P_k(\mathbb{R}^3)$ for the space of kth degree polynomials on $\mathbb{R}^3$. Assume that the map

$$\Delta : P_k(\mathbb{R}^3) \longrightarrow P_{k-2}(\mathbb{R}^3) \tag{5.85}$$

is onto. Use the rank-nullity theorem to show that $\dim \mathcal{H}_l(\mathbb{R}^3) = 2l+1$. If you need help you might consult Example 3.24, as well as (3.89).

(b) To complete the proof we need to show that (5.85) is onto. I have not seen a clean or particularly enlightening proof of this fact, so I don't heartily recommend this part of the problem, but if you really want to show it you might try showing (inductively on k) that the two-dimensional Laplacian

$$\Delta : P_k(\mathbb{R}^2) \longrightarrow P_{k-2}(\mathbb{R}^2)$$

is onto, and then use this to show (inductively on k) that (5.85) is onto as well.

5-11. Let T and U be *commuting* linear operators on a complex vector space V, so that $[T, U] = 0$. We will show that T and U have a simultaneous eigenvector.
Using the standard argument we employed in the proof of Proposition 5.5, show that T has at least one eigenvector. Denote that vector by v_a and its eigenvalue by a. Let V_a denote the span of *all* eigenvectors of T with eigenvalue a; V_a may just be the one-dimensional subspace spanned by v_a, or it may be bigger if there are other eigenvectors that also have eigenvalue a. Use the fact that U commutes with T to show that V_a is invariant under U (i.e., $U(v) \in V_a$ whenever $v \in V_a$). We can then restrict U to V_a to get a linear operator

$$U|_{V_a} \equiv U_a \in \mathcal{L}(V_a).$$

Then use the standard argument again to show that U_a has an eigenvector $v_0 \in V_a$. Show that this is a simultaneous eigenvector of T and U.

5-12. In this problem we'd like to show that the tensor product "distributes" over direct sums, in the sense that for any vector spaces V, W, and Z, there exists a vector space isomorphism

$$\phi : (V \oplus W) \otimes Z \to (V \otimes Z) \oplus (W \otimes Z).$$

Define such a map on decomposable elements by

$$\phi : (v, w) \otimes z \mapsto (v \otimes z, w \otimes z)$$

and extend linearly. Show that ϕ is linear (you'll have to be careful about how addition works in the direct sum spaces and the tensor product spaces) as well as one-to-one and onto, so that it is a vector space isomorphism.

5-13. (*) In this problem we'll study the $O(3,1)$ representation $(\Lambda^2\Pi, \Lambda^2\mathbb{C}^4)$ and its decomposition into $(\mathbf{1},\mathbf{0}) \oplus (\mathbf{0},\mathbf{1})$ under $SO(3,1)_o$ and $\mathfrak{so}(3,1)$.

(a) Define the ***star*** operator $*$ on $\Lambda^2\mathbb{C}^4$ by

$$\begin{aligned} * : \Lambda^2\mathbb{C}^4 &\to \Lambda^2\mathbb{C}^4 \\ e_i \wedge e_j &\mapsto \frac{1}{2}\epsilon_{ijkl}\eta^{km}\eta^{ln} e_m \wedge e_n \end{aligned}$$

and extending linearly. Here η^{ln} are the components of the inverse of the Minkowski metric and ϵ_{ijkl} are the components of the usual epsilon tensor on $\mathbb{R}^4$. Show that $*$ is an intertwiner between $\Lambda^2\mathbb{C}^4$ and itself when viewed as an $SO(3,1)_o$ representation, but not as an $O(3,1)$ representation. You will need (4.19) as well as your results from part (a) of Problem 3-4.

(b) Compute the action of $*$ on the basis vectors f_i, $i = 1, \ldots, 6$ defined in Example 5.24. Use the f_i to construct eigenvectors of $*$, and show that $*$ is diagonalizable with eigenvalues $\pm i$, so that $\Lambda^2\mathbb{C}^4$ decomposes into $V_{+i} \oplus V_{-i}$ where V_{+i} is the eigenspace of $*$ in which every vector is an eigenvector with eigenvalue $+i$ and likewise for V_{-i}. Then use Schur's lemma to conclude that any $SO(3,1)_o$-irreducible subspace must lie entirely in V_{+i} or V_{-i}. In particular, this means that $\Lambda^2\mathbb{C}^4$ is not irreducible as an $SO(3,1)_o$ or $\mathfrak{so}(3,1)$ representation.

(c) A convenient basis for V_{+i} which you may have discovered above is

$$\begin{aligned} v_{+1} &\equiv f_1 + i f_4 \\ v_{+2} &\equiv f_2 + i f_5 \\ v_{+3} &\equiv f_3 + i f_6. \end{aligned}$$

Likewise, for V_{-i} we have the basis

$$\begin{aligned} v_{-1} &\equiv f_1 - i f_4 \\ v_{-2} &\equiv f_2 - i f_5 \\ v_{-3} &\equiv f_3 - i f_6. \end{aligned}$$

Now consider the vectors

$$v_{0,0} \equiv v_{+1} + i v_{+2}$$
$$w_{0,0} \equiv v_{-1} + i v_{-2}.$$

Show that these are highest weight vectors for $(\mathbf{0}, \mathbf{1})$ and $(\mathbf{1}, \mathbf{0})$ respectively. Then count dimensions to show that as $\mathfrak{so}(3,1)$ representations, $V_{+i} \simeq (\mathbf{0}, \mathbf{1})$ and $V_{-i} \simeq (\mathbf{1}, \mathbf{0})$.

(d) Show directly that $\Lambda^2\Pi(P)(v_{+j}) = v_{-j}$, $j = 1, 2, 3$ and likewise for T, so that P and T interchange V_{+i} and V_{-i}, as expected.

5-14. Let $F_{\mu\nu}$ be the electromagnetic field tensor from Example 3.16. Restrict the $*$ operator from the previous problem to $\Lambda^2\mathbb{R}^4$ and apply it to F to obtain another antisymmetric $(2, 0)$ tensor, known as ***dual field tensor***. This may be familiar from relativistic treatments of electromagnetism, such as found in Griffiths [10].

Chapter 6
The Representation Operator and Its Applications

In this chapter we'll introduce the somewhat abstract notion of a *representation operator*, which is absent from much of the physics literature but allows for a unified treatment of several topics of interest in quantum mechanics, including tensor operators, spherical tensors, quantum-mechanical selection rules, and the Wigner–Eckart theorem. As usual, we begin with a heuristic introduction in which we'll try to dispel some of the widespread confusion about spherical tensors and the Wigner–Eckart theorem, as well as motivate the definition of a representation operator.

6.1 Invitation: Tensor Operators, Spherical Tensors, and Wigner–Eckart

Consider an $SO(3)$ representation $(\Pi, \mathcal{H})$, so that $\Pi : SO(3) \to GL(\mathcal{H})$. We will assume that $\mathcal{H}$ is a complex Hilbert space, since we have quantum-mechanical applications in mind. The representation Π on $\mathcal{H}$ induces a tensor product representation $(\Pi^1_1, \mathcal{L}(\mathcal{H}))$ as per (5.29):

$$\Pi^1_1(R)\,T = \Pi(R)\,T\,\Pi(R)^{-1} \qquad \forall\, R \in SO(3),\ T \in \mathcal{L}(\mathcal{H}).$$

We emphasize that for $(\Pi^1_1, \mathcal{L}(\mathcal{H}))$, the vector space being acted upon is the space of *operators* on $\mathcal{H}$, and that $R \in SO(3)$ acts on these operators via similarity transformations by the operators $\Pi(R)$.

Now, as an $SO(3)$ representation, $\mathcal{L}(\mathcal{H})$ typically contains nontrivial invariant subspaces. This is not as unfamiliar as it may sound, as the next example will show.

N. Jeevanjee, *An Introduction to Tensors and Group Theory for Physicists*,
DOI 10.1007/978-3-319-14794-9_6

Example 6.1. *Vector operators as invariant subspaces of $\mathcal{L}(L^2(\mathbb{R}^3))$*

Let $\mathcal{H} = L^2(\mathbb{R}^3)$. What are some $SO(3)$-invariant subspaces of $\mathcal{L}(\mathcal{H})$? If we consider the subspace

$$\text{Span}\{\hat{p}_x, \hat{p}_y, \hat{p}_z\} \subset \mathcal{L}(\mathcal{H})$$

then this is clearly an invariant subspace of $\mathcal{L}(\mathcal{H})$, since

$$\Pi^1_1(R)\hat{p}_j = \Pi(R)\hat{p}_j\Pi(R)^{-1} = \sum_j R_{ij}\,\hat{p}_i \tag{6.1}$$

as you will show in Exercise 6.1. The same is true for $\text{Span}\{\hat{x}, \hat{y}, \hat{z}\}$. Furthermore, it should be clear that as (complex) $SO(3)$ representations,

$$\text{Span}\{\hat{p}_x, \hat{p}_y, \hat{p}_z\} \simeq \text{Span}\{\hat{x}, \hat{y}, \hat{z}\} \simeq \mathbb{C}^3.$$

Since the sets of operators $\{\hat{p}_x, \hat{p}_y, \hat{p}_z\}$ and $\{\hat{x}, \hat{y}, \hat{z}\}$ each span a space equivalent to the (complex) vector representation of $SO(3)$, these sets are known as ***vector operators*** (see Exercise 6.2 for the equivalence of this definition with that from Sect. 3.7). Note that by (6.1), the elements of the vector operators transform under Π^1_1 like the *basis vectors* e_i of $\mathbb{R}^3$.

Exercise 6.1. Prove the second equality in (6.1). Do this by letting both sides act on a $\hat{\mathbf{p}}$ eigenket $|\mathbf{p}\rangle$, where $\Pi(R)\,|\mathbf{p}\rangle = |R\mathbf{p}\rangle$.

Exercise 6.2. Prove that this definition of a vector operator is equivalent to the old one. Do this by letting $R = e^{tL_i}$ in (6.1), differentiating at $t = 0$, and identifying $i\pi(L_i)$ with what we've called the total angular momentum operator J_i. This should yield the definition (3.57).

Example 6.2. *2nd rank tensor operators as invariant subspaces of $\mathcal{L}(L^2(\mathbb{R}^3))$*

Vector operators aren't the only kind of invariant subspaces lurking around in $\mathcal{L}(\mathcal{H})$. Consider also the subspace

$$\text{Span}\{\hat{x}_i\,\hat{p}_j\} \subset \mathcal{L}(\mathcal{H})$$

(in what follows Roman indices run from 1 to 3 unless otherwise noted). These operators transform under $R \in SO(3)$ as

$$\begin{aligned}
\Pi^1_1(R)(\hat{x}_i\,\hat{p}_j) &= \Pi(R)\hat{x}_i\,\hat{p}_j\,\Pi(R)^{-1} \\
&= (\,\Pi(R)\,\hat{x}_i\,\Pi(R)^{-1})(\,\Pi(R)\,\hat{p}_j\,\Pi(R)^{-1}) \\
&= \sum_{k,l} R_{ki}\,\hat{x}_k\,R_{lj}\,\hat{p}_l \\
&= \sum_{k,l} R_{ki}\,R_{lj}\,\hat{x}_k\,\hat{p}_l,
\end{aligned}$$

which is of course just the transformation law for a 2nd rank tensor! Thus, the set $\{\hat{x}_i \hat{p}_j\}$ is known as a ***2nd rank tensor operator***. Again, the elements $\hat{x}_i \hat{p}_j$ of the tensor operator transform like the basis vectors $e_i \otimes e_j$ of the (complex) 2nd rank tensor representation $\mathbb{C}^3 \otimes \mathbb{C}^3$. □

We can generalize the previous two examples with the following definition: A ***(r,s) tensor operator*** on a complex $SO(3)$ representation $(\Pi, \mathcal{H})$ is a *set* of 3^{r+s} operators $\{T^{i_1,\cdots,i_r}{}_{j_1,\cdots,j_2}\} \subset \mathcal{L}(\mathcal{H})$ that transform under Π^1_1 like the basis vectors $\{e^{i_1} \otimes \cdots \otimes e^{i_r} \otimes e_{j_1} \otimes \cdots \otimes e_{j_s}\}$ of $\mathcal{T}^r_s(\mathbb{C}^3)$. Put another way,

$$\text{Span}\left\{T^{i_1,\cdots,i_r}{}_{j_1,\cdots,j_2}\right\} \simeq \mathcal{T}^r_s(\mathbb{C}^3)$$

as $SO(3)$ representations.

Now, recall that $\mathcal{T}^r_s(\mathbb{C}^3)$ is typically decomposable, and so the span of a tensor operator should be decomposable also. We illustrate this in the next example.

Example 6.3. *Decomposition of tensor operators into spherical tensors*

Consider again the tensor operator $\{\hat{x}_i \hat{p}_j\}$. We saw that the span of these operators forms an $SO(3)$-invariant subspace equivalent to $\mathcal{T}^0_2(\mathbb{C}^3)$, so by Example 5.33 we must have a decomposition into irreducibles as

$$\text{Span}\{\hat{x}_i \hat{p}_j\} \simeq V_0 \oplus V_1 \oplus V_2. \tag{6.2}$$

Here V_0, V_1, and V_2 should correspond to $\mathbb{C}g$, $\Lambda^2(\mathbb{C}^3)$, and $S^{2'}(\mathbb{C}^3)$ as per (5.54). But what does this decomposition look like, explicitly?

Since $g = \sum_i e_i \otimes e_i$, it's clear that V_0 should be given by

$$V_0 = \text{Span}\left\{\sum_i \hat{x}_i \hat{p}_i\right\} = \text{Span}\{\underbrace{\hat{\mathbf{x}} \cdot \hat{\mathbf{p}}}_{T^{(0)}_0}\}. \tag{6.3}$$

The appearance of the dot product here shouldn't be a surprise, as we know V_0 must be spanned by a scalar. The $T^{(0)}_0$ will be explained momentarily.

As for the vector representation V_1, we know it should correspond to $\Lambda^2(\mathbb{C}^3) \subset \mathcal{T}^0_2(\mathbb{C}^3)$. This means we're looking for some antisymmetric combinations of the $\hat{x}_i$ and $\hat{p}_j$ which transform like a vector; these, of course, are nothing but the angular momentum operators $\hat{L}_i = \sum_{j,k} \epsilon_{ijk} \hat{x}_j \hat{p}_k$! Thus

$$V_1 = \text{Span}\{\hat{L}_i\} = \text{Span}\left\{\underbrace{\frac{1}{\sqrt{2}}(\hat{L}_x + i\hat{L}_y)}_{T^{(1)}_1},\ \underbrace{\hat{L}_z}_{T^{(1)}_0},\ \underbrace{-\frac{1}{\sqrt{2}}(\hat{L}_x - i\hat{L}_y)}_{T^{(1)}_{-1}}\right\}. \tag{6.4}$$

In the last expression we switched to (normalized) J_z eigenvectors satisfying (5.65), and again the $T_j^{(1)}$ will be explained momentarily.

For V_2, the simplest approach is to recall that $\text{Span}\{\hat{x}_i \hat{p}_j\} \simeq \mathcal{T}_2^0(\mathbb{C}^3) \simeq V_1 \otimes V_1$, and so our preferred J_z eigenbasis is just given by the analogs in $\text{Span}\{\hat{x}_i \hat{p}_j\}$ of the vectors in (3.60). Translating from the Dirac notation there, using the J_z eigenbasis for vector operators given in (6.4), and also normalizing gives

$$V_2 = \text{Span}\left\{\begin{array}{r} \frac{1}{2}(\hat{x}+i\hat{y})(\hat{p}_x+i\hat{p}_y) \equiv T_2^{(2)}, \\ \frac{1}{2}\left(\hat{z}(\hat{p}_x+i\hat{p}_y)+(\hat{x}+i\hat{y})\hat{p}_z\right) \equiv T_1^{(2)}, \\ \frac{1}{\sqrt{6}}\left(2\hat{z}\hat{p}_z-\hat{x}\hat{p}_x-\hat{y}\hat{p}_y\right) \equiv T_0^{(2)}, \\ -\frac{1}{2}\left(\hat{z}(\hat{p}_x-i\hat{p}_y)+(\hat{x}-i\hat{y})\hat{p}_z\right) \equiv T_{-1}^{(2)}, \\ \frac{1}{2}(\hat{x}-i\hat{y})(\hat{p}_x-i\hat{p}_y) \equiv T_{-2}^{(2)} \end{array}\right\}. \tag{6.5}$$

In (6.4) and (6.5) one can see a strong analogy with the $l = 1, 2$ spherical harmonics (cf. Exercise 2.2). In fact, what we have done is decompose the tensor operator $\{\hat{x}_i \hat{p}_j\}$ into three subsets

$$\mathbf{T}^{(0)} \equiv \{T_0^{(0)}\}, \quad \mathbf{T}^{(1)} \equiv \{T_m^{(1)}\}_{m=-1,0,1}, \quad \mathbf{T}^{(2)} \equiv \{T_m^{(2)}\}_{m=-2,\ldots,2}$$

composed of the operators $T_m^{(l)}$ on the right-hand sides of (6.3)–(6.5).[1] By construction each $\mathbf{T}^{(l)}$ is a set of $2l+1$ operators which transform like the v_k from (5.65), or alternatively like the spherical harmonics Y_m^l. Accordingly, each $\mathbf{T}^{(l)}$ is said to be a ***spherical tensor operator*** (or just ***spherical tensor***) of degree l. We can thus say that

A spherical tensor of degree l is a collection of $(2l+1)$ operators $\mathbf{T}^{(l)} \equiv \{T_m^{(l)}\}_{m=-l,\ldots,l}$ whose elements $T_m^{(l)}$ transform under $SO(3)$ similarity transformations like the spherical harmonics Y_m^l.

Returning to our original question, we can then decompose $\text{Span}\{\hat{x}_i \hat{p}_j\}$ as

$$\text{Span}\{\hat{x}_i \hat{p}_j\} \simeq \text{Span}\,\mathbf{T}^{(0)} \oplus \text{Span}\,\mathbf{T}^{(1)} \oplus \text{Span}\,\mathbf{T}^{(2)},$$

which we might call a "decomposition of $\{\hat{x}_i \hat{p}_j\}$ into spherical tensors." □

[1]Note that the 'l' in the superscript of $T_m^{(l)}$ is in parentheses; this is because the l isn't really an active index, but just serves to remind us which $SO(3)$ representation we're dealing with.

With an understanding of spherical tensors now in place, we can give a quick, heuristic overview of angular momentum selection rules and the Wigner–Eckart theorem. A more precise treatment will be given in the following sections. Let us switch to Dirac notation, and consider the product $T^{(q)}_{m_q}\,|l,m_l\rangle$ of a J_z, $\mathbf{J}^2$ eigenket $|l,m_l\rangle$ and degree q spherical tensor operator $T^{(q)}_{m_q}$. Can the $SO(3)$ transformation properties of the factors tell us anything about the transformation properties of the product? The answer is yes, and in fact

$$\begin{aligned}\Pi(R)\,T^{(q)}_{m_q}\,|l,m_l\rangle &= \Pi(R)T^{(q)}_{m_q}\Pi(R)^{-1}\Pi(R)\,|l,m_l\rangle \\ &= \Pi_q(R)^{n_q}_{m_q}\,T^{(q)}_{n_q}\,\Pi_l(R)^{n_l}_{m_l}\,|l,n_l\rangle \end{aligned} \tag{6.6}$$

where $\Pi_q(R)^{n_q}_{m_q}$ are just the matrix elements of R in the preferred basis for V_q, and similarly for $\Pi_l(R)^{n_l}_{m_l}$.[2] The appearance of these matrix elements in (6.6) tells us that:

$$\boldsymbol{T^{(q)}_{m_q}\,|l,m_l\rangle}\ \textbf{lives in a subspace equivalent to}\ \boldsymbol{V_q\otimes V_l}.$$

Colloquially, when an operator that lives in V_q acts on a vector that lives in V_l, the result behaves like an element of $V_q \otimes V_l$!

This unsurprising fact actually has far-reaching consequences. Take another angular momentum eigenket $|j,m_j\rangle$ and consider the inner product $\langle j,m_j|\,T^{(q)}_{m_q}\,|l,m_l\rangle$. Since $T^{(q)}_{m_q}\,|l,m_l\rangle$ transforms like an element of $V_q\otimes V_l$, this inner product must vanish unless $V_j\subset V_q\otimes V_l$. This, combined with (5.56), immediately gives the very useful angular momentum selection rule

$$\langle j,m_j|\,T^{(q)}_{m_q}\,|l,m_l\rangle = 0 \quad \text{unless } |l-q|\le j\le l+q.$$

Now suppose that indeed $V_j\subset V_q\otimes V_l$, so that the above inner product is nonzero. Then a glance back at Example 3.21 shows that this inner product bears a strong resemblance to the Clebsch–Gordan coefficients $\langle j,m_j|m_q,m_l\rangle$ from (3.59). The content of the Wigner–Eckart theorem is that this is more than just a resemblance, but is in fact a strict proportionality:

Proposition 6.1 (Wigner–Eckart I). *Let* $(\Pi,\mathcal{H})$ *be a representation of* $SO(3)$ *on a quantum-mechanical Hilbert space* $\mathcal{H}$*. Let* $\mathbf{T}^{(q)}\equiv\{T^{(q)}_{m_q}\}$ *be a spherical tensor of degree* q*, and let* $|l,m_l\rangle$, $|j,m_j\rangle$ *be* J_z, $\mathbf{J}^2$ *eigenkets. Then*

$$\langle j,m_j|\,T^{(q)}_{m_q}\,|l,m_l\rangle = c\langle j,m_j|m_q,m_l\rangle$$

where c *is independent of* m_l, m_q*, and* m_j*.*

We will state this more generally, precisely, and with proof in the next section.

[2] These are known as ***Wigner functions*** or ***Wigner D-matrices***.

6.2 Representation Operators, Selection Rules, and the Wigner–Eckart Theorem

You may have noticed that the definitions of vector, tensor, and spherical tensor operators from the last section were somewhat unwieldy; in particular, they relied on an equivalence between the transformation properties of *specific* operators and that of *specific* basis vectors for the V_l. It would be preferable to have a basis-independent definition which subsumes the previous definitions. To that end, we introduce in this section the notion of a *representation operator*, which generalizes the notions of vector and tensor operators as well as spherical tensors. We'll then use representation operators to derive a fundamental quantum-mechanical selection rule, which lays the foundation for the various selection rules one encounters in standard quantum mechanics courses. Representation operators will also play a key role in the Wigner–Eckart theorem, as we'll see.

Given a representation (Π_0, V_0) of a group G on some auxiliary vector space V_0, as well as a unitary representation $(\Pi, \mathcal{H})$ of G on some Hilbert space $\mathcal{H}$, we define a ***representation operator*** to simply be an intertwiner between V_0 and $\mathcal{L}(\mathcal{H})$, or in other words a linear map $\rho : V_0 \to \mathcal{L}(\mathcal{H})$ satisfying

$$\boxed{\rho(\Pi_0(g)v) = \Pi(g)\rho(v)\Pi(g)^{-1} \quad \forall\, g \in G, v \in V_0.} \tag{6.7}$$

Note that in terms of maps between V_0 and $\mathcal{L}(\mathcal{H})$, this just says

$$\rho \circ \Pi_0(g) = \Pi_1^1(g) \circ \rho \qquad \forall\, g \in G.$$

What this definition is saying, roughly, is that we have a subspace $\rho(V_0) \subset \mathcal{L}(\mathcal{H})$ which, even though it's composed of operators acting on $\mathcal{H}$, *actually transforms like the space V_0 under similarity transformations by the operators Π_g*. Note that, strictly speaking, the representation operator ρ is not itself an operator, but an intertwiner between representations.

How does this definition subsume the previous ones? Let V_0 have a basis $\{e_i\}$; then since ρ is linear, it's completely determined by its action on this basis. Plugging a basis vector e_i into (6.7) then yields

$$\Pi(g)\rho(e_i)\Pi(g)^{-1} = (\Pi_0(g))_i{}^j \rho(e_j).$$

If we set $\rho(e_i) \equiv B_i$, this becomes

$$\Pi(g)B_i\Pi(g)^{-1} = (\Pi_0(g))_i{}^j B_j. \tag{6.8}$$

This, of course, just says that under the representation Π_1^1, the B_i transform like basis vectors of the representation (Π_0, V_0), which was how we defined vector operators, tensor operators, and spherical tensors in the first place! In fact, if we take $G = SO(3)$ and $V_0 = \mathbb{R}^3$, then (6.8) becomes (writing the matrices of the fundamental representation with both indices down)

$$\Pi(R)B_i\Pi(R)^{-1} = \sum_j R_{ji}B_j$$

which is of course just (6.1). To get tensor operators or spherical operators, we just take $V_0 = \mathcal{T}_s^r(\mathbb{R}^3)$ or $V_0 = \mathcal{H}_l(\mathbb{R}^3)$, respectively.

Now that we have defined representation operators, we can go on to formulate the fundamental selection rule from which the usual quantum-mechanical selection rules can be derived. Then we can state and prove the Wigner–Eckart theorem, which is a kind of complement to the angular momentum selection rules. First, though, we need the following fact, the proof of which we only sketch. The details are deferred to the problems referenced below.

Proposition 6.2. *Let W_1 and W_2 be finite-dimensional inequivalent irreducible subspaces of a unitary representation $(\Pi, \mathcal{H})$ equipped with an inner product $(\cdot\,|\,\cdot)$. Then W_1 is orthogonal to W_2.*

Proof sketch. Define the ***orthogonal projection operator*** $P : \mathcal{H} \to W_2$ to be the map which sends $v \in \mathcal{H}$ to the *unique* vector $P(v) \in W_2$ satisfying

$$(P(v)|w) = (v|w) \quad \forall\, w \in W_2.$$

This is depicted schematically in Fig. 6.1. You will check in Problem 6-1 that such a vector $P(v)$ exists and is in fact unique. If we now restrict P to W_1, we get $P|_{W_1} : W_1 \to W_2$, and using the unitarity of Π and the invariance of the W_i one can show that $P|_{W_1}$ is an intertwiner. One can then use Schur's lemma to conclude that $P|_{W_1} = 0$, which then implies that W_1 is orthogonal to W_2. To fill out the details, see Problem 6-2. □

With this in hand, we can now state and prove

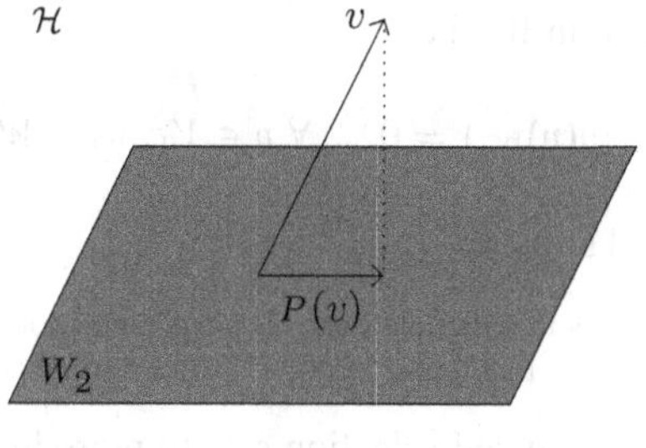

Fig. 6.1 Action of the orthogonal projection operator $P : \mathcal{H} \to W_2$ on a vector $v \in \mathcal{H}$

Proposition 6.3 (Selection Rule). *Let G be a semi-simple group, and let W_1 and W_2 be finite-dimensional inequivalent irreducible subspaces of a unitary representation $(\Pi, \mathcal{H})$ of G. Also let $\rho : V_0 \to \mathcal{L}(\mathcal{H})$ be a representation operator, where (Π_0, V_0) is some auxiliary representation. Then*

$$(w_1|\rho(v)w_2) = 0 \qquad \forall\, v \in V_0,\ w_i \in W_i$$

unless the decomposition of $V_0 \otimes W_2$ into irreducibles contains a representation equivalent to W_1.

Before proving this, note that this proposition just formalizes what we observed in the last section: namely, that a vector of the form $\rho(v)w_2$ is a kind of "product" of elements of V_0 and W_2, and thus transforms like something in $V_0 \otimes W_2 = U_1 \oplus \cdots \oplus U_k$. Thus, for there to be any overlap with an irreducible subspace W_1, W_1 must then be equivalent to one of the U_i.

Proof of proposition. Define a map by

$$\begin{aligned} T : V_0 \otimes W_2 &\to \mathcal{H} \\ v \otimes w_2 &\mapsto \rho(v)w_2 \end{aligned} \tag{6.9}$$

and extend linearly to arbitrary elements of $V_0 \otimes W_2$. You can check that this is a linear map between vector spaces, and so the image $T(V_0 \otimes W_2) \equiv D \subset \mathcal{H}$ is a vector subspace of $\mathcal{H}$. Now, since G is semi-simple we can decompose $V_0 \otimes W_2$ into irreducibles as

$$V_0 \otimes W_2 = U_1 \oplus \cdots \oplus U_k$$

for some irreps U_k. As you will show below, the fact that ρ is a representation operator implies that T is in fact an intertwiner, and this further implies that the kernel of T (cf. Exercise 4.19) is an invariant subspace of $V_0 \otimes W_2$. This means that (with a possible relabeling of the U_i) we can write the kernel of T as $U_1 \oplus \cdots \oplus U_m$ for some $m \leq k$, which then implies that D is equivalent to $U_{m+1} \oplus \cdots \oplus U_k$. If none of the U_i are equivalent to W_1, then by Proposition 6.2, every vector in D is orthogonal to every vector in W_1, i.e.

$$(w_1|\rho(v)w_2) = 0 \qquad \forall\, v \in V_0,\ w_i \in W_i.$$

which is what we wanted to prove. □

Exercise 6.3. Quickly show that the fact that ρ is a representation operator implies that T is an intertwiner. Show further that the kernel of T is an invariant subspace of $V_0 \otimes W_2$.

We'll now use this generalized selection rule to reproduce some of the familiar selection rules from quantum mechanics.

Example 6.4. *Parity Selection Rules*

Let $(\Pi, \mathcal{H})$ be a complex unitary representation of the two-element group $\mathbb{Z}_2 = \{I, P\}$ where P is the parity operator. Let $v_\alpha,\ v_\beta \in \mathcal{H}$ be parity eigenstates with eigenvalues $c_\alpha,\ c_\beta$. If $c_\alpha = 1$, then v_α spans the one-dimensional trivial representation of $\mathbb{Z}_2$, and if $c_\alpha = -1$ then v_α spans the alternating representation. Likewise for c_β. Now let V_0 be another one-dimensional irrep of $\mathbb{Z}_2$ with parity eigenvalue c_0, and let $B \in \rho(V_0)$, where ρ is a representation operator. Then a little thought shows that the selection rule implies

$$(v_\beta | B v_\alpha) = 0 \ \text{ unless } \ c_\beta = c_0 c_\alpha .$$

Thus if is B is parity-odd ($c_0 = -1$) then it can only connect states of opposite parity, and if it is parity even ($c_0 = +1$) then it can only connect states of the same parity. If we were looking at dipolar radiative transitions (which emit a photon) between electronic states of an atom or molecule, the relevant operators are the components of the dipole operator $\mathbf{p}$ which is parity odd (since it is a vector operator). The above then tells us that dipolar radiative transitions can only occur between electronic states of opposite parity.

Example 6.5. *Angular Momentum Selection Rules*

Let $(\Pi, \mathcal{H})$ be a complex unitary representation of $SU(2)$, with two subspaces W_l and W_j equivalent to V_l and V_j respectively, $2l, 2j \in \mathbb{Z}$. Also suppose we have a $SU(2)$ representation operator $\mathbf{A} : V_q \to \mathcal{L}(\mathcal{H})$, $q \in \mathbb{Z}$ (in other words, a spherical tensor). Then for any $v \in W_l$, $v' \in W_j$, $A \in \mathbf{A}(V_q)$, the selection rule tells us that

$$(v'|Av) = 0 \quad \text{unless} \quad |l - q| \le j \le l + q.$$

If we again consider a dipolar radiative transition between electronic states of an atom or molecule, the relevant operator is still the dipole $\mathbf{p}$ whose components p_i live in $\rho(V_1)$, and so we find that a dipolar radiative transition between states with angular momentum j and l can only occur if $l - 1 \le j \le l + 1$. □

The famous Wigner–Eckart theorem can be seen as a kind of complement to the angular momentum selection rule above. In the notation of the previous example, the Wigner–Eckart theorem says (roughly) that when $(v'|Av)$ is *not* equal to zero, it is still tightly constrained and is in fact determined up to a constant by the fact that $\mathbf{A}$ is a representation operator. The precise statement is as follows:

Proposition 6.4 (Wigner–Eckart II). *Let $(\Pi, \mathcal{H})$ be a complex unitary representation of $SU(2)$, with two subspaces W_l and W_j equivalent to V_l and V_j respectively, $2l, 2j \in \mathbb{Z}$. Also suppose we have two $SU(2)$ representation operators $\mathbf{A}, \mathbf{B} : V_q \to \mathcal{L}(\mathcal{H})$, $q \in \mathbb{Z}$, which yield two spherical tensors with components $A_k \equiv \mathbf{A}(v_k)$, $B_k \equiv \mathbf{B}(v_k)$, $0 \le k \le 2q + 1$. Finally, assume that*

$$(v'|A_k v) \ne 0 \ \textit{ for some } k \textit{ and } \ v \in W_l, v' \in W_j . \tag{6.10}$$

Then for all k and $w \in W_l, w' \in W_j$, we have

$$(w'|B_k w) = c(w'|A_k w) \tag{6.11}$$

for some constant $c \in \mathbb{C}$ which is independent of k, w, and w'.

Proof. Let $\mathcal{L}(W_l, W_j)$ denote the vector space of all linear maps from W_l to W_j. By restricting to W_l and using the orthogonal projection operator $P_j : \mathcal{H} \to W_j$, we can turn $\mathbf{A} : V_q \to \mathcal{L}(\mathcal{H})$ into a map

$$\begin{aligned} \tilde{\mathbf{A}} : V_q &\to \mathcal{L}(W_l, W_j) \\ v &\mapsto P_j \circ A(v)|_{W_l}. \end{aligned}$$

All we've done here is take the linear operator $A(v) \in \mathcal{L}(\mathcal{H})$ and restrict it to W_l and then project onto W_j. Now, since W_l, W_j are $SU(2)$-invariant subspaces, the vector space $\mathcal{L}(W_l, W_j)$ actually furnishes a representation $(\tilde{\Pi}_1^1, \mathcal{L}(W_l, W_j))$ of $SU(2)$ by

$$\tilde{\Pi}_1^1(g)T \equiv \Pi(g)T\Pi(g)^{-1} \quad T \in \mathcal{L}(W_l, W_j),\ g \in SU(2).$$

You will check below that this is a bona fide representation, and is in fact equivalent to the tensor product representation $W_l^* \otimes W_j \simeq W_l \otimes W_j$! Furthermore, since the action of $SU(2)$ on $\mathcal{H}$ commutes with restriction and projection (cf. Problem 6-2), it's not hard to see that $\tilde{\mathbf{A}}$ is an intertwiner. From (6.10) we know that $\tilde{\mathbf{A}}$ is not zero, and so from Schur's lemma we conclude that $\mathcal{L}(W_l, W_j) \simeq W_l \otimes W_j$ has a subspace U_q equivalent to V_q, and that $\tilde{\mathbf{A}}$ is a vector space isomorphism from V_q to U_q.

Now, we can also use our second representation operator $\mathbf{B}$ to construct a second intertwiner $\tilde{\mathbf{B}} : V_q \to U_q \subset \mathcal{L}(W_l, W_j)$. Then we invoke the corollary of Schur's lemma that you proved in Exercise 5.34 to conclude that $\tilde{\mathbf{B}} = c\tilde{\mathbf{A}}$. But this then means that

$$\begin{aligned} \tilde{\mathbf{A}}(v_k) &= c\,\tilde{\mathbf{B}}(v_k) \qquad \forall\, k \\ \Rightarrow (w'|\tilde{\mathbf{A}}(v_k)w) &= c(w'|\tilde{\mathbf{B}}(v_k)w) \ \forall\, k \text{ and } w \in W_l,\ w' \in W_j \\ \Rightarrow \quad (w'|A_k w) &= c(w'|B_k w) \quad \forall\, k \text{ and } w \in W_l,\ w' \in W_j \end{aligned}$$

and so we are done. We here used the definition of $\tilde{\mathbf{A}}$, the definition of the orthogonal projection operator P_j, and the definitions $A_k = A(v_k)$, $B_k = B(v_k)$. □

Exercise 6.4. Show that if $T \in \mathcal{L}(W_l, W_j)$ then so is $\Pi(g)T\Pi(g)^{-1}$, so that $(\tilde{\Pi}_1^1, \mathcal{L}(W_l, W_j))$ really is a representation; in fact, it is an invariant subspace of $(\Pi_1^1, \mathcal{L}(\mathcal{H}))$. In analogy to the equivalence between $V^* \otimes V$ and $\mathcal{L}(V)$, show that $(\tilde{\Pi}_1^1, \mathcal{L}(W_l, W_j))$ is equivalent to $W_l^* \otimes W_j$, which by Problem 5-4 is equivalent to $W_l \otimes W_j$.

You may have noticed that this is not the way we stated the Wigner–Eckart theorem earlier, which may also be familiar from advanced quantum mechanics texts like Sakurai [17]. To make the connection, consider the intertwiner

$$\begin{aligned} T : V_q \otimes W_l &\to W_j \\ v \otimes w &\mapsto P_j(\mathbf{A}(v)w). \end{aligned}$$

If we work with standard bases $\{v_m\}_{m=0-2q}$, $\{w_n\}_{n=0-2l}$, $\{w'_p\}_{p=0-2j}$ for V_q, W_l, and W_j, then this map has components

$$\begin{aligned} T_{mn}{}^{p} &= T(v_m, w_n, w'^{p}) \\ &= w'^{p}(P_j(\mathbf{A}(v_m)w_n)) \\ &= \frac{1}{||w'_p||}(w'_p|P_j(A_m w_n)) \\ &= \frac{1}{||w'_p||}(w'_p|A_m w_n). \end{aligned} \tag{6.12}$$

Now let's switch to the orthonormal bases familiar from quantum mechanics, which look like

$$\begin{aligned} |q, m_q\rangle &\in V_q \qquad -q \le m_q \le q \\ |l, m_l\rangle &\in W_l \qquad -l \le m_l \le l \\ |j, m_j\rangle' &\in W_j \qquad -j \le m_j \le j \end{aligned}$$

(notice the prime on the last set of vectors, which will distinguish it from vectors in $V_q \otimes W_l$ with the same quantum numbers). With this basis and notation, the components (6.12) of T become the matrix elements $\langle j, m_j|' A_{m_q} |l, m_l\rangle$.

What do these matrix elements have to do with Clebsch–Gordan coefficients? Recall that $V_q \otimes W_l$ has two convenient sets of orthonormal basis vectors:

$$\begin{aligned} |q, m_q\rangle \otimes |l, m_l\rangle \equiv |ql; m_q, m_l\rangle &\in V_q \otimes W_l \qquad -q \le m_q \le q,\ -l \le m_l \le l \\ |l', m_{l'}\rangle &\in V_q \otimes W_l \qquad |l-q| \le l' \le l+q,\ -l' \le m_{l'} \le l'. \end{aligned}$$

The Clebsch–Gordan coefficients are just the inner products $\langle l', m_{l'}|ql; k, m_l\rangle$ of these basis vectors. Since $T : V_q \otimes W_l \to W_j$ is nonzero, $V_q \otimes W_l$ must contain a subspace U_j equivalent to W_j and so we can consider the orthogonal projection operator $P : V_q \otimes W_l \to U_j$. By Problem 6-1 this is given by

$$\begin{aligned} P : V_q \otimes W_l &\to U_j \\ |ql; m_q, m_l\rangle &\mapsto \sum_{-j \le m_j \le j} \langle j, m_j|ql; m_q, m_l\rangle\, |j, m_j\rangle \end{aligned}$$

and is an intertwiner by Problem 6-2. By then making the obvious identification of U_j with W_j and hence $|j, m_j\rangle \mapsto |j, m_j\rangle'$, we get the intertwiner

$$\begin{aligned} P' : V_q \otimes W_l &\to W_j \\ |ql; m_q, m_l\rangle &\mapsto \sum_{-j \le m_j \le j} \langle j, m_j \, |ql; m_q, m_l\rangle \, |j, m_j\rangle' \end{aligned}$$

whose components are nothing but the Clebsch–Gordan coefficients! By Wigner–Eckart, though, this intertwiner must be proportional to T, and so its components must be proportional to those of T. We thus have the following component version of the Wigner–Eckart theorem:

Proposition 6.5 (Wigner-Eckart III). *Let $(\Pi, \mathcal{H})$ be a complex unitary representation of $SU(2)$, with two subspaces W_l and W_j equivalent to V_l and V_j respectively, $2l, 2j \in \mathbb{Z}$. Also suppose we have a degree q spherical tensor $\mathbf{A} = \{A_{m_q}\}$. Then*

$$\langle j, m_j | \, A_{m_q} \, |l, m_l\rangle = c \, \langle j, m_j \, |ql; m_q, m_l\rangle$$

where c is a constant independent of m_l, m_q, and m_j.

6.3 Gamma Matrices and Dirac Bilinears

We conclude this short chapter with what is essentially an extended example, which involves both the representation operators we've met in this chapter and the $O(3,1)$ representation theory we developed in the last. This section relies on a familiarity with the theory of the Dirac electron; if you have not seen this material, and in particular are unfamiliar with gamma matrices and Dirac bilinears, then this section can be skipped. This section borrows heavily from Frankel [6].

Let $(D \equiv \Pi \oplus \bar{\Pi}, \mathbb{C}^4)$ be the Dirac spinor representation of $SL(2, \mathbb{C})$, so that

$$\begin{aligned} D : SL(2, \mathbb{C}) &\to GL(4, \mathbb{C}) \\ A &\mapsto D(A) \equiv \begin{pmatrix} A & 0 \\ 0 & A^{\dagger -1} \end{pmatrix}. \end{aligned}$$

We claim that the gamma matrices γ_μ can be seen as the components of a representation operator $\gamma : \mathbb{R}^4 \to \mathcal{L}(\mathbb{C}^4) = M_4(\mathbb{C})$, where $\mathbb{R}^4$ is the Minkowski four-vector representation of $SL(2, \mathbb{C})$. To define γ, we need the following two identifications of $\mathbb{R}^4$ with $H_2(\mathbb{C})$, where $X = (\mathbf{x}, t) \in \mathbb{R}^4$ and $\boldsymbol{\sigma} \equiv (\sigma_x, \sigma_y, \sigma_z)$:

$$\mathbb{R}^4 \longleftrightarrow H_2(\mathbb{C})$$

$$X \longleftrightarrow X_* \equiv \mathbf{x} \cdot \boldsymbol{\sigma} + tI \tag{6.13a}$$

$$X \longleftrightarrow X^* \equiv \mathbf{x} \cdot \boldsymbol{\sigma} - tI. \tag{6.13b}$$

Note that (6.13a) is just the identification we used in Examples 4.23 and 5.5 in defining the four-vector representation of $SL(2,\mathbb{C})$, and (6.13b) is just a slight variation of that. Now, using the well-known property of the sigma matrices that

$$\sigma_i\sigma_j + \sigma_j\sigma_i = 2\delta_{ij}, \tag{6.14}$$

you will verify below that

$$X^* = \eta(X,X)X_*^{-1}. \tag{6.15}$$

Also, if $\rho : SL(2,\mathbb{C}) \to SO(3,1)_o$ is the homomorphism from Example 4.23 which defines the four-vector representation, then by the definition of ρ we have

$$(\rho(A)X)_* = AX_*A^\dagger \tag{6.16}$$

which when combined with (6.15) yields

$$(\rho(A)X)^* = A^{\dagger-1}X^*A^{-1} \tag{6.17}$$

as you will also show below. If we then define a map γ by

$$\begin{aligned}\gamma : \mathbb{R}^4 &\to M_4(\mathbb{C})\\ X &\mapsto \begin{pmatrix} 0 & X_* \\ X^* & 0 \end{pmatrix}\end{aligned}$$

you can then use (6.16) and (6.17) to check that γ is a representation operator, i.e. that

$$\gamma(\rho(A)X) = D(A)\gamma(X)D(A)^{-1}. \tag{6.18}$$

If we define the gamma matrices as $\gamma_\mu \equiv \gamma(e_\mu)$, then we find that

$$\begin{aligned}\gamma_i &= \begin{pmatrix} 0 & \sigma_i \\ \sigma_i & 0 \end{pmatrix} \quad i = 1,2,3\\ \gamma_4 &= \begin{pmatrix} 0 & I \\ -I & 0 \end{pmatrix}\\ \gamma_5 \equiv \gamma_1\gamma_2\gamma_3\gamma_4 &= \begin{pmatrix} -I & 0 \\ 0 & I \end{pmatrix}.\end{aligned} \tag{6.19}$$

Up to a few minus signs that have to do with our choice of signature for the Minkowski metric, as well as the fact that most texts write γ_0 instead of γ_4, this is the familiar ***chiral*** representation of the gamma matrices (the chiral representation is the one in which γ_5 is diagonal). If we let $\Lambda_\mu{}^\nu$ be the components of $\rho(A)$, we can take $X = e_\mu$ in (6.18) to get

$$\Lambda_\mu{}^\nu \gamma_\nu = D(A)\gamma_\mu D(A)^{-1} \tag{6.20}$$

which is how (6.18) usually appears in physics texts. Another important property is that the gamma matrices satisfy the fundamental anticommutation relation

$$\gamma_\mu \gamma_\nu + \gamma_\nu \gamma_\mu = 2\eta_{\mu\nu} \tag{6.21}$$

which you can easily verify. This relationship is the starting point for ***Clifford algebras***, which we also mentioned at the very end of Chap. 5.[3]

Exercise 6.5. Verify (6.15), (6.17), (6.18), and (6.21).

Now we'd like to construct the Dirac bilinears. Denote an element of $\mathbb{C}^4$ by

$$\psi = \begin{pmatrix} \psi_L \\ \psi_R \end{pmatrix}$$

where ψ_L and ψ_R are two-component spinors living in $\left(\frac{1}{2}, \mathbf{0}\right)$ and $\left(\mathbf{0}, \frac{1}{2}\right)$ respectively. Also define the row vector $\bar{\psi}$ by

$$\bar{\psi} \equiv (\psi_R^*, \psi_L^*)$$

where the "*" denotes complex conjugation (note that the positions of the right- and left-handed spinors are switched here). You can easily check that if ψ transforms like a Dirac spinor, i.e. $\psi \mapsto D(A)\psi$, then $\bar{\psi}$ transforms like

$$\bar{\psi} \longmapsto \bar{\psi} \begin{pmatrix} A^{-1} & 0 \\ 0 & A^\dagger \end{pmatrix} = \bar{\psi} D(A)^{-1}. \tag{6.22}$$

If we consider the associated column vector $\bar{\psi}^T$, it transforms like

$$\bar{\psi}^T \longmapsto \begin{pmatrix} A^{-1T} & 0 \\ 0 & A^* \end{pmatrix} \bar{\psi}^T$$

which you should recognize as the representation dual to the Dirac spinor, since the matrices of dual representations are just the inverse transpose of the original matrices. However, we know from Problem 5-3 that $\left(\frac{1}{2}, \mathbf{0}\right)$ and $\left(\mathbf{0}, \frac{1}{2}\right)$ are equivalent to their duals, so we conclude that $\bar{\psi}$ transforms like a Dirac spinor as well.

---

[3] Also, note the analogy between (6.21) and (6.14); in fact, the Pauli matrices can be thought of as a lower-dimensional analog of the gamma matrices, and it's no coincidence that the gamma matrices γ_i in (6.19) are built out of the Pauli matrices! For more on this see Frankel [6].

With $\bar{\psi}$ in hand, we can now define the ***Dirac bilinears***

$\bar{\psi}\psi$	scalar
$\bar{\psi}\gamma_\mu\psi$	vector
$\bar{\psi}\gamma_\mu\gamma_\nu\psi, \quad \mu \neq \nu$	antisymmetric 2-tensor
$\bar{\psi}\gamma_\mu\gamma_\nu\gamma_\rho\psi, \quad \mu \neq \nu \neq \rho$	pseudovector
$\bar{\psi}\gamma_5\psi$	pseudoscalar

Each of these contains a product of two Dirac spinors, and can be seen as components of tensors living in $\mathbb{C}^4 \otimes \mathbb{C}^4$. Note the similarities between the names of the Dirac bilinears and the first five entries of the table from Example 5.41. This makes it seem like the Dirac bilinears should transform like the components of antisymmetric tensors (of ranks 0 through 4). Is this true? Well, using (6.22), we find that the scalar transforms like

$$\bar{\psi} \longmapsto \bar{\psi} D(A)^{-1} D(A)\psi = \bar{\psi}\psi$$

and so really does transform like a scalar. Similarly, using (6.20), we find that the vector transforms like

$$\begin{aligned}\bar{\psi}\gamma_\mu\psi &\longmapsto \bar{\psi} D(A)^{-1}\gamma_\mu D(A)\psi \\ &= {\Lambda^{-1}}_\mu{}^\nu \bar{\psi}\gamma_\nu\psi\end{aligned}$$

and so really does transform like a vector. Using the anticommutation relation (6.21), you can similarly verify that the antisymmetric 2-tensor, pseudovector, and pseudoscalar transform like antisymmetric tensors of ranks 2, 3, and 4 respectively. If we let $\Lambda^*(\mathbb{R}^4)$ denote the set of *all* antisymmetric tensor products of $\mathbb{R}^4$, i.e.

$$\Lambda^*\mathbb{R}^4 \equiv \bigoplus_{k=0}^{4} \Lambda^k\mathbb{R}^4,$$

and let $\Lambda^*\mathbb{R}^4_{\mathbb{C}}$ denote its complexification, then $\mathbb{C}^4 \otimes \mathbb{C}^4$ contains a subspace equivalent to $\Lambda^*\mathbb{R}^4_{\mathbb{C}}$. However, one can check that both spaces have (complex) dimension 16, so as $\mathfrak{sl}(2,\mathbb{C})_{\mathbb{R}}$ representations

$$\mathbb{C}^4 \otimes \mathbb{C}^4 \simeq \Lambda^*\mathbb{R}^4_{\mathbb{C}}.$$

In other words. **the tensor product of the Dirac spinor representation with itself is equivalent to the space of all antisymmetric tensors**!

Another way to obtain this same result is to use our tensor product decomposition (5.76). The tensor product $\mathbb{C}^4 \otimes \mathbb{C}^4$ of two Dirac spinors is given by

$$\begin{aligned}
&[(\tfrac{1}{2},\mathbf{0}) \oplus (\mathbf{0},\tfrac{1}{2})] \otimes [(\tfrac{1}{2},\mathbf{0}) \oplus (\mathbf{0},\tfrac{1}{2})] = \\
&\quad [(\tfrac{1}{2},\mathbf{0}) \otimes (\tfrac{1}{2},\mathbf{0})] \oplus [(\tfrac{1}{2},\mathbf{0}) \otimes (\mathbf{0},\tfrac{1}{2})] \oplus [(\mathbf{0},\tfrac{1}{2}) \otimes (\tfrac{1}{2},\mathbf{0})] \oplus [(\mathbf{0},\tfrac{1}{2}) \otimes (\mathbf{0},\tfrac{1}{2})] \\
&= (\mathbf{1},\mathbf{0}) \oplus (\mathbf{0},\mathbf{0}) \oplus (\tfrac{1}{2},\tfrac{1}{2}) \oplus (\tfrac{1}{2},\tfrac{1}{2}) \oplus (\mathbf{0},\mathbf{1}) \oplus (\mathbf{0},\mathbf{0}) \\
&= (\mathbf{0},\mathbf{0}) \oplus (\tfrac{1}{2},\tfrac{1}{2}) \oplus (\mathbf{1},\mathbf{0}) \oplus (\mathbf{0},\mathbf{1}) \oplus (\tfrac{1}{2},\tfrac{1}{2}) \oplus (\mathbf{0},\mathbf{0}) \qquad (6.23)
\end{aligned}$$

where in the first equality we used the fact that the tensor product distributes over direct sums (see Problem 5-12), in the second equality we used (5.76), and in the third equality we just rearranged the summands. However, from Example 5.41 we know that as an $\mathfrak{so}(3,1)$ representation, $\Lambda^*\mathbb{R}^4_{\mathbb{C}}$ decomposes as

$$\Lambda^*\mathbb{R}^4_{\mathbb{C}} \simeq (\mathbf{0},\mathbf{0}) \oplus (\tfrac{1}{2},\tfrac{1}{2}) \oplus (\mathbf{1},\mathbf{0}) \oplus (\mathbf{0},\mathbf{1}) \oplus (\tfrac{1}{2},\tfrac{1}{2}) \oplus (\mathbf{0},\mathbf{0})$$

which is just (6.23)!

Exercise 6.6. Verify (6.22). Also verify that the antisymmetric 2-tensor, pseudovector and pseudoscalar bilinears transform like the components of antisymmetric tensors of rank 2, 3, and 4.

Chapter 6 Problems

Note: Problems marked with an "$*$" tend to be longer, and/or more difficult, and/or more geared towards the completion of proofs and the tying up of loose ends. Though these problems are still worthwhile, they can be skipped on a first reading.

6-1. In this problem we'll establish a couple of the basic properties of the orthogonal projection operator. To this end, let $\mathcal{H}$ be a Hilbert space with inner product $(\cdot\,|\,\cdot)$ and let W be a finite-dimensional subspace of $\mathcal{H}$.

(a) Show that for any $v \in \mathcal{H}$, there exists a *unique* vector $P(v) \in W$ such that

$$(P(v)|w') = (v|w') \quad \forall\, w' \in W.$$

This defines the ***orthogonal projection*** map $P : \mathcal{H} \to W$ which projects $\mathcal{H}$ onto W. (Hint: there are a few ways to show that $P(v)$ exists and is unique. One route is to consider the map $L : \mathcal{H} \to \mathcal{H}^*$ given by $L(v) = (v|\cdot)$ and then play around with restrictions to W.)

(b) Quickly show that $P(w) = w$ for all $w \in W$, and hence that $P^2 = P$.

(c) Let $\{e_i\}$ be a (possibly infinite) orthonormal basis for $\mathcal{H}$ where the first k vectors e_i, $i = 1, \ldots, k$ are a basis for W. Show that if we expand an arbitrary $v \in \mathcal{H}$ as $v = v^i e_i$ where the implied sum is over *all* i (and where the sum may be infinite), then P takes the simple form

$$P(v) = P(v^i e_i) = \sum_{i=1}^{k} v^i e_i.$$

Thus P can be thought of as "projecting out all the components orthogonal to W."

(d) Let $\{e_{j'}\}$ be an *arbitrary* orthonormal basis for $\mathcal{H}$. Show that in this case the action of P is given by

$$P(e_{j'}) = \sum_{i=1}^{k} (e_i | e_{j'}) e_i.$$

6-2. Let $(\Pi, \mathcal{H})$ be a unitary representation of a group G on a Hilbert space $\mathcal{H}$. In this problem we'll show that inequivalent irreducible subspaces of $\mathcal{H}$ are orthogonal.

(a) Let $W \subset \mathcal{H}$ be a finite-dimensional irreducible subspace and $P : \mathcal{H} \to W$ the orthogonal projection operator onto W. Use the defining property of P, the unitarity of Π, and the invariance of W to show that P is an intertwiner.

(b) Let $V \subset \mathcal{H}$ be another irreducible subspace inequivalent to W. Restrict P to V to get $P|_V : V \to W$ which is still an intertwiner. Use Schur's lemma to deduce that $P|_V = 0$, and conclude that V and W are orthogonal.

Appendix A
Complexifications of Real Lie Algebras and the Tensor Product Decomposition of $\mathfrak{sl}(2,\mathbb{C})_{\mathbb{R}}$ Representations

The goal of this appendix is to prove Proposition 5.9 about the tensor product decomposition of two $\mathfrak{sl}(2,\mathbb{C})_{\mathbb{R}}$ representations. The proof is long but will introduce some useful notions, like the direct sum and complexification of a Lie algebra. We'll use these notions to show that the representations of $\mathfrak{sl}(2,\mathbb{C})_{\mathbb{R}}$ are in 1-1 correspondence with certain representations of the *complex* Lie algebra $\mathfrak{sl}(2,\mathbb{C}) \oplus \mathfrak{sl}(2,\mathbb{C})$. That this complex Lie algebra is a direct sum will imply certain properties about its representations, which in turn will allow us to prove Proposition 5.9.

A.1 Direct Sums and Complexifications of Lie Algebras

In this text we have dealt only with *real* Lie algebras, as that is the case of greatest interest for physicists. From a more mathematical point of view, however, it actually simplifies matters to focus on the complex case, and we will need that approach to prove Proposition 5.9. With that in mind, we make the following definition (in total analogy to the real case): A ***complex Lie algebra*** is a complex vector space $\mathfrak{g}$ equipped with a complex-linear Lie bracket $[\cdot,\cdot] : \mathfrak{g} \times \mathfrak{g} \to \mathfrak{g}$ which satisfies the usual axioms of antisymmetry

$$[X,Y] = -[Y,X] \;\; \forall\, X,Y \in \mathfrak{g},$$

and the Jacobi identity

$$[[X,Y],Z] + [[Y,Z],X] + [[Z,X],Y] = 0 \;\; \forall\, X,Y,Z \in \mathfrak{g}.$$

Examples of complex Lie algebras are $M_n(\mathbb{C}) = \mathfrak{gl}(n,\mathbb{C})$, the set of all complex $n \times n$ matrices, and $\mathfrak{sl}(n,\mathbb{C})$, the set of all complex, traceless $n \times n$ matrices. In both cases the bracket is just given by the commutator of matrices.

N. Jeevanjee, *An Introduction to Tensors and Group Theory for Physicists*,
DOI 10.1007/978-3-319-14794-9

For our application we'll be interested in turning real Lie algebras into complex Lie algebras. We already know how to complexify vector spaces, so to turn a real Lie algebra into a complex one we just have to extend the Lie bracket to the complexified vector space. This is done in the obvious way: given a real Lie algebra $\mathfrak{g}$ with bracket $[\cdot,\cdot]$, we define its ***complexification*** to be the complexified vector space $\mathfrak{g}_{\mathbb{C}} = \mathbb{C}\otimes\mathfrak{g}$ with Lie bracket $[\cdot,\cdot]_{\mathbb{C}}$ defined by

$$\begin{aligned}[1\otimes X_1 + i\otimes X_2, 1\otimes Y_1 + i\otimes Y_2]_{\mathbb{C}} &\equiv 1\otimes[X_1,Y_1] - 1\otimes[X_2,Y_2] \\ &+ i\otimes[X_1,Y_2] + i\otimes[X_2,Y_1]\end{aligned}$$

where $X_i, Y_i \in \mathfrak{g}$, $i = 1,2$. If we abbreviate $i\otimes X$ as iX and $1\otimes X$ as X, this tidies up and becomes

$$[X_1 + iX_2, Y_1 + iY_2]_{\mathbb{C}} \equiv [X_1,Y_1] - [X_2,Y_2] + i\left([X_1,Y_2] + [X_2,Y_1]\right).$$

This formula defines $[\cdot,\cdot]_{\mathbb{C}}$ in terms of $[\cdot,\cdot]$, and is also exactly what you'd get by naively using complex linearity to expand the left-hand side.

What does this process yield in familiar cases? For $\mathfrak{su}(2)$ we define a map

$$\begin{aligned}\phi : \mathfrak{su}(2)_{\mathbb{C}} &\to \mathfrak{sl}(2,\mathbb{C}) \\ 1\otimes X_1 + i\otimes X_2 &\mapsto X_1 + \begin{pmatrix} i & 0 \\ 0 & i\end{pmatrix} X_2, \quad X_1, X_2 \in \mathfrak{su}(2).\end{aligned} \tag{A.1}$$

You will show below that this is a Lie algebra isomorphism, and hence $\mathfrak{su}(2)$ complexifies to become $\mathfrak{sl}(2,\mathbb{C})$. You will also use similar maps to show that $\mathfrak{u}(n)_{\mathbb{C}}\simeq\mathfrak{gl}(n,\mathbb{R})_{\mathbb{C}}\simeq\mathfrak{gl}(n,\mathbb{C})$. If we complexify the real Lie algebra $\mathfrak{sl}(2,\mathbb{C})_{\mathbb{R}}$, we also get something nice. The complexified Lie algebra $(\mathfrak{sl}(2,\mathbb{C})_{\mathbb{R}})_{\mathbb{C}}$ has complex basis

$$\begin{aligned}\mathcal{M}_i &\equiv \tfrac{1}{2}\left(1\otimes S_i - i\otimes\tilde{K}_i\right), \quad i = 1,2,3 \\ \mathcal{N}_i &\equiv \tfrac{1}{2}\left(1\otimes S_i + i\otimes\tilde{K}_i\right), \quad i = 1,2,3\end{aligned} \tag{A.2}$$

which we can again abbreviate[1] as

$$\begin{aligned}\mathcal{M}_i &\equiv \tfrac{1}{2}\left(S_i - i\tilde{K}_i\right), \quad i = 1,2,3 \\ \mathcal{N}_i &\equiv \tfrac{1}{2}\left(S_i + i\tilde{K}_i\right), \quad i = 1,2,3.\end{aligned}$$

[1]Careful here! When we write $i\tilde{K}_i$, this is *not* to be interpreted as i times the matrix $\tilde{K}_i$, as this would make the $\mathcal{N}_i$ identically zero (check!); it is merely shorthand for $i\otimes\tilde{K}_i$.

These expressions are very similar to ones found in our discussion of $\mathfrak{sl}(2,\mathbb{C})_\mathbb{R}$ representations, and using the bracket on $(\mathfrak{sl}(2,\mathbb{C})_\mathbb{R})_\mathbb{C}$ one can verify that the analogs of Eq. (5.72) hold, i.e. that

$$[\mathcal{M}_i, \mathcal{N}_j]_\mathbb{C} = 0 \tag{A.3}$$

$$[\mathcal{M}_i, \mathcal{M}_j]_\mathbb{C} = \sum_{k=1}^{3} \epsilon_{ijk}\mathcal{M}_k$$

$$[\mathcal{N}_i, \mathcal{N}_j]_\mathbb{C} = \sum_{k=1}^{3} \epsilon_{ijk}\mathcal{N}_k.$$

Notice that both Span$\{\mathcal{M}_i\}$ and Span$\{\mathcal{N}_i\}$ (over the complex numbers) are Lie subalgebras of $(\mathfrak{sl}(2,\mathbb{C})_\mathbb{R})_\mathbb{C}$ and that both are isomorphic to $\mathfrak{sl}(2,\mathbb{C})$, since

$$\text{Span}\{\mathcal{M}_i\} \simeq \text{Span}\{\mathcal{N}_i\} \simeq \mathfrak{su}(2)_\mathbb{C} \simeq \mathfrak{sl}(2,\mathbb{C}). \tag{A.4}$$

Furthermore, the bracket between an element of Span$\{\mathcal{M}_i\}$ and an element of Span$\{\mathcal{N}_i\}$ is 0, by (A.3). Also, as a (complex) vector space $(\mathfrak{sl}(2,\mathbb{C})_\mathbb{R})_\mathbb{C}$ is the direct sum Span$\{\mathcal{M}_i\}\oplus$ Span$\{\mathcal{N}_i\}$. When a Lie algebra $\mathfrak{g}$ can be written as a direct sum of subspaces W_1 and W_2, where the W_i are each subalgebras *and* $[w_1, w_2] = 0$ for all $w_1 \in W_1$, $w_2 \in W_2$, we say that the original Lie algebra $\mathfrak{g}$ is a ***Lie algebra direct sum*** of W_1 and W_2, and we write $\mathfrak{g} = W_1 \oplus W_2$.[2] Thus, we have the Lie algebra direct sum decomposition

$$(\mathfrak{sl}(2,\mathbb{C})_\mathbb{R})_\mathbb{C} \simeq \mathfrak{sl}(2,\mathbb{C}) \oplus \mathfrak{sl}(2,\mathbb{C}). \tag{A.5}$$

This decomposition will be crucial in our proof of Proposition 5.9.

Exercise A.1. Prove that (A.1) is a Lie algebra isomorphism. Remember, this consists of showing that ϕ is a vector space isomorphism, and then showing that ϕ preserves brackets. Then find similar Lie algebra isomorphisms to prove that

$$\mathfrak{u}(n)_\mathbb{C} \simeq \mathfrak{gl}(n,\mathbb{C})$$

$$\mathfrak{gl}(n,\mathbb{R})_\mathbb{C} \simeq \mathfrak{gl}(n,\mathbb{C}).$$

[2]Notice that this notation is ambiguous, since it could mean either that $\mathfrak{g}$ is the direct sum of W_1 and W_2 *as vector spaces* (which would then tell you nothing about how the direct sum decomposition interacts with the Lie bracket), or it could mean Lie algebra direct sum. We'll be explicit if there is any possibility of confusion.

A.2 Representations of Complexified Lie Algebras and the Tensor Product Decomposition of $\mathfrak{sl}(2,\mathbb{C})_{\mathbb{R}}$ Representations

In order for (A.5) to be of any use, we must know how the representations of a real Lie algebra relate to the representations of its complexification. First off, we should clarify that when we speak of a representation of a complex Lie algebra $\mathfrak{g}$ we are ignoring the complex vector space structure of $\mathfrak{g}$; in particular, the Lie algebra homomorphism $\pi : \mathfrak{g} \to \mathfrak{gl}(V)$ is only required to be ***real-linear***, in the sense that $\pi(cX) = c\pi(X)$ for all *real* numbers c. If $\mathfrak{g}$ is a complex Lie algebra, the representation space V is complex, *and* $\pi(cX) = c\pi(X)$ for all *complex* numbers c, then we say that π is ***complex-linear***. Not all complex representations of complex Lie algebras are complex-linear. For instance, all of the $\mathfrak{sl}(2,\mathbb{C})_{\mathbb{R}}$ representations described in Sect. 5.11 can be thought of as representations of the complex Lie algebra $\mathfrak{sl}(2,\mathbb{C})$, but only those of the form $(\mathbf{j},\mathbf{0})$ are complex-linear, as you will show below.

Exercise A.2. Consider all the representations of $\mathfrak{sl}(2,\mathbb{C})_{\mathbb{R}}$ as representations of $\mathfrak{sl}(2,\mathbb{C})$ as well. Show directly that the fundamental representation $\left(\frac{1}{2},\mathbf{0}\right)$ of $\mathfrak{sl}(2,\mathbb{C})$ is complex-linear, and use this to prove that all $\mathfrak{sl}(2,\mathbb{C})$ representations of the form $(\mathbf{j},\mathbf{0})$ are complex-linear. Furthermore, by considering the operators N_i defined in (5.71), show that these are the *only* complex-linear representations of $\mathfrak{sl}(2,\mathbb{C})$.

Now let $\mathfrak{g}$ be a *real* Lie algebra and let (π, V) be a complex representation of $\mathfrak{g}$. Then we can extend (π, V) to a complex-linear representation of the complexification $\mathfrak{g}_{\mathbb{C}}$ in the obvious way, by setting

$$\pi(X_1 + iX_2) \equiv \pi(X_1) + i\pi(X_2) \quad X_1, X_2 \in \mathfrak{g}.$$

(Notice that this representation is complex-linear by definition, and that the operator $i\pi(X_2)$ is only well defined because V is a *complex* vector space.) Furthermore, this extension operation is reversible: that is, given a complex-linear representation (π, V) of $\mathfrak{g}_{\mathbb{C}}$, we can get a representation of $\mathfrak{g}$ by simply restricting π to the subspace $\{1 \otimes X + i \otimes 0 \mid X \in \mathfrak{g}\} \simeq \mathfrak{g}$, and this restriction reverses the extension just defined. Furthermore, you will show below that (π, V) is an irrep of $\mathfrak{g}_{\mathbb{C}}$ if and only if it corresponds to an irrep of $\mathfrak{g}$. We thus have

Proposition A.1. *The irreducible complex representations of a real Lie algebra $\mathfrak{g}$ are in one-to-one correspondence with the irreducible complex-linear representations of its complexification $\mathfrak{g}_{\mathbb{C}}$.*

This means that we can identify the irreducible complex representations of $\mathfrak{g}$ with the irreducible complex-linear representations of $\mathfrak{g}_{\mathbb{C}}$, and we will freely make this identification from now on. Note the contrast between what we're doing here and what we did in Sect. 5.10; there, we complexified real representations to get complex representations which we could then classify; here, the representation space is fixed (and is always complex!) and we are complexifying the *Lie algebra itself*, to get a representation of a complex Lie algebra on the same representation space we started with.

Exercise A.3. Let (π, V) be a complex representation of a real Lie algebra $\mathfrak{g}$, and extend it to a complex-linear representation of $\mathfrak{g}_{\mathbb{C}}$. Show that (π, V) is irreducible as a representation of $\mathfrak{g}$ if and only if it is irreducible as a representation of $\mathfrak{g}_{\mathbb{C}}$.

Example A.1. *The complex-linear irreducible representations of* $\mathfrak{sl}(2,\mathbb{C})$

As a first application of Proposition A.1, consider the complex Lie algebra $\mathfrak{sl}(2,\mathbb{C})$. Since $\mathfrak{sl}(2,\mathbb{C}) \simeq \mathfrak{su}(2)_{\mathbb{C}}$, we conclude that its complex-linear irreps are just the irreps (π_j, V_j) of $\mathfrak{su}(2)$! In fact, you can easily show directly that the complex-linear $\mathfrak{sl}(2,\mathbb{C})$ representation corresponding to (π_j, V_j) is just $(\mathbf{j}, \mathbf{0})$. □

As a second application, note that by (A.5) the complex-linear irreps of $\mathfrak{sl}(2,\mathbb{C}) \oplus \mathfrak{sl}(2,\mathbb{C})$ are just the representations $(\mathbf{j_1}, \mathbf{j_2})$ coming from $\mathfrak{sl}(2,\mathbb{C})_{\mathbb{R}}$. Since $\mathfrak{sl}(2,\mathbb{C}) \oplus \mathfrak{sl}(2,\mathbb{C})$ is a direct sum, however, there is another way to construct complex-linear irreps. Take two complex-linear irreps of $\mathfrak{sl}(2,\mathbb{C})$, say (π_{j_1}, V_{j_1}) and (π_{j_2}, V_{j_2}). We can take a modified tensor product of these representations such that the resulting representation is not of $\mathfrak{sl}(2,\mathbb{C})$ but rather of $\mathfrak{sl}(2,\mathbb{C}) \oplus \mathfrak{sl}(2,\mathbb{C})$. We denote this representation by $(\pi_{j_1} \bar{\otimes} \pi_{j_2}, V_{j_1} \otimes V_{j_2})$ and define it by

$$(\pi_{j_1} \bar{\otimes} \pi_{j_2})(X_1, X_2) \equiv \pi_{j_1}(X_1) \otimes I + I \otimes \pi_{j_2}(X_2) \in \mathcal{L}(V_1 \otimes V_2), \quad X_1, X_2 \in \mathfrak{sl}(2,\mathbb{C}). \tag{A.6}$$

Note that we have written the tensor product in $\pi_{j_1} \bar{\otimes} \pi_{j_2}$ as "$\bar{\otimes}$" rather than "$\otimes$"; this is to distinguish this tensor product of representations from the tensor product of representations defined in Sect. 5.4. In the earlier definition, we took a tensor product of two $\mathfrak{g}$ representations and produced a third $\mathfrak{g}$ representation given by $(\pi_1 \otimes \pi_2)(X) = \pi_1(X) \otimes I + I \otimes \pi_2(X)$, where the *same* element $X \in \mathfrak{g}$ gets fed into both π_1 and π_2; here, we take a tensor product of two $\mathfrak{g}$ representations and produce a representation of $\mathfrak{g} \oplus \mathfrak{g}$, where two *different* elements $X_1, X_2 \in \mathfrak{g}$ get fed into π_1 and π_2.

Now, one might wonder if the representation $(\pi_{j_1} \bar{\otimes} \pi_{j_2}, V_{j_1} \otimes V_{j_2})$ of $\mathfrak{sl}(2,\mathbb{C}) \oplus \mathfrak{sl}(2,\mathbb{C})$ defined above is equivalent to $(\mathbf{j_1}, \mathbf{j_2})$; this is in fact the case! To prove this, recall the following notation: the representation space $V_{j_1} \otimes V_{j_2}$ is spanned by vectors of the form $v_{k_1} \otimes v_{k_2}$, $k_i = 0, \ldots, 2j_i$, $i = 1, 2$ where the v_{k_i} are characterized by (5.65). Similarly, the representation space $V_{(j_1, j_2)}$ of $(\mathbf{j_1}, \mathbf{j_2})$ is spanned by vectors of the form v_{k_1, k_2}, $k_i = 0, \ldots, 2j_i$, $i = 1, 2$, where these vectors are characterized by (5.74). We can thus define the obvious intertwiner

$$\begin{aligned} \phi : V_{j_1} \otimes V_{j_2} &\to V_{(j_1, j_2)} \\ v_{k_1} \otimes v_{k_2} &\mapsto v_{k_1, k_2}. \end{aligned}$$

Of course, we must check that this map actually is an intertwiner, i.e. that

$$\phi \circ [(\pi_{j_1} \bar{\otimes} \pi_{j_2})(X_1, X_2)] = \pi_{(j_1, j_2)}(X_1, X_2) \circ \phi \;\; \forall\, X_1, X_2 \in \mathfrak{sl}(2,\mathbb{C}). \tag{A.7}$$

Since $\mathfrak{sl}(2,\mathbb{C}) \oplus \mathfrak{sl}(2,\mathbb{C})$ is of (complex) dimension six, it suffices to check this for the six basis vectors

$$\begin{aligned}
&(i\mathcal{M}_z, 0),\\
&(0, i\mathcal{N}_z),\\
&(\mathcal{M}_+, 0) \equiv (i\mathcal{M}_x - \mathcal{M}_y, 0),\\
&(\mathcal{M}_-, 0) \equiv (i\mathcal{M}_x + \mathcal{M}_y, 0),\\
&(0, \mathcal{N}_+) \equiv (0, i\mathcal{N}_x - \mathcal{N}_y),\\
&(0, \mathcal{N}_-) \equiv (0, i\mathcal{N}_x + \mathcal{N}_y),
\end{aligned}$$

where the $\mathcal{M}_i$ and $\mathcal{N}_i$ were defined in (A.2). We now check (A.7) for $(i\mathcal{M}_z, 0)$ on an arbitrary basis vector $v_{k_1} \otimes v_{k_2}$, and leave the verification for the other five $\mathfrak{sl}(2,\mathbb{C}) \oplus \mathfrak{sl}(2,\mathbb{C})$ basis vectors to you. The left hand of (A.7) gives (careful with all the parentheses!)

$$\begin{aligned}
\phi((\pi_{j_1} \bar{\otimes} \pi_{j_2})(i\mathcal{M}_z, 0)(v_{k_1} \otimes v_{k_2})) &= \phi((J_z \otimes I)(v_{k_1} \otimes v_{k_2}))\\
&= \phi((j_1 - k_1)v_{k_1} \otimes v_{k_2})\\
&= (j_1 - k_1)v_{k_1,k_2}
\end{aligned}$$

where in the first equality we identified $\mathcal{M}_z$ with $S_z \in \mathfrak{su}(2)_{\mathbb{C}}$ as per (A.4). Meanwhile, viewing $\mathcal{M}_z$ as an element of $(\mathfrak{sl}(2,\mathbb{C})_{\mathbb{R}})_{\mathbb{C}}$, the right-hand side of (A.7) is

$$\begin{aligned}
\pi_{(j_1,j_2)}(i\mathcal{M}_z, 0)(\phi(v_{k_1} \otimes v_{k_2})) &= (\pi_{(j_1,j_2)}(i\mathcal{M}_z, 0))(v_{k_1,k_2})\\
&= i M_z\, v_{k_1,k_2}\\
&= (j_1 - k_1)v_{k_1,k_2}
\end{aligned}$$

and so the two sides agree. The verification for the other five $\mathfrak{sl}(2,\mathbb{C}) \oplus \mathfrak{sl}(2,\mathbb{C})$ basis vectors proceeds similarly. This proves that

Proposition A.2. *Let* $(\mathbf{j_1}, \mathbf{j_2})$ *denote both the usual* $\mathfrak{sl}(2,\mathbb{C})_{\mathbb{R}}$ *irrep and its extension to a complex-linear irrep of* $(\mathfrak{sl}(2,\mathbb{C})_{\mathbb{R}})_{\mathbb{C}} \simeq \mathfrak{sl}(2,\mathbb{C}) \oplus \mathfrak{sl}(2,\mathbb{C})$. *Let* $(\pi_{j_1} \bar{\otimes} \pi_{j_2}, V_{j_1} \otimes V_{j_2})$ *be the representation of* $\mathfrak{sl}(2,\mathbb{C}) \oplus \mathfrak{sl}(2,\mathbb{C})$ *defined in (A.6). Then*

$$(\mathbf{j_1}, \mathbf{j_2}) \simeq (\pi_{j_1} \bar{\otimes} \pi_{j_2}, V_{j_1} \otimes V_{j_2}). \tag{A.8}$$

Exercise A.4. Verify (A.7) for the other five $\mathfrak{sl}(2,\mathbb{C}) \oplus \mathfrak{sl}(2,\mathbb{C})$ basis vectors, and thus complete the proof of Proposition A.2.

We will now use Proposition A.2 to compute tensor products of the $(\mathfrak{sl}(2,\mathbb{C})_{\mathbb{R}})_{\mathbb{C}}$ irreps $(\mathbf{j_1},\mathbf{j_2})$, which by Proposition A.1 will give us the tensor products of the $(\mathbf{j_1},\mathbf{j_2})$ as $\mathfrak{sl}(2,\mathbb{C})_{\mathbb{R}}$ irreps, which was what we wanted! In the following computation we will use the fact that

$$(\pi_{j_1}\bar{\otimes}\pi_{j_2}) \otimes (\pi_{k_1}\bar{\otimes}\pi_{k_2}) \simeq (\pi_{j_1} \otimes \pi_{k_1})\bar{\otimes}(\pi_{j_2} \otimes \pi_{k_2}), \tag{A.9}$$

which you will prove below (note which tensor product symbols are "barred" and which are not). With this in hand, and using the $\mathfrak{su}(2)_{\mathbb{C}}\simeq\mathfrak{sl}(2,\mathbb{C})$ tensor product decomposition (5.56), we have (omitting the representation spaces in the computation)

$$\begin{aligned}
(\mathbf{j_1},\mathbf{j_2}) \otimes (\mathbf{k_1},\mathbf{k_2}) &\simeq (\pi_{j_1}\bar{\otimes}\pi_{j_2}) \otimes (\pi_{k_1}\bar{\otimes}\pi_{k_2}) \\
&\simeq (\pi_{j_1} \otimes \pi_{k_1})\bar{\otimes}(\pi_{j_2}\bar{\otimes}\pi_{k_2}) \\
&\simeq \left(\bigoplus_{l_1=|j_1-k_1|}^{j_1+k_1} \pi_{l_1}\right)\bar{\otimes}\left(\bigoplus_{l_2=|j_2-k_2|}^{j_2+k_2} \pi_{l_2}\right) \\
&\simeq \bigoplus_{(l_1,l_2)} \pi_{l_1}\bar{\otimes}\pi_{l_2} \quad \text{where } |j_i - k_i| \le l_i \le j_i + k_i,\ i = 1,2 \\
&\simeq \bigoplus_{(l_1,l_2)} (\mathbf{l_1},\mathbf{l_2}) \quad \text{where } |j_i - k_i| \le l_i \le j_i + k_i,\ i = 1,2.
\end{aligned}$$

This gives the tensor product decomposition of complex-linear $(\mathfrak{sl}(2,\mathbb{C})_{\mathbb{R}})_{\mathbb{C}}$ irreps. However, these irreps are just the extensions of the irreps $(\mathbf{j_1},\mathbf{j_2})$ of $\mathfrak{sl}(2,\mathbb{C})_{\mathbb{R}}$, and you can check that the process of extending an irrep to the complexification of a Lie algebra commutes with taking tensor products, so that the extension of a product is the product of an extension. From this we conclude that

Proposition A.3. *The decomposition into irreps of the tensor product of two* $\mathfrak{sl}(2,\mathbb{C})_{\mathbb{R}}$ *irreps* $(\mathbf{j_1},\mathbf{j_2})$ *and* $(\mathbf{k_1},\mathbf{k_2})$ *is given by*

$$\begin{aligned}
(\mathbf{j_1},\mathbf{j_2}) \otimes (\mathbf{k_1},\mathbf{k_2}) = \bigoplus (\mathbf{l_1},\mathbf{l_2}) \ \ \textit{where } &|j_1 - k_1| \le l_1 \le j_1 + k_1, \\
&|j_2 - k_2| \le l_2 \le j_2 + k_2
\end{aligned}$$

which is just Proposition 5.9.

Exercise A.5. Prove (A.9) by referring to the definitions of both kinds of tensor product representations and by evaluating both sides of the equation on an arbitrary vector $(X_1, X_2) \in \mathfrak{sl}(2,\mathbb{C}) \oplus \mathfrak{sl}(2,\mathbb{C})$.

References

1. V.I. Arnold, *Mathematical Methods of Classical Mechanics*, 2nd edn. (Springer, New York, 1989)
2. L. Ballentine, *Quantum Mechanics: A Modern Development* (World Scientific, Singapore, 1998)
3. F. Battaglia, T.F. George, Tensors: a guide for undergraduate students. Am. J. Phys. **81**(7), 498–511 (2013)
4. A.C. Da Silva, *Lectures on Symplectic Geometry*. Lecture Notes in Mathematics, vol. 1764 (Springer, Berlin, 2001)
5. D. Fleisch, *A Student's Guide to Vectors and Tensors* (Cambridge University Press, Cambridge, 2011)
6. T. Frankel, *The Geometry of Physics*, 1st edn. (Cambridge University Press, Cambridge, 1997)
7. S. Gasiorowicz, *Quantum Physics*, 2nd edn. (Wiley, New York, 1996)
8. H. Goldstein, *Classical Mechanics*, 2nd edn. (Addison-Wesley, Massachusetts, 1980)
9. M. Göckeler, T. Schücker, *Differential Geometry, Gauge Theories, and Gravity*. Cambridge Monographs on Mathematical Physics (Cambridge University Press, Cambridge, 1987)
10. D. Griffiths, *Introduction to Electrodynamics*, 3rd edn. (Prentice Hall, Upper Saddle River, 1999)
11. B. Hall, *Lie Groups, Lie Algebras and Representations: An Elementary Introduction* (Springer, New York, 2003)
12. I. Herstein, *Topics in Algebra*, 2nd edn. (Wiley, New York, 1975)
13. K. Hoffman, D. Kunze, *Linear Algebra*, 2nd edn. (Prentice Hall, Englewood Cliffs, 1971)
14. A.L. Onishchik, *Lectures on Real Semisimple Lie Algebras and Their Representations*. ESI Lectures in Mathematics and Physics (European Mathematical Society, Zürich, 2004)
15. M. Reed, B. Simon, *Methods of Modern Mathematical Physics I: Functional Analysis* (Academic, San Diego, 1972)
16. W. Rudin, *Principles of Mathematical Analysis*, 3rd edn. (McGraw Hill, New York, 1976)
17. J.J. Sakurai, *Modern Quantum Mechanics*, 2nd edn. (Addison Wesley Longman, Reading, MA, 1994)
18. B. Schutz, *Geometrical Methods of Mathematical Physics* (Cambridge University Press, Cambridge, 1980)
19. S. Sternberg, *Group Theory and Physics* (Princeton University Press, Princeton, 1994)
20. V.S. Varadarajan, *Lie Groups, Lie Algebras and Their Representations* (Springer, New York, 1984)
21. F. Warner, *Foundations of Differentiable Manifolds and Lie Groups* (Springer, New York, 1979)
22. N.M.J. Woodhouse, *Geometric Quantization* (Clarendon Press, Oxford, 1980)

N. Jeevanjee, *An Introduction to Tensors and Group Theory for Physicists*,
DOI 10.1007/978-3-319-14794-9

23. C. Zachos, Deformation quantization: quantum mechanics lives and works in phase-space. Int. J. Mod. Phys. A **17**(3), 297–316 (2002)
24. A. Zee, *Quantum Field Theory in a Nutshell* (Princeton University Press, Princeton, 2010)

Index

Symbols
$\mathbb{C}^n$, 13
  as complexification of $\mathbb{R}^n$, 249
$C(P)$, 167, 207
ϵ. *See* Levi–Civita tensor
$\mathfrak{gl}$ (n, $\mathbb{C}$), 150
GL (n, $\mathbb{R}$). *See* General linear group
$\mathfrak{gl}$(V), 166
$GL(V)$. *See* General linear group
$H_n(\mathbb{C})$. *See* Hermitian matrices
$\mathcal{H}_l(\mathbb{R}^3)$. *See* Harmonic polynomials
$\tilde{\mathcal{H}}_l$, *See* Spherical harmonics
J_+, 242, 246, 247, 249
J_-, 242, 246, 247
J_z, 242, 246, 247, 249
L^2([−a, a]), 15, 21, 25, 33, 37, 43, 77
$L^2(\mathbb{R})$, 15, 77
$L^2(\mathbb{R}^3)$, 81, 202, 210
$L^2(S^2)$, 205, 235
$\mathcal{L}(V)$. *See* Operators, linear
$M_n(\mathbb{C})$, 289
  as complexification of $M_n(\mathbb{R})$, 249
$M_n(\mathbb{R})$, $M_n(\mathbb{C})$, 13, 19
$O(3)$, 126
$O(3, 1)$, 132, 181
$O(n)$. *See* Orthogonal matrices, orthogonal group
$\mathfrak{o}(n)$, 151
$\mathfrak{o}(n-1, 1)$, 153
$O(n-1, 1)$. *See* Lorentz group
$\hat{p}$, 25, 77, 79, 169
$P_l(\mathbb{C}^2)$, 200, 238, 245
$P_l(\mathbb{R}^3)$, 15
q, 169
S_n. *See* Permutation group
$S^l(\mathbb{C}^2)$, 216, 246, 254, 255
$\mathfrak{sl}(n, \mathbb{C})$, 289
$SL(2, \mathbb{C})$, 134, 199
  relationship to $SO(3, 1)_o$, 141, 177
$\mathfrak{sl}(2, \mathbb{C})_{\mathbb{R}}$, 164
  relationship to $\mathfrak{so}(3, 1)$, 177
$SO(2)$, 111, 125
$SO(3)$, 125, 138, 168
  infinitesimal elements of, 147
$SO(3, 1)_o$, 130, 181
$SU(2)$, 127
  relationship to $SO(3)$, 138, 176, 179
$SU(3)$, 225
$\mathfrak{so}(n)$, 154
$\mathfrak{so}(2)$, 158
$\mathfrak{so}(3)$, 113, 158, 168
$\mathfrak{so}(3, 1)$, 113, 158, 168
SO(n). *See* special orthogonal group
$\mathfrak{su}(n)$, 154
$\mathfrak{su}(2)$, 138, 162
  isomorphic to $\mathfrak{so}(3)$, 176
SU(n). *See* special unitary group
2-sphere, 129
3-sphere, 129
$\mathfrak{u}$(n), 151
$U(n)$. *See* Unitary group
$U(1)$, 137
V^*. *See* Dual space
$\hat{x}$, 25, 77
Y^l_m. *See* Spherical harmonics
$\mathbb{Z}_2$, 123, 197
  and parity, time-reversal, 142
  and sgn homomorphism, 143
  irreducible representations of, 241
  is semi-simple, 241, 267

N. Jeevanjee, *An Introduction to Tensors and Group Theory for Physicists*,
DOI 10.1007/978-3-319-14794-9

A
Abelian (group), 116
Active transformation, 66, 103
Addition of angular momentum, 82, 83, 234
Additive quantum number, 209–211
Ad homomorphism, 172, 178
Adjoint, 120
Adjoint representation, 26
of Lorentz group, 196, 227
of orthogonal group, 196, 227
Alternating tensor, 88
Angular momentum, 9, 168
commutation relations, 83, 115, 160
Angular velocity vector, 9, 99, 114, 161
Anti-Hermitian matrices, 151
Anti-Hermitian operators, Lie algebra of, 166, 172
Antisymmetric
matrices, 19, 151
2nd rank tensor representation, 196
tensor, 88
tensors and Lie algebras of isometry groups, 229
Axial vector, 96
Axis of rotation, 95, 181

B
BAC-CAB rule, 97
Baker-Campbell-Hausdorff formula, 157
Basis (for a vector space), 18
Bilinear, 34
Bivector, 96
Body axes, 23
Body frame, 99
Boost, 130, 163, 182
Bosons, 89, 145
additivity of, 164
generators, 153
Bra, 41

C
Canonical
quantization, 167, 189
transformation, 168
variables, 170
Cartesian product, xv
Chiral representation (of gamma matrices), 283
Clebsch-Gordan coefficients, 84, 234, 281
Clifford algebra, 263, 284
Closed (group), 146
Cofactor expansion of determinant, 93
Commutator, 155
Completely reducible, 232
Completeness (of vector space), 37
Complexification
of Lie algebra, 289
of representation, 248
of vector space, 248
Complex Lie algebra, 289
Complex-linear, 292
Complex plane, 18
Complex vector space, 13
Components
contravariant, 42, 61
covariant, 42, 61
of dual vector, 31
of matrix, 27
of Minkowski metric, 36
of tensor, 3, 6, 7, 53, 72
of vector, 21
Conjugate-linear, 34
Conjugate representation. *See* Representation, conjugate
Connected group, 127
Contraction, 73
Cross product, 55, 97
Cyclic permutation, 92

D
Decomposable, 232
Determinant, 92, 104–105
as group homomorphism, 137
of a matrix exponential, 153
and oriented volume, 94, 105
and sgn homomorphism, 144
and trace, 180
Diagonalizable matrix, 153
Dimension
complex, 19
real, 19
of a vector space, 18
Dirac bilinear, 284
Dirac delta functional, 33, 43, 78
Dirac notation, xvi, 41, 54, 74, 78, 84
Dirac spinor, 194, 199, 236, 261, 282, 284
and antisymmetric tensors, 285
Direct sum, 230, 236
external, 236
internal, 236
of Lie algebras, 291
representation, 236
Direct sum decomposition
of $\Lambda^2\mathbb{R}^4$, 234
of $L^2(S)^2$, 235
of $M_n(C)$, 231

of M_n(R), 231
of tensor product representations, 233
Disconnected group, 127
Doppler shift, 132
Dot product, 35
Double cover
of $SO(3)$ by $SU(2)$, 140
of $SO(3,1)_o$ by $SL(2, \mathbb{C})$, 141
Double dual, 48
Dual
basis, 32
metric, 49
representation, 211, 212
space, 31
vector (*see* Linear functional)
Dummy index, 51

E
Einstein summation convention, 51, 61
Einstein velocity addition law, 164
Electromagnetic field tensor, 77
dual, 270
transformation properties of, 196, 234, 258
Entanglement, 85
Equivalence of representations, 220
Euler angles, 125
Euler's theorem, 181
Exponential of matrix, 148
Exterior derivative, 77

F
Faithful representation, 266
Fermions, 89, 145
Fourier coefficients, 25
Fourier series, 25
Fourier transform, 79
Four-vector, 193
Four-vector representation. *See* Representation, four-vector
Free index, 51

G
Gamma matrices, 282–284
chiral representation of, 283
General linear group, 17, 118, 150
Generator. *See* Infinitesimal generator
Gram-Schmidt, 34
Group, 102, 111
axioms, 115
definition of, 115

H
Hamilton's equations, 208
Harmonic polynomials, 15, 20, 29, 46, 64, 86, 205, 247
Heisenberg algebra, 169, 206, 210
Heisenberg equation of motion, 26
Heisenberg picture, 68
Hermite polynomials, 39
Hermitian adjoint, 48
Hermitian matrices, 14, 20
Hermitian scalar product, 35
Hermiticity, 33
Highest weight vector, 243, 252
Hilbert-Schmidt norm (on $M_n(\mathbb{C})$), 157
Hilbert space, 21, 37
Homomorphism
group, 135
Lie algebra, 171

I
Identical particles, 89, 197
Indices
offsetting, 54
raising and lowering, 42, 54
Infinitesimal generator, 111, 145, 149
Inner product, 34
on $\mathbb{C}^n$, 35
on $L^2([-a, a])$, 37
on $M_n(\mathbb{C})$, 35
on $P(\mathbb{R})$, 38, 50
on $\mathbb{R}^n$, 35
Inner product space, 34
Intertwiner, 220
Intertwining map. *See* Intertwiner
Invariant subspace, 231
nontrivial, 231
Inversion, 95, 127
Invertibility, 26, 105
Irreducible representation, 232, 237–238
of abelian group, 241
of double-cover of $O(3, 1)$, 260
of $O(3,1)$, 261–263
of a real Lie algebra and its complexification, 292
of $\mathfrak{sl}(2, \mathbb{C})$, 293
of $\mathfrak{sl}(2, \mathbb{C})_{\mathbb{R}}$, 251–256
of $\mathfrak{sl}(2, \mathbb{C}) \bigoplus \mathfrak{sl}(2, \mathbb{C})$, 293–294
of $SL(2, \mathbb{C})$, 256
of $SO(3)$, 247
of $SO(3, 1)_o$, 256
of $\mathfrak{su}(2)$, 242–244
of $\mathbb{Z}_2$, 241

Irrep. *See* Irreducible representation
Isometry, 118, 135, 150, 166, 172
Isomorphism
  group, 135
  Lie algebra, 171

J
Jacobi identity, 155, 173

K
Kernel, 137
Ket, 41
Killing form, 179, 186
Kinetic energy of rigid body, 9

L
Laguerre polynomials, 50
Left-handed spinor. *See* Representation, spinor
Legendre polynomials, 39
Levi–Civita symbol. *See* Levi–Civita tensor
Levi–Civita tensor, 4–6, 54, 91, 144
Lie algebra, 115, 145, 150–170
  abstract, 166
  complex, 289
  definition, 150
  direct sum, 291
  geometric interpretation of, 149
  physicist's definition, 152
Lie algebra homomorphism, 171
  induced, 173
Lie algebra representation, 192
Lie bracket
  and similarity transformation, 179
Lie group, 111, 145–147
  dimension of, 146
Lie product formula, 156, 175
Lie, Sophus, 145
Lie subalgebra, 167
Linear combinations, 17
Linear dependence, independence, 18
Linear functional, 30
Linearity, 25
Linear map, 136
Linear operator
  definition of, 25
  Lie algebra of, 166
  as (1,1) tensor, 52, 74
Lorentz force law, 77
Lorentz group, 121, 130–133
Lorentz transformations, 65, 130
  decomposition into boost and rotation, 182

M
Majorana representation. *See* Majorana spinor
Majorana spinor, 237, 266
Matrix
  diagonalizable, 153
  exponential, 148
  of linear operator, 28
  logarithm, 185
  of Minkowski metric, 36
Matrix Lie group, 111, 146
Maxwell stress tensor, 76
Metric, 34
  as intertwiner between V and $V*$, 225
Metric dual, 40
Minkowski metric, 36, 121
Moment of inertia tensor, 9, 55, 75
Momentum representation, 79
Multilinearity, 3, 5, 52
Multiplicative quantum number, 209–211
Multipole moments, 56, 86

N
Noether's theorem, 168
Non-abelian (group), 116
Non-commutative (group), 116
Non-degeneracy, 33, 34
Non-degenerate Hermitian form, 33
Norm, 34
Normal subgroup, 137
Nullity, 138
Null space, 138

O
Observables, 167
One-parameter subgroup, 148
One-to-one, 26
Onto, 26
Operators
  adjoint of, 48
  angular momentum, 26, 29, 190–191
  exponential of, 47
  Hermitian, 49
  invertible, 26
  linear (*see* Linear operator)
  matrix representation of, 28
  self-adjoint, 49
  symmetric, 49

Orbital angular momentum, 189
Orientation, 95, 105
Orthogonal
 complement, 267
 matrices, 63, 102
 projection operator, 277, 286
 set, 35
Orthogonal group, 119, 125–127
 proof that it's a matrix Lie group, 146
Orthonormal basis, 34
 for hilbert space, 37

P
Parity, 132, 210, 242
 and Dirac spinors, 237
 and $\mathbb{Z}_2$, 142
Passive transformation, 66, 103
Pauli matrices, 14
Permutation(s). *See* also Permutation group
 even, odd, 89, 144
Permutation group, 123, 197
 relation to $\mathbb{Z}_2$, 143
Permutation operator, 90
Phase space, 167
Pin groups, 263
Poisson bracket, 167
 formulation of mechanics, 208
Polar decomposition theorem, 182
Polynomials, 15
 harmonic (*see* Harmonic polynomials)
 Hermite, 39
 Laguerre, 50
 Legendre, 39
 real, 38, 50
Positive-definite, 34
Positive matrix, 182
Principal axes, 62
Principal minors, 183
Principal moments of inertia, 62
Product
 of matrices, 28
 state, 85
Pseudoscalar, 218, 219
Pseudovector, 96, 196, 218, 219, 234

Q
Quark, antiquark, 225

R
Rank (of map), 138
Rank (of tensor), 3, 54
Rank-nullity theorem, 138
Rapidity, 130
Real-linear, 292
Real numbers (as additive group), 117
Real vector space, 13
Representation, 167, 187, 188, 192
 adjoint, 195
 adjoint of $C(P)$, 207
 alternating, 197
 complex, 192
 conjugate, 265
 dual, 211, 212, 225
 equivalence of, 220
 faithful, 266
 four-vector, 193, 199, 257
 on function spaces, 199
 fundamental, 193
 of Heisenberg algebra on $L^2(\mathbb{R})$, 206
 irreducible, 232
 on linear operators, 214
 pseudoscalar, 219
 pseudovector, 219
 real, 192
 of S_n , 197
 scalar, 217
 2nd rank antisymmetric tensor, 196, 227, 258
 2nd rank antisymmetric tensor and adjoint of O(n), 217
 sgn, 197
 spin-one, 190, 250
 $SO(2)$ on $\mathbb{R}^2$, 241, 250
 $SO(3)$ on $\mathcal{H}_l(\mathbb{R}^3)$ $\mathrm{H}_{\updownarrow}$ and $L^2(S^2)$, 205
 $SO(3)$on $L^2(\mathbb{R}^3)$, 202
 space, 192
 spinor, 194, 199, 226, 253, 254, 261, 282
 spin s, 190, 200, 210
 spin-two, 250
 $SU(2)$ on $P_l(\mathbb{C}^2)$, 200
 on symmetric and antisymmetric tensors, 215–220
 symmetric traceless tensors, 250
 on $\mathcal{T}^r_s(V)$, 211
 tensor product, 208
 trivial, 193
 unitary, 188, 192

Representation, (*cont.*)
vector, 193
and "spin-one", 249
of $\mathbb{Z}_2$, 197
Representation operator, 276
Right-handed spinor. *See* Representation, spinor
Rigid body motion, 23, 68, 100
Rotation, 95, 111, 125–126, 168, 180
generators, 112, 113, 151, 153, 158, 160
improper, 127
proper, 127

S
Scalar representation. *See* Representation, scalar
Scalars, 12
Schouten convention, 58
Schrodinger picture, 68
Schur's lemma, 239
Selection rule, 277
angular momentum, 275, 279
parity, 279
Self-adjoint, 49
Semi-simple (group or algebra), 233
Separable state, 85
sgn homomorphism, 143
sgn representation, 197
Sigma matrices. *See* Pauli matrices
Similarity transformation, 63, 214
Space axes, 23
Space frame, 99
Span (of a set of vectors), 17
Special linear group, 134
Special orthogonal group, 122, 180
Special unitary group, 122, 127
Spectral theorem, 206
Spherical harmonics, 15, 20, 29, 46, 191, 205, 235
Spherical Laplacian, 16, 206
Spherical tensor, 274
Spin, 13, 82, 189
Spin angular momentum, 83
Spinor. *See* Representation, spinor
Square-integrable, 15, 205
Star operator, 269
Stern-Gerlach experiment, 24
Stone-von Neumann theorem, 189
Structure constants, 171
Subgroup, 118
normal, 137
Subspace, 14
Symmetric
form, 34
matrices, 19
tensors, 85
Symmetric group. *See* Permutation group
Symmetrization postulate, 90, 145, 197
Symmetry and degeneracy, 241

T
$[T]$, 28
T^r_s. *See* Tensors, of type (r, s)
Tensor operator, 272, 273
Tensor product
as addition of degrees of freedom, 81–85, 209
of operators, 80
representation, 208
of $\mathfrak{sl}(2, \mathbb{C})_{\mathbb{R}}$ irreps, 257, 295
of $\mathfrak{su}(2)$ irreps, 234
of vectors, 70
of vector spaces, 70
Tensors
alternating, 88
antisymmetric, 88
basis for vector space of, 72
components of, 3, 6, 7, 53, 72
contraction of, 73
definition of, 52
linear operators, 8, 52, 74
and matrices, 8
rank of, 54
symmetric, 85
as tensor product space, 72
transformation law, 8, 57–69
type (r, s), 52, 72
Time-reversal, 132
and $\mathbb{Z}_2$, 142
Trace, 32, 63
cyclic property of, 156
and determinant, 180
interpretation of, 154
Transformation law
of linear operators, 63, 214
of metric tensors, 64
of vectors and dual vectors, 60, 212
Translations, 96, 170
Transpose, 48
Transposition, 85, 143
Trivial representation. *See* Representation, trivial

U
Unitary
    group, 120
    matrices, 63
    operator, 120
    representation, 192

V
Vector operators, 80, 271–272
Vector representation. *See* Representation, vector
Vector space
    as additive group, 118
    axioms, 12
    complex, 13
    definition of, 12
    isomorphism, 136
    real, 13

W
Wedge product, 88, 144
Weight function, 39
Weyl spinor, 194
Wigner D-matrix, 275
Wigner-Eckart theorem, 275, 279–282
Wigner function, 275

Printed in the United States
By Bookmasters